Method and Tactics
in Cognitive Science

METHOD AND TACTICS IN COGNITIVE SCIENCE

Edited by

WALTER KINTSCH
JAMES R. MILLER
PETER G. POLSON
University of Colorado

LEA LAWRENCE ERLBAUM ASSOCIATES, PUBLISHERS
1984 Hillsdale, New Jersey   London

Lawrence Erlbaum Associates, Inc., Publishers
365 Broadway
Hillsdale, New Jersey 07642

Library of Congress Cataloging in Publication Data
Main entry under title:

Method and tactics in cognitive science.

Bibliography: p.
Includes index.
1. Cognition—Congresses. 2. Artificial intelligence
—Congresses. I. Kintsch, Walter, 1932–
II. Miller, James R. III. Polson, Peter G.
BF311.M4494 1984   153   83-25489
ISBN 0-89859-327-1

Printed in the United States of America
10 9 8 7 6 5 4 3 2 1

Contents

Preface

It is never easy to do good research, but it is even harder to do good interdisciplinary research. The editors of this volume are three psychologists whose work has brought them into frequent contact with other disciplines within cognitive science, particularly linguistics and artificial intelligence. There is the excitement of discovering new possibilities, of finding novel contexts for one's work. A new literature with new concepts and fresh methods deals with the familiar old problems and is discovered with enthusiasm. It is of course not the case that the researcher with interdisciplinary interests fancies him or herself a professional in the discipline he or she is interested in, nor even do most of us intend to become real experts in it. We want to stay within psychology, philosophy, AI, or linguistics, whatever the case may be. But we think we can do better by paying close attention to what is going on in related fields. In all too many cases this interdisciplinary enthusiasm is dampened soon: troubles build up, we find ourselves not being taken seriously by the pros in the new field, we can't get them to listen to our concerns, and worse, they seem to be doing everything wrong.

Effective cooperation across disciplines takes even more patience, tolerance, and good sense than scientific cooperation always does. Even when concerned with the same phenomenon, scientists from different disciplines will look at it differently, ask different questions, are qualified to deal with different aspects of the problem. But that does not mean that they cannot profit from each other. Cognitive science is based on the belief that crossing the boundaries of the traditional disciplines is not merely possible, but indeed essential in the study of cognition. Without abandoning our own scientific identity, we must learn to take advantage of the results and insights obtained by researchers in other disciplines in order to progress more rapidly in the study of our exceedingly complex and difficult subject matter.

The purpose of this volume is to facilitate this interaction among the disciplines that constitute cognitive science. We want to learn to talk to each other, and we want to show others how that can be achieved, and what the problems and limits are of such interactions. We are trying to do this not by an abstract discussion of methodological issues, but by concrete example. We take some highly respected, and in some ways typical, work from several of the sub-disciplines of cognitive science, and watch how they do it, how they pose their questions, what they accept as answers. Then we have someone from another discipline who is interested and competent in the subject under discussion in his own field comment on this work from the perspective of another discipline. Thus, Lehnert and Clancy describe their work in artificial intelligence, and the philosopher Haugeland comments on it. Bresnan & Kaplan and Givón describe their approaches to linguistics, and Clark and Malt react to them as psychologists. Cognitive psychology is represented here by Swinney and VanLehn, Brown, & Greeno and discussed by Charniak from an artificial intelligence perspective. Finally, Suppes and Mandler present an overview and evaluation of the various issues that were raised here. Thus, being anchored in concrete problems all of the time facilitates communication and helps to maintain a common focus. The problem is to get people to listen to each other, to develop some appreciation for the other's problems and concerns, thereby learning something about one's own limitations.

The Sloan Foundation graciously agreed to support a conference of this nature, that met in the summer of 1981 in Boulder, Colorado. Discussions were lively and satisfying. All participants had handed in complete drafts of their papers beforehand, so that the conference could be devoted to discussion rather than to the reading of papers. Subsequently, all papers were extensively revised. The chapters presented here are the result of this rather intensive interaction. The editors cooperated equally in all phases of this project and the various permutations of our names that we used here are arbitrary.

The editors would like to thank the authors for their cooperation throughout the preparation of this volume. It was their willingness to enter into the spirit of the enterprise we had proposed that created this book. We also thank the Sloan Foundation for their financial support that made the entire project possible. And finally, we thank our secretary P. Bochert whose competent help with the project is much appreciated.

Walter Kintsch
James R. Miller
Peter G. Polson

Contributors

DR. JOAN BRESNAN
Department of Linguistics
Stanford University
Stanford, California 94305

DR. JOHN SEELY BROWN
Xerox Palo Alto Research Center
333 Coyote Hill Road
Palo Alto, California 94303

DR. EUGENE CHARNIAK
Department of Computer Science
Brown University
Providence, Rhode Island 02912

DR. WILLIAM CLANCEY
Department of Computer Science
Stanford University
Stanford, California 94305

DR. HERBERT H. CLARK
Department of Psychology
Stanford University
Stanford, California 94305

DR. TALMY GIVÓN
Department of Linguistics
University of Oregon
Eugene, Oregon 97403

DR. JAMES GREENO
Learning Research and Development
  Center
University of Pittsburgh
Pittsburgh, Pennsylvania 15260

DR. JOHN HAUGELAND
Department of Philosophy
University of Pittsburgh
Pittsburgh, Pennsylvania 15260

DR. RONALD KAPLAN
Xerox Palo Alto Research Center
333 Coyote Hill Road
Palo Alto, California 94303

DR. WALTER KINTSCH
Department of Psychology
University of Colorado
Boulder, Colorado 80309

DR. WENDY LEHNERT
Department of Computer Science
University of Massachusetts
Amherst, Massachusetts 01003

DR. BARBARA MALT
Institute of Human Learning
Cognitive Science Program
University of California
Berkeley, California 94720

DR. GEORGE MANDLER
Department of Psychology, C-009
University of California, San Diego
La Jolla, California 92093

DR. JAMES R. MILLER
Computer Thought Corp.
1721 W. Plano Pkwy.
Plano, Texas 75075

DR. PETER POLSON
Department of Psychology
University of Colorado
Boulder, Colorado 80309

DR. PATRICK SUPPES
Department of Philosophy
Stanford University
Stanford, California 94305

DR. DAVID SWINNEY
Department of Psychology
Tufts University
Medford, Massachusetts 02155

DR. KURT VAN LEHN
Xerox Palo Alto Research Center
333 Coyote Hill Road
Palo Alto, California 94303

1 Problems of Methodology in Cognitive Science,

James R. Miller
University of Colorado

Peter G. Polson
University of Colorado

Walter Kintsch
University of Colorado

The conglomerate that makes up cognitive science includes portions of psychology, artificial intelligence, and linguistics, as well as anthropology, philosophy, and neuroscience. There can be no doubt that, in some of these disciplines, considerable progress has occurred during the last decade and that, therefore, cognitive science is a field of genuine promise today. But in what sense does cognitive science transcend these individual disciplines? At the moment, attitudinal differences in methodology sometimes preclude productive interactions. Of course, each discipline should have its own research style, and there can be no question of imposing some common methodology on cognitive science as a whole. But we must aim for a better understanding of the goals and methods employed by other disciplines and an appreciation of the strengths and limitations of these methods. After all, cognitive science is built on the premise that these methods complement each other, thereby permitting us to advance research in cognition beyond our own discipline's boundaries.

But when, in fact, is cognitive science research really interdisciplinary? Suppose a linguist makes an excellent case for the proposition that anaphora rules should apply to indexed noun phrases at a quasi-syntactic level intermediate between the autonomous syntactic part of the grammar and the logical formalism; he then shows that this is not the case in several current AI language comprehension programs, and dismisses them. Is this cognitive science? Alter-

1

natively, consider a program for text comprehension that makes bridging inferences when there are gaps in the text. The author points out that people can make such inferences, too, and therefore asserts the "psychological reality" of the program. Is this cognitive science? Or, if a psychologist shows that reaction times to passive questions are longer than reaction times to active questions, does this establish the "linguistic reality" of the passive transformation? These examples are not real ones, but neither are they fanciful. Is this limited extension of research from one domain to another what we mean by interdisciplinary work?

Before addressing what we perceive as potentially serious methodological problems, we need to remark on the rather broad view of methodology taken here. We are concerned as much with methods of observation, data collection, and analysis, as with problems of theory construction and evaluation. Indeed, it seems that the latter are primary; once you have decided what sort of theory you want, the empirical methods simply follow.

We first make a few brief observations about the history of cognitive science, because there are some problems we have inherited and have not yet come to terms with. We then enumerate a number of issues that we think deserve attention and that seem to spring from long-standing, unresolved historical conflicts. These sections comprise our diagnosis of what is wrong with cognitive science. We then suggest some treatments to set things right again. We explore how cognitive science could make more effective use of experimental methods and we discuss the opportunities and dangers that stem from the heavy reliance on introspective reports within cognitive science. Most importantly, we take a look at the nature of theory construction in cognitive science, and consider some of its puzzles and contradictions. It is certainly not the case that we have all the answers, but we like to raise questions, even at times in an intentionally pejorative manner, in the hope that the discussions will help us to find some of the answers.

The Dual Origins of Cognitive Science

Cognitive science, as it is practiced today, has two distinct historical roots. It derives from one scientific tradition that emphasizes objectivity and the study of behavior from the outside, and from another that is subjectively oriented and that has proposed to study mental life from the inside (Aebli, 1980). The information processing tradition of cognitive psychology on the one hand, and action theory and purposive or intentional descriptions of behavior on the other, are representative examples of these incompatible trends with cognitive science. Essentially, it is the same conflict that was acted out in a more extreme form earlier in this century by behaviorism and phenomenology.

Some of cognitive science is based on self-observation. We observe a process in ourselves, analyze it, then recognize it again when we see it in others. We understand what happens in others because we know what happens with ourselves. Cognitive theory, thus, is an externalization of processes gained through self-knowledge. The data to be accounted for are direct, simple observations; complicated operationalizations and statistical manipulations would only obscure this straightforward characterization. It is cognitive science from the inside.

The antithesis is the black-box approach that characterizes much of current information-processing psychology. The organism is a black-box, whose input and output can be studied, and we must infer often complex information-processing mechanisms in order to relate the two. These mechanisms are conceived as fixed, interacting components, typically described in rather mechanistic terms. For example, our information-processor might have a short-term memory with a capacity of 7 ± 2 units, various pattern matching devices, a high sensory threshold, and so on.

This is very different from a person who has goals, purposes, and a conscious mind. The contrast is perhaps best expressed by their opposing views of verbal communication. Central to the information-processing approach is the information transmission metaphor. On one end, we have a sender, on the other the receiver, each of which may be decomposed into complex information-processing components; between the two, there is a channel with certain properties, through which information flows from the sender to the receiver. The alternative conceptualization holds that there are two conscious minds, each in a certain state; one is trying to influence the other, that is to change its state, by providing it with certain cues (spoken words, gestures), from which the receiver will infer what the sender meant. Here there is no flow of information anywhere (Hörmann, 1981).

We argue that the problem for cognitive science is to find the right synthesis of these approaches. Neither is satisfactory in itself, but their synthesis must be something other than thoughtless confusion.

METHODOLOGICAL QUESTIONS

The Rejection of Intuition

Despite the growth of cognitive science over the last 10 years, many experimental psychologists are still puzzled by some of the goals of cognitive science. Particularly bothersome is the fact that a theory in cognitive science often amounts to an explication and externalization of intuitions. An anecdote may help to focus this issue. Suppose, like some of us, you are working on a model of text comprehension. We are all able to comprehend text and we have fairly good intuitions about some aspects of the comprehension process. What is the function of a model under these circumstances? In part, it is to make these intuitions explicit, in part to fill out these components of the process where our intuitions fail. For instance, one component where intuitions have proven to be reliable is the following: If you read a text that you can comprehend well, you can predict quite accurately which portions of that text will be recalled later and which will be forgotten. One of the feats of the model we have been developing (Miller & Kintsch, 1980), is that it is also able to make such predictions, just about as successfully as people do. Recently, an experimental psychologist reviewing a paper describing such recall predictions rejected them as being trivial: All they show is that people recall the important passages and forget the irrelevant detail, as everyone knows and as alternative theories also predict. (Indeed, any theory that wouldn't predict this trend should be rejected.)

A fundamental problem revealed by this criticism is the collision between an approach that views behavior from the *outside* with one whose goal is to specify and explicate behavior on the *inside,* including peoples' intuitions about comprehension. However, such general intuitive knowledge about comprehension is implicit, vague, and cannot be used reliably to troubleshoot comprehension problems nor can it improve the readability of a textbook chapter. The model discussed above is not trivial simply because it makes some predictions anybody can make, just as the linguistic analysis of a sentence is not useless simply because it parses the sentence into constituents any naive speaker could identify. Explaining intuitions is not a bad goal for a theory.

On Mistaking Intuitions for Theory

Self-observation and intuition, and task analyses based on them, are important tools in cognitive science. But it is important to realize that they provide a starting point for research, not an ultimate goal. Intuitions are not in themselves a cognitive theory; rather, they need to be explained. Cognitive scientists do not always seem to appreciate this point. Let us invent an anecdote to illustrate this point. Imagine a psychologist delivering a paper entitled, "Everyday planning techniques: How to bake a cake." He relates how he personally planned this procedure, and then, after reporting protocols of 15 other subjects, identifies three major cake-baking strategies, through which he claims 85% of the statements in the protocols can be explained. Among traditional psychologists, such a paper would be greeted with derision. Some cognitive scientists might applaud. Both reactions miss the point: The paper is inadequate because it offers no theory; it fails to explain, because it merely restates intuitions, and offers them as explanations. On the other hand, such a paper might provide a valuable task analysis, which is a necessary step in formulating a theory of planning strategies, that could account for the behaviors in question.

Describing intuitions is the first step toward explaining them, but no more. What does constitute an explanation? What is the difference between the model that explains prose recall and the planning strategies for baking a cake that merely describe what people do without explaining that process? The comprehension model is an information-processing model: It identifies certain processes and mechanisms (a short-term memory buffer, cyclical processing, memory retrieval) that interact to produce comprehension. As a result of this interaction, certain parts of the text will be recalled later while others will not be recalled. In a sense we are not directly explicating intuitions in this case at all. Instead, what we have are two ways of understanding a phenomenon like text recall: One is by intuitions that we as scientists share with every literate adult; the other is by explaining behavior as a result of particular information-processing operations that are specified in our model.

The point of this example is not that all explanations must involve information-processing notions. Any kind of general theoretical framework, not just that of information-processing psychology, could, in principle, provide an adequate scientific explanation. That is, our cake-baking strategies could have been related to any concrete, specific theory of planning—whatever the basic terms of that theory. In the cake-baking example it is the atheoretical treatment of the task that we object to, not the teleological terminology.

Are theories at different levels—a strategy explanation versus an information-processing account—real alternatives that can be compared and tested against each other, or are they merely different ways of talking about the same thing? A somewhat idealized description of a recent dissertation at Colorado illustrates the dilemma (Caccamise, 1981). When subjects are asked to generate ideas relevant to a given topic, they occasionally repeat ideas they have mentioned before—more frequently under some experimental conditions than under others. The question of interest concerns the nature of these repetitions. Consider first an information-processing model of idea generation from long-term memory, that gives a reasonably accurate account of these repetitions. In such a model, repetitions indicate an interference process: once an idea is generated, its accessibility in memory increases, and the retrieval process is more likely to follow this same route because of its increased accessibility. Thus, idea repetitions are explained as consequences of the retrieval dynamics.

An interesting alternative views the same phenomenon as a planning task. Ideas are repeated because the subject wants to do so, and not because of interference effects over which he has only partial control. The subject plans the repetition because it might serve as a starting point for a process of elaboration that did not occur when the idea was mentioned the first time. This is certainly a very reasonable intuition, namely, that a repeated idea may lead to a chain of associations that were not produced the first time.

Thus, the information-processing model accounts for precisely the same observations as the planning model. Which model is right? The appealing intuition or the mechanistic model? The mechanistic model clearly "explains" the data; does the strategy model? If so, how do we tell which explanation is better? Are they really alternatives, or is the one a description of processes, while the other describes our intuitions? It can't be that easy, because there is a clear incompatibility between an interference process and a planned elaboration! Can you have both—one to describe the dynamics of an automatic retrieval process and the other to describe a consciously controlled planning strategy that makes use of that retrieval process? We have returned to the conflict between cognition as mechanistic information-processing and cognition as intentional behavior.

Conceptual Confusions, New Insights, and a Latter Day Tower of Babel

One thing that can enormously upset the more meticulous observers of the cognitive science scene is the terminological inflation and confusion. If only words were explanations, especially fuzzy words that can mean anything and everything! We have only to invent three more terms as all-encompassing as "schema" and cognitive science will be ripe for Nobel prizes. But young sciences are notoriously hard to pin down to terminological purity, and terms like "schema" obviously have their uses.

More interesting, and perhaps more serious, is the confusion between purposive and mechanistic language that characterizes much of the writing in cognitive science. As if it were the most natural thing in the world, purposive terminology has been imported into an information-processing framework: subgoals are stored in short-term memory; unconscious expectations are processed in parallel; opinions are represented propositionally; the mind contains schemata. Is it a sign of conceptual weakness, or merely an excusable sloppiness in the use of language? Or is it not confusion at all, but a new synthesis that cognitive science has achieved? Can the methodology of artificial intelligence overcome the distinctions between information-processing terms and purposive concepts? It is not the same to say that we now can talk about intentionality in information-processing terms (a claim we believe to be valid) and to simply disregard the distinction! We need to explore whether or not the distinction has any merit today. If the answer is no, we need to be able to tell people why; if the answer is yes, the current practice in cognitive science is badly in need of a change.

But even if we were much more careful about our language than we are, cognitive science has a serious communication problem. At one time theories were eminently describable, because they were simply described verbally. The spectre of fuzzy-headed thinking was always present, of course, and substantial amounts of handwaving could be (and often was) concealed by a carefully designed paragraph. The precision of mathematical models helped to remove some of this fuzziness. One needed more than a passing familiarity with mathematics to understand these models, but this was nothing that some time spent with a couple of books on probability theory and Markov chains could not cure. As the topics of interest to psychologists grew increasingly complex and knowledge-laden, these abstract mathematical techniques became less useful, and simulation models, based on the technology (if not the methodology) of artificial intelligence, have come to dominate this aspect of cognitive psychology. As powerful as these models are, and as hopeful as the outlook on understanding the true nature of cognitive processes has become as a result of these models, significant communication problems have appeared.

The communication problem that characterizes computer models cannot be solved as easily as that of mathematical models. It is very difficult to understand how a simulation model works without gaining experience in constructing models of this type, and this requires access to extremely expensive computer systems, as well as training in a particularly technical area. This problem is compounded by the multiple and incompatible versions of LISP or other languages that are convenient for designing AI systems: It is very difficult for programs to be exchanged between researchers at different institutions that use different computers, or even different operating systems running on the same computer. Even with appropriate tools and experience, the difficulty of understanding another person's program is well known among programmers of any computer language. It may be possible for a program to be a reasonable instantiation of a theory, but it would be completely unreasonable to publish a copy of the program's code and rely upon the reader's ability to infer the properties of the theory from this listing. Practically speaking, this is rarely done anyway; instead we have returned to describing these models verbally, and the opportunities for hand waving and fuzzy thinking (or at least fuzzy descriptions) have returned as well. In fact, there is an even greater opportunity now for hand waving, because potentially significant aspects of a model can be implemented by programming tricks that bear no relation to what is known about cognition. Has a new Tower of Babel arisen through our own doing?

So far, we have only complained about what is wrong with cognitive science, trying to trace the reasons that lead some of our colleagues (in all disciplines) to complain that cognitive science is in fact something less than a true science. Now, we shall try to offer some constructive suggestions—about the role of experimentation in cognitive science, about the development of a methodology for taking introspective reports, and most important of all, about the nature of theorizing in cognitive science.

METHODOLOGICAL SUGGESTIONS

Sometimes An Experiment Might Just Be The Right Thing to Do!

Insofar as cognitive psychology is a part of experimental psychology, it shares with the latter its preference for the use of the experimental method. In linguistics and AI, the psychologist's preference for experimental data seems to be regarded as perverse, sometimes with benevolent tolerance, sometimes with ridicule. One of the questions we want to raise here concerns the use of the experimental method—when is it proper, when is it necessary, what does it have to offer to cognitive science? What are its limitations?

The first point to be established is that introspection alone is not enough. For many psychologists, the traumatic collapse of introspective psychology at the beginning of this century is unforgettable, and they hardly need to be convinced. Linguistics, AI, and philosophy could learn many arguments against introspection from the history of psychology, but learning from history is a surprisingly rare occurrence. If people can easily be mistaken about their own states of pain (Dennett, 1978), it is hardly advisable to rely completely on introspection. Furthermore, many of the processes cognitive scientists are interested in are unconscious ones, such as comprehension and motor behavior. Similarly, the elementary laws of learning and forgetting simply do not lend themselves to introspective study, and we are forced to use experimental methods for their exploration.

The question then becomes when and how to use experimental methods. The attitude of some experimentalists, that nothing is proven unless there are some experimental results about it, is surely unacceptable, and probably one of the reasons why so many non-psychologists reject experimentation altogether. There are, indeed, too many trivial experiments, but that is a poor reason for neglecting a powerful methodological tool. Experimentation is the potential contribution that psychology can make to cognitive science.

Experiments can be used for two somewhat different reasons: for exploration and for hypothesis testing. Experimental psychologists, traditionally have been taught to use experiments almost exclusively for the latter purpose. One formulates a hypothesis, sets up an experiment to test it, and then decides whether or not the results of the experiment permit the hypothesis to be rejected. Very sophisticated procedures are available for designing experiments properly, and there exists an extensive inventory of statistical techniques to analyze their results efficiently and effectively. There are numerous experiments of this type that have made substantial contributions to our knowledge. But the hypothesis testing method has been greatly oversold to psychologists. Many hypotheses tested with great skill and effort are not worth testing: they are uninteresting, trivial, isolated from any theory, or tenuously related to the interesting aspects of the theory, or they are simply superfluous because we already know the outcome.

The use of experiments as systematic exploration procedures, where introspection or field observation fail, is less standard. Sometimes, it is useful to set up "interesting" experimental conditions, just "to see what happens." Sometimes, an investigator may want to explore systematically the boundary conditions of a phenomenon. Sometimes, he is merely concerned with demonstrating control over a phenomenon. The machinery of hypothesis testing is irrelevant in such cases (though not infrequently pseudo-hypotheses are invented by the successful investigator after the fact). But experimental exploration is becoming more and more important as cognitive science enters fields where the problems are unsuitable for introspection, the first crop of problems that lent themselves to introspection having been exhausted.

What needs to be developed more systematically is an experimental methodology for exploration. The hypothesis testing paradigm is well established, but its reflexive use by some psychologists has contributed to divorcing that discipline from important scientific developments. On the other hand, experimental methodology is underdeveloped in precisely those areas where it is needed most: the systematic exploration of new research issues, and the global evaluation of theories. More will be said about this later.

Protocol Analysis: Disciplined Introspection

In parallel with the further development of experimental methodology, a great deal of attention should be devoted to introspective methods. How introspection is used makes a difference. It is possible to be misled, as the history of psychology shows. At this point, some very promising beginnings for a post-behavioristic introspective methodology have been made, as described by Ericsson and Simon (1980).

Verbal reports are data in the same sense as key presses, response latencies, patterns of eye fixations, and sequences of hand motions. Thus, verbal reports are to be explained by correspondences between observation performance and predictions of a model just like any other class of data. In addition, some investigators argue that they are a preferred source of information about cognitive processes because of the richness and the density of information that can be obtained from verbal reports concurrent with the performance of a complex task.

Ericsson & Simon's classification scheme identifies two dimensions of verbal reports. The first is the time of verbalization. This dimension distinguishes information that is reported while in the focus of attention (i.e., while being held in short-term memory) versus information reported after completion of the experimental procedure. The other dimension concerns the relationship between the attended to or retrieved information and the content of the requested verbal report. This relationship can range from a direct report of the information to reports requiring complex or ill-specified transformations of this information.

Ericsson and Simon provide a theoretical analysis of the processes involved in generating various kinds of verbal reports, making use of widely accepted assumptions concerning the structure and dynamics of a human information-processing system, in particular, short- and long-term memory. Data take on meaning in the context of a particular theory of the process under study. Relationships between measurements taken in the laboratory and entities specified by the theory are defined by that same body of theoretical ideas. What Ericsson and Simon have shown is that our current body of theoretical knowledge in cognitive psychology is powerful enough to put our use of verbal reports as data on a sound theoretical and logical foundation.

When we talk about introspective methodology, the main concern is when and how far we can trust protocols. But, as we have just seen, some progress is being made in that respect. A more serious problem is the matter we discussed earlier, the occasional tendency of investigators to confuse the content of a collection of protocols with a model of the processes that generated the protocols. What is wrong with such confusions is that they lead to bad theories. Intuitive descriptions of cognitive processes, whether derived from self-observation or the transcribed protocols of experts, are not necessarily contributions to our understanding of these processes. The objective of cognitive science is the development of a rigorous body of theoretical knowledge about such intuitions. We cannot let the intuitions substitute for principled theoretical explanations of the observed phenomena.

Ecological Validity: Why "Artificial" Is Not Always Bad

Perhaps the main reason why nonpsychologists (and some psychologists as well) question the usefulness of the experimental method has to do with the apparent lack of ecological validity of experiments. Indeed, it is not much of an exaggeration if we say that any good experiment is unnatural and lacks ecological face validity, for in order to be good, an experiment must be well controlled, and that simply can't be done without imposing artificial, unnatural constraints. This point has been made so often from within psychology (Brunswick, 1956; Neisser, 1976) and without (Schank & Abelson, 1977), that it hardly needs elaboration.

There is no question that it has some merit. It is easy to point to experimental investigations of simple laboratory tasks that are concerned almost entirely with task–specific strategies that the experimental subjects generated in response to the peculiar demands of this particular laboratory task. Outside the laboratory, in more natural contexts, these strategies may play no role whatever. The results of such experiments are indeed of questionable significance, because the task that was studied lacks interest in itself, and the strategies that were discovered are not generalizable.

Nevertheless, this criticism can easily be overdone. On the one hand we must point out that there has been a strong trend within cognitive psychology in recent years to bring tasks that are of intrinsic interest into the laboratory (e.g., Kieras & Polson, 1982). On the other hand, and perhaps more importantly, it is not necessarily the case that ecologically invalid, artificial experiments cannot contribute to ecologically valid theories. The question is really what is studied in the simplified, artificial experiment: some general principle of information-processing, or some specialized strategy that the subject develops to deal with the task at hand. (Strategies are important to study in ecologically valid situations, but they are of little interest in artificial, trivial environments.) It is certainly not the case that experiments must yield useless (or at best esoteric) results. We shall try to make this point with an example.

Consider current knowledge about reading as a perceptual decoding process. Understanding in this field has reached a high level (Perfetti & Lesgold, 1977);

theories are well-worked out and are having a considerable practical impact (Resnick & Weaver, 1979). Indeed, there are claims that we know about as much as we need to know about teaching reading in the first few grades and that the problem is now one of putting this knowledge into practice (Bateman, 1979). Where did that successful theory come from? It is based upon an almost 100-year-old, rich tradition of laboratory experiments, employing for the most part a highly unnatural device, namely the tachistoscope. It is indeed surprising that this research has anything at all to do with 'real' reading. As Carr (1981) points out, even the time scale is off: in word identification experiments the activation of a word's meaning requires only about 30–40 msec, whereas the normal duration of fixation in reading is about 200–300 msec. How easy would it have been to reject a priori any relevance of these experiments to reading! And where would we be now without these 100 years of experimental work? That during these 100 years quite a few unnecessary experiments were done (from our perspective, enjoying the benefits of hindsight) is not important: no scientific method can guarantee steady progress along a straight line.

What makes an experiment ecologically valid or not is a more complex issue than the detractors of the experimental method admit. It is certainly one that must be raised and examined carefully. But we can't simply reject an experiment because it is unnatural. A frequent horror story that is told about the experimental method involves the verbal learning research employing nonsense syllables and lists of unrelated words, that supposedly was all in vain. One can argue, however, that the memory theory that resulted from that work is quite adequate, though limited in scope, and superior to current attempts to intuit the principles of memory (Schank, 1980). Specifically (Kintsch, 1982), classical memory theory, based for the most part on list learning experiments, provides a fairly good account of memory for text; it is not at all the case that we have to forget "that old theory" (Jenkins, 1974) and start again from scratch when dealing with the more complex, ecologically significant phenomenon of discourse memory.

What has been said here about ecological validity with respect to experiments holds as well for other scientific methods, for example, observational procedures. In the case of language, which should we observe: a complex, ecologically valid phenomena (a real discussion between two mathematicians) or a set of highly selected, artificial examples ("John was hit by Mary")? As Chomsky (1980) points out, other sciences did rather well by paying close attention to some quite unnatural paradigms. The issue is an important (and divisive) one within linguistics, where we have on the one side a deep concern for ecological validity, exhibited by studies of actual conversations in their social context (Franck, 1980), and, on the other side, a narrow focus on certain selected language phenomena, for which the work of Chomsky is the primary example. The principal argument against studying ecologically valid situations is that it can't be done (for Chomsky's pessimism, see Chomsky, 1975). That argument is plainly wrong. In and out of linguistics, complex, ecologically significant situations have been studied with considerable success. Although to some (e.g., Dresher & Hornstein, 1976), these results are not of true scientific significance because they have no bearing on their favorite theoretical issues, the problem there may be more with their choice of theory in the first place.

On the other side, the main argument against studying selected, special phenomena is that, in doing so, elegant but misleading theories are obtained, for in context everything is changed. As we have pointed out in our discussion of the ecological significance of experiments, this is a dangerous argument to make: while it is sometimes right, it would lead us at other times to reject very powerful simplifications. The trouble is that it is difficult to tell beforehand which simplification is the right one, and which will be misleading. Whether looking only at the syntactic form of sentences is a good or a poor research strategy is an empirical issue: we know it leads to elegant and "deep" theory, but whether it bears on ecologically significant issues remains to be seen.

THEORY CONSTRUCTION IN COGNITIVE SCIENCE

The Interaction Between Theory Construction and Experimentation

Discussions of the interaction between theories and experiments meant to validate those theories generally reduce to the question, "When should a model be tested?" Although this is not a bad question, it may inappropriately imply that experiments are run only to determine the validity of a theory. There are in fact several ways in which experimentation can guide the development of a complex theory, in ways short of absolute acceptance or rejection.

Consider the development of a model from its beginning as an idea to the ultimate evaluation of the model as an accurate psychological explanation of its phenomenon. There are three general stages through which a model passes, and, correspondingly, three ways in which experiments can be useful.

Initial Theory Formulation.   When a theory is first formulated—when it is little more than an idea—experiments can be used to comment on the general structure of the idea, and determine, at a global level, whether or not it is configured in a psychologically plausible way. These experiments can do little more than suggest that the idea appears to be on the right (or wrong) track. It is probably not appropriate to say that they "test the theory," simply because "testing" implies that the theory is sufficiently detailed to make specific predictions that can be compared to data, that may not be the case at this stage in the theory's development.

In practice, this stage has typically been the place for "experiments by example," that are sometimes indistiguishable from introspeetion. Examples play an important role in AI; one often hears that "the best kind of experiments are those that you don't even have to run, because you know how they'll come out just by thinking about them." While these "thought-experiments" should be used with caution, a good set of examples can be a useful tool in the construction of a theory, by identifying the phenomena with which the researcher is concerned and isolating for study a manageable part of a complex domain.

Explicit Process Models.   This is the final stage of the theory construction process (the middle stage of this process will be discussed next). Here, an explicit psychological model of some process exists, one that is designed in such a way that quantitative predictions about data can be made. This is where traditional experimental psychology shines, by providing elegant techniques for fitting a model to data, or using data to decide from among a number of models. The times when these techniques are unambiguously applicable are somewhat limited: enough psychological research needs to have been done to have produced some consensus about what concepts and findings are critical to the phenomena, so that the goals of the evaluation procedure are well-defined.

However, the presence of an explicit model does not necessarily restrict one to quantitative evaluation: not every aspect of a model *needs* to be tested. This is one of the more powerful aspects of simulation modeling. In some of our own research on prose comprehension (Miller & Kintsch, 1980), certain parts of the experiments' recall and reading time data were not fit by the model's predictions as well as others. The fine-grained way in which the process of prose comprehension was described by this model allowed us to determine the causes of these shortcomings without recourse to further experiments. This should be contrasted with a more abstract theory of comprehension, where the processes being studied can be diagnosed only through several layers of generalization and through experiments that reach beneath this generalization to reveal the flaws in the theory.

The Middle Ground.   Here, one has a theory in which structures and processes have been specified in sufficient detail to support both the instantiation of the theory as an executable computer program and qualitative experimental studies of the theory. Bower, Black, and Turner's (1979) study of script theory is a good example of this use of experimentation. When these experiments were run, the early frame and script proposals had been sufficiently fleshed out that specific qualitative predictions could be made and tested. The fact that experiments may take place at the same time that the theory is being instantiated as a program does not mean that the experiments must be an evaluation of a particular instantiation. There are often good reasons for tying the experiments to the original theory, especially the difficulties of deciding whether a particular implementation is a good instantiation of the theory, and whether a certain part of a program is theoretically important or a trivial implementation detail. Note that Bower et al.'s experiments followed this course, studying not the details of existing script-based story-understanding programs, but more abstract proposals, such as the sharing of parts of representations by related scripts and the ordering of events within scripts.

This middle ground is the right place for the AI researcher to consider the correspondence between his model and well-established psychological constraints. Some examples of this can be seen in the relationship between the limited capacity of of human working memory and two AI projects: the story understanding system SAM (Cullingford, 1978) and the speech understanding system HEARSAY (Newell et al., 1973). When SAM generated inferences from a story and its scriptal knowledge, the lack of constraints on its inference generation process meant that many of these inferences were irrelevant to the comprehension task at hand. Similarly, the early versions of HEARSAY generated a vast number of hypotheses about the content of an utterance, far more than it could efficiently evaluate. What both of these systems needed—and what both systems or their descendants ultimately acquired—was a knowledge-driven way to focus their attention on the particularly important parts of the problem: in effect, a set of sophisticated strategies for the control of working memory. In retrospect, both projects would have benefited by considering this psychological principle from the start.

This middle stage is also particularly important for psychologists, because the properties of a model in this stage of development are generally detailed enough to serve as at least a possible explanation of how people perform this task. However, psychologists should be wary of the temptation to rush such a model off to the laboratory for rigorous quantitative evaluation. Not only may the time not be right for such work, but such efforts may obscure the fact that these models can often be of interest and importance to psychologists *regardless* of whether they can be demonstrated to be psychologically valid. Consider, in this light, Winograd's (1972) language understanding system. Psychologists interested in language had much to gain by examining this program, but comparing the system's comprehension times (in terms of cpu cycles or some other measure) to the comparable times for people would have served little purpose. Psychologists need to remember that a program can be an experiment in the same way as can bringing a group of people into a room and collecting data: it may not be meant to discover any universal truth, but merely to find out whether some process works the way you think it does.

Because research in many parts of cognitive science lies in this middle ground, we should ask whether there is anything that can be done to facilitate the interaction among the member disciplines, and to further a joint understanding of their methodologies. Understanding a foreign methodology is not an easy task, but it is one that is critical to cognitive science, as many of the arguments in this field can be traced to such misunderstandings.

Formal Logic, Or Is The Brain A Kludge?

Cognitive scientists differ widely in the tools that are preferred for use in theory construction. On the one hand are scientists who believe that theories must be elegant, and must exploit the powerful formalisms of mathematics and logic as much as possible. On the other are theorists who feel that these techniques are not suited for the kind of problems studied by cognitive science, and that we have to develop new formalisms capable of serving our purposes.

The conflict between logic-based theories and alternative approaches, (e.g., schema theories) goes through all branches of cognitive science. For AI, Kolata (1982) has highlighted this issue by contrasting the approaches of Minsky and McCarthy. McCarthy places his trust in first order mathematical logic. He agrees that ordinary mathematical reasoning differs from common sense reasoning, but sees his task as one of modifying the ordinary system to expand it so that it is able to deal with common sense reasoning. This expansion is obviously not a trivial task, but McCarthy prefers its intricacies to the idea of abandoning such a proven, powerful instrument as mathematical logic.

Minsky proposes exactly such an abandoment because he believes that the concern of logic with 'facts' and 'truth values' is simply inappropriate for most problems in AI. Instead of an elegant, precise formalism, he suggests that "the human brain is a kludge," that cognition is inherently messy, and must deal with many different techniques and methods.

As another example, consider the field of language comprehension. The formal semanticists, with their elegant, sophisticated models, and their modest problems and restricted perspective, stand in contrast to another group of researchers from various disciplines, who choose to study a wider range of linguistic phenomena with "home-made" formalisms. There can be no question that these formalisms are inferior to logic or semantics as formal systems, but are they at least suitable for cognitive science research? Or must we wait until the logicians can extend their methods so that they can deal with the problems we are interested in? If Minsky is right and the brain is indeed a kludge, that day may never come.

Clearly, there is no hope of obtaining agreement among cognitive scientists on an issue like this. Nevertheless, we should raise such issues, and try to clarify the various positions as well as we can. It is not a matter of legislating what is right or wrong, but of knowing what we are doing, and why we are doing it.

ON THE POSSIBILITY OF COGNITIVE SCIENCE

At the most general level, there is some agreement on what cognitive science should be:

> Cognitive science is the study of the principles by which intelligent entities interact with their environments [Walker 1978, p. 3].

> to define cognitive science as the domain of inquiry that seeks to understand intelligent systems and the nature of intelligence . . . Intelligence systems exhibit their intelligence by achieving goals (e.g., meeting their needs for survival) in the face of difficult and challenging environments [Simon, 1980, p. 35].

Beyond this very general definition, there is very little, if any, consensus concerning a set of more specific goals and metatheoretical assumptions that could define a coherent field of inquiry. Walker (1978) continues, ''. . . To discover the representational and computational capacities of the mind and their structural and functional representations in the brain [p. 6].'' Simon (1980) continues ''We have learned that intelligence is not a matter of substance—whether protoplasm or glass or wire—but the forms that that substance take and the processes it undergoes [p. 35].'' Thus, even the initial attempts to refine and give substance to top-level goals lead to conflicts about the role of neuroscience in cognitive science as a whole.

The differences expressed in these quotes are not trivial. They are in fact quite representative. Every cognitive scientist tends to take the presuppositions of his or her own work for granted, and views them as expressions of the new synthesis that is supposed to define cognitive science as a whole.

A less imperialistic view of cognitive science emerges if one treats it merely as the collection of several pairwise intersections among anthropology, computer science, linguistics, neuroscience, philosophy, and psychology. Walker (1978) lists 11 subfields formed in this way, ranging from neurolinguistics to cognitive anthropology. Each of these fields is already well established, many with their own journals and conferences. Membership in one of these subdisciplines by some definitions, makes one a cognitive scientist.

Is that all there is to cognitive science? It is surely not the case that if you are a psycholinguist then you are, ipso facto, a cognitive scientist, simply because many psycholinguists would reject that label. What is missing here is a consideration of the autonomy of the various subfields. There exist linguists who insist on studying language as an object in itself, divorced from the realities of language use. Whatever a psycholinguist would tell them about how people deal with anaphora is of no interest to them—that is, of no professional interest; they might very well agree that it was an interesting bit of psycholinguistics. For them, their field of study is autonomous, with its own questions, methodology, and criteria for evaluation. Indeed, even within traditional academic disciplines such insistence on the autonomy of subfields is not uncommon. Within linguistics, to continue with our example, many have postulated the autonomy of syntax.

The cognitive scientist is to the psycholinguist as is the latter to the linguist.

The difference is in what he considers relevant to his inquiry. For the grammarian the object of inquiry is syntax, and complex problems that go beyond that point are not to be asked. The semanticist argues that the problem thus narrowly defined is intractable and must be broadened. The psycholinguist adds another dimension of complexity. The study of language is now embedded into questions about language use. This embedding does not necessarily negate the autonomy of the more narrowly defined discipline—for example, there may still be an autonomous syntax that is more or less directly related (and relevant) to psycholinguistics. Cognitive science represents another step in this broadening of the problem definition. The psycholinguist is a cognitive scientist insofar as he or she explicitly accepts the relevance or problems and results from fields other than linguistics and psychology (e.g., a theory of anaphora use may be rejected because of unacceptable implications derived from computational linguistics). Once more, the question about the autonomy of the more specialized fields need not be answered unequivocally. The cognitive scientist is ready to work in his complex, highly interactive discipline, but it would be presumptious to prescribe to others their choice of problem definition.

If, as suggested here, cognitive science is a collection of autonomous subfields, then considerations of methodology become of paramount importance because we need to communicate and understand each other. If cognitive science is to succeed, we have to develop criteria for the evaluation of work as cognitive science. Right now an unfortunate amount of research that is not good enough for AI, linguistics, or psychology audiences tends to get passed off as cognitive science. We have tried to explore the criteria for a sound cognitive science. In terms of experimental methodology, protocol analysis, and principles of theory construction, adequate criteria exist. We hope that this volume will help to explore them more fully, and to make cognitive scientists more aware of the importance of these criteria.

REFERENCES

Aebli, H. *Denken: Das Ordnen des Tuns.* Stuttgart: Klett, 1980.

Bateman, B. Teaching reading to learning disabled and other hard-to-teach children. In L. B. Resnick & P. A. Weaver (Eds.), *Theory and practice of early reading.* Vol. 1. Hillsdale, N.J.: Lawrence Erlbaum Associates, 1979.

Bower, G. H., Black, J. B., & Turner, T. J. Scripts in memory for text. *Cognitive Psychology,* 1979, *11,* 177–220.

Brunswick, E. *Perception and the representative design of psychological experiments.* Berkeley: University of California Press, 1956.

Caccamise, D. J. *Cognitive processes in writing: Idea generation and integration.* Ph.D. Thesis, University of Colorado, 1981.

Carr, T. H. Research on reading: Meaning, context effects, and comprehension. *Journal of Experimental Psychology: Human Perception and Performance,* 1981, *7,* 592–603.

Chomsky, N. *Reflections on language.* New York: Pantheon, 1975.

Chomsky, N. Rules and representations. *The Behavioral and Brain Sciences,* 1980, *3,* 1–61.

Cullingford, R. E. *Script application.* Technical Report No. 116, Computer Science Department, Yale University, 1978.

Dennett, D. C. *Brainstorms.* Montgomery, Vt: Bradford, 1978.

Dresher, B. E., & Hornstein, N. On some supposed contributions of artificial intelligence to the scientific study of language. *Cognition,* 1976, *4,* 321–398.

Ericsson, K. A., & Simon, H. A. Verbal reports as data. *Psychological Review,* 1980, *87,* 215–251.

Franck, D. *Grammatik und Konversation.* Königstein, TS: Scriptor, 1980.

Hörmann, H. *To mean—to understand.* Berlin: Springer, 1981.

Jenkins, J. J. Remember that old theory of memory? Well, forget it! *American Psychologist,* 1974, *29,* 785–795.

Kieras, D. E., & Polson, P. G. *An approach to the formal analysis of user complexity.* Project on User Complexity of Devices and Systems Working Paper No. 2. 1982.

Kintsch, W. Memory for text. In A. Flammer & W. Kintsch (Eds.), *Discourse processing.* Amsterdam: North Holland, 1982.

Kolata, G. How can computers get common sense? *Science* 1982, *217,* 1237–1238.

Miller, J. R., & Kintsch, W. Readability and recall of short prose passages: A theoretical analysis. *Journal of Experimental Psychology: Human Learning and Memory,* 1980, *6,* 335–354.

Neisser, U. *Cognition and Reality.* San Francisco: W. H. Freeman and Company, 1976.

Newell, A., Barnett, J., Forgie, J., Green, C., Klatt, D. H., Licklider, J. C. R., Munson, J., Reddy, D. R., & Woods, W. A. *Speech understanding systems: Final report of a study group.* Amsterdam: North-Holland, 1973.

Perfetti, C. A., & Lesgold, A. M. Discourse comprehension and individual differences. In P. Carpenter & M. Just (Eds.), *Cognitive processes in comprehension.* Hillsdale, N.J.: Lawrence Erlbaum Associates, 1977.

Resnick, L. B., & Weaver, P. A. *Theory and practice of early reading* (Vol. 1). Hillsdale, N.J.: Lawrence Erlbaum Associates, 1979.

Schank, R. C. Language and memory. *Cognitive Science,* 1980, *4,* 243–284.

Schank, R. C., & Abelson, R. P. *Scripts, plans, goals, and understanding.* Hillsdale, N.J.: Lawrence Erlbaum Associates, 1977.

Simon, H. A. Cognitive science: The newest science of the artificial. *Cognitive Science,* 1980, *4,* 33–46.

Walker, E. (Ed.) *Cognitive Science, 1978:* Report of the State of the Art Committee to the Advisors of the Alfred P. Sloan Foundation, October 1, 1978.

Winograd, T. *Understanding natural language.* New York: Academic Press, 1972.

ARTIFICIAL INTELLIGENCE

Ever since its beginnings, research in artificial intelligence has been intertwined with that of psychology. This interaction was one of the prominent forces in the establishment of cognitive science, but it has not always been a peaceful one. This has largely been due to some confusion about precisely what AI is, and whether it is possible to do AI without also doing psychology.

Some highly constrained domains exist in which seemingly "intelligent" behavior can be obtained through computer algorithms that bear no resemblance to human cognitive processes. Today's most successful chess programs achieve their success not so much by simulating the knowledge-based strategies of chess masters, but by utilizing specialized algorithms that evaluate as many moves as possible in as little time as possible; the more moves a program can consider, the more likely it is to win. However, these algorithms are less effective in domains such as prose comprehension and problem solving, where knowledge about the domain is the component that is critical to successful performance. These domains can be so unconstrained and dependent upon the manipulation of domain knowledge that the author of a successful AI program may have no alternative but to model his program after the cognitive structures and processes of a skilled human working in the same domain.

This modeling of human processes can just be a matter of expediency—people offer a plausible set of strategies that can be exploited by the AI researcher and mixed with other techniques that are simply good computer science to yield a program that achieves the researcher's goals. Alternatively, the researcher may believe that the program offers a sufficiently accurate portrayal of the cognitive strategies people apply to this task so that the resulting program can be considered a psychological model of that process. The chapters in this section are primarily concerned with these issues, especially those detailing when and how the conditions of psychological validity are met, and what advantages are offered by these different approaches to the study of cognition.

The paper by Wendy Lehnert focuses on two aspects of the relation between AI and psychology: how the construction of an AI-based model can be compared to psychological experiments that are concerned with similar questions, and what breadth and depth of cognitive processes must be represented in a competent model of some domain. Given the relevance of domain knowledge to the understanding of complex cognitive processes (whether implemented as computer programs or not), William Clancey discusses the different kinds of knowledge that are needed for a particular task: what kinds of knowledge are relevant, how they might be represented, and what techniques might be used to discover them. Finally, John Haugeland summarizes these two chapters by describing some of the theoretical and metatheoretical contrasts that appear throughout them, and throughout the field of cognitive science as a whole. Consequently, these first three chapters address many of the themes discussed in this volume's introduction, (Chapter 1) particularly those concerned with the distinctions between programs, models, and theories, and the questions of what cognitive science really is and how research in this multidisciplinary field can best be done.

2 Paradigmatic Issues in Cognitive Science

Wendy G. Lehnert
University of Massachusetts

CONCEPTS OF SCIENTIFIC INVESTIGATION

I recall from my early school days a daily routine that was divided among subjects such as English, social studies, math, and science. The "science" hours moved through disconnected cycles: 3 weeks on astronomy, 1 week of geology, 6 weeks devoted to biology, and so on. At that stage I conceptualized the scientific disciplines in terms of their topics. Astronomy dealt with heavenly bodies, geology was concerned with rock formations, and biology took care of living things. If I thought at all about the investigative tools involved, it was only to notice that telescopes were sort of like microscopes, and all the sciences appeared to require some type of equipment.

I bring this all up because I assume we have all been educated under similar circumstances, and I suspect that our first impressions of science die hard. For example, most of us think of mathematics as a "special" sort of science (if a science at all) because it requires such modest tools. Anyone with paper and pencil can participate. Mathematics is also thought to be less "experimental" for this reason as well; we naturally associate experiments with tools and physical observations.

A Search for Universals

Many of us (including professional scientists) never progress far beyond this "topics and tools" view of science. Scientific disciplines are characterized by the topics they investigate and the tools these investigations require. In fact, topics and tools can be viewed as "universals" in the world of science. We all

21

assume that such universals exist, although we may never think about them per se. The very term "scientific standards" suggests that all the sciences share some universal methodology. For example, it is standard to assume that the results of a valid experiment are reproducible. "Scientific premises" constitute another universal, although specific premises vary across disciplines. Most sciences share the premise that time is unidirectional, but a cosmologist who assumes the irreversibility of time is making a big mistake.

Scientists are extremely conscious of their topics, tools, methodologies, and premises. These universals are taught in courses and discussed at length in seminars. But there is another scientific universal that is rarely examined in the classroom (unless perhaps one is taking a course in the philosophy of science). This is the notion of a scientific "paradigm"—by which I mean nothing less than special ways of approaching and solving problems.

The word paradigm comes up rarely in the conversation of scientists who share common topics, tools, methodologies, and premises. But it appears frequently in interdisciplinary dialogues, as in "this makes sense *only* if you're working in *that* paradigm." It has come to be sort of a catch-all explanation for a suspiciously large number of communication failures.

Paradigm Conflicts

The most interesting paradigm discussions arise when it is not clear which tools, methodologies, and premises should be utilized for a given topic of inquiry. Consider the topic of schizophrenia. In the paradigm of neurophysiology, schizophrenia is assumed to be a problem of brain chemistry that results in characteristic behavioral patterns. But in the paradigm of clinical psychology, schizophrenia can be viewed as a purely psychological phenomenon, that happens to involve some characteristic pathology of brain chemistry. Until the question of causality is answered, it is impossible to know which paradigm holds the proper premise about schizophrenia. Perhaps both premises are correct: many researchers now believe that schizophrenia occurs when a physiological predisposition is combined with a psychological problem. If this holistic view turns out to be correct, past paradigm conflicts over schizophrenia should evaporate into a united attack on the problem.

In the meantime, some thought-provoking debates are taking place about the nature of mental diseases, techniques of behavioral modification, the relationship between a disease and its symptoms, and our usual notions of mind and brain. Paradigm conflicts often kick up a lot of dust around ideas we would otherwise take for granted. On a philosophical level, these questions about disease and disease symptoms will remain worthy of discussion long after the specifics of schizophrenia are worked out. But the "front-line" fighters, the neurophysiologists and the clinical psychologists, are less likely to participate in such discussions after the dust has settled and their favored topic has been pigeon-holed once and for all to everyone's satisfaction. Most of the heat underlying professional debates is a territorial reaction about who owns what problems: when the threatening territorial forces dissipate, discussions cool off considerably. Yet it would be a mistake to reduce all of our complex interdisciplinary dialogues to sociobiological interactions. It sometimes helps to discuss the lessons we think we're learning, in the hope that others might learn less slowly.

My own personal lessons tend to focus on the boundaries between cognitive psychology and artificial intelligence (AI). Having spent some time in recent years on both sides of this particular fence, I have discovered some natural tensions that serve to separate these two areas in spite of their common concerns. If we limit our observations to research in natural language, it would seem that psychology and AI share a common topic (how language works), although differing with respect to investigative tools, methodologies, premises, and paradigms. Although the tools (human subjects for psychology versus computer programs for AI) might at first seem to be the likely source of all other differences, this chapter shows that the boundaries between psychology and AI are more accurately drawn in terms of paradigmatic differences. The tools by themselves are little more than a red herring.

Experimental Methodologies

When the topic of research is natural language processing, the ostensible goals of psychology and AI seem to be quite distinct: (1) psychologists want to study people as information-processing systems to explicate human language processing abilities; and (2) AI workers want to design adequate algorithms for computational language processing systems. Those who would identify themselves as "cognitive scientists" are liable to argue that these two goals subsume each other, but that is a premise that will not concern us at this time. For the moment we will restrict our attention to the methodologies that both groups are practicing in service of their research goals. These two fields foster seemingly different methods of empirical investigation, yet their basic strategies for testing a theory have striking similarities.

First, consider the methodology of an experimental psychologist. Given some information-processing phenomenon as the object of investigation, a psychologist moves through the following steps:

1. Propose a theory to explain the phenomenon.
2. Design an experiment to test the theory.
3. Run the experiment.
4. Analyze the experimental data.

One of two things can happen at the completion of Step 4. Either the data confirms the theory and the experiment is "successful," or, the data is not consistent with the theory. In the latter case, it is appropriate to go back to either: a) Step 1 to revise or reformulate the proposed theory; or b) Step 2 to revise a faulty experimental design. (It may also be appropriate to revise a faulty design if the experiment is successful, but design faults do not force themselves on an experimenter when his hypothesis appears to have been confirmed.)

Now consider the research methodology of artificial intelligence. Given some information-processing phenomenon as the object of investigation, a researcher in AI moves through a similar set of steps:

1. Propose a theory to explain the phenomenon.
2. Implement the theory in a computer program designed to simulate the phenomenon.
3. Run the program.
4. Analyze the program's output.

As before, one of two things can happen with the completion of Step 4. The program output may simulate the target phenomenon and the program is "successful," or, the program's behavior is not consistent with the target phenomenon. In the latter case, a program may fail in one of two ways. If the program fails to execute because of programming errors—"bugs" in the program—we have a technical failure. Such failures are of no theoretical interest and can be fixed. But if the program fails to execute because the theory is inadequate, or executes but fails to simulate the target phenomenon, then we have a failure of theoretical significance. In this case it is appropriate to go back to Step 1, and revise or reformulate the proposed theory. Bugs are fixed by returning to Step 2 and ironing out the programming errors.

The parallels are obvious. Computer programs confirm or contradict hypotheses just as experimental data confirm or contradict hypotheses. Technical problems at the level of getting a program to run correspond to technical problems at the level of experimental design. But most important is the procedural loop: programs that fail in an interesting way are perfectly analogous to experiments that fail in an interesting way. In each case the experimenter learns something about his hypothesis that would not have surfaced without running the experiment (program).

Given this view of the two methodologies, it may appear that the only difference between psychology and AI are the tools used to test a theory: psychology uses subjects in experiments, whereas AI uses programs on computers. But the tools alone cannot account for all the differences we see in these two research areas. For example, a computer can be advantageously employed to test descriptive theories of a stochastic nature. But such undertakings are normally proscribed by AI practitioners[1] for reasons that are all too often lost on psychol-

---

[1] I am talking about natural language work here—there is plenty of non-descriptive statistical analysis in other areas of AI (e.g., vision processing).

ogists. Not all computer programs are valid research tools in AI but the "topic-tool" approach to science is not refined enough to explain what is and is not an AI program.

Another interesting difference concerns end results. In psychology, a successful experiment stimulates ideas for one or two or ten more experiments. But when a psychology experiment fails, that failure does not normally shed light on the theoretical problem being investigated (except to possibly rule out the hypothesis being tested). One either improves the design of the experimemt, or rejects the hypothesis altogether. Yet in AI, the "theoretical" failure of a computer program is extremely instructive. In fact, AI workers habitually look to programs that fail in order to learn more about the theoretical ideas they are investigating. While a successful program undeniably stimulates interest outside AI, and AI worker will always be watching for the things that the successful program still can't do. These limitations and failures are at the heart of the AI enterprise.

At least in the case of natural language-processing, psychology and AI share a common topic and somewhat similar methodological procedures. Many advocates of cognitive science have noticed this and have suggested that the two fields unite to develop a deeper understanding of cognitive processes by sharing their investigative tools: psychologists should write computer programs and AI workers should run experiments. Being one of these people (indeed—having tried it myself), I can appreciate how demanding this path is. On a very mundane level, time is a limiting factor. It takes a lot of time to write programs and design experiments, let alone manage both. But more importantly, the lessons we need to share will not surface if the psychologist writes a program that is "not AI" or the AI worker gathers experimental data that cannot be analyzed. Some appreciation of differing premises and paradigms is in order here, and to achieve this appreciation we must get beyond the simplistic "topic-and-tools" view of psychology and AI.

THE ROLE OF TASK ORIENTATIONS

As we saw in the last section, the methodologies of psychology and AI can be characterized by similar discovery procedures. Whereas both fields utilize a hypothesis-test-revise loop as a standard mode of operation, they each bring a separate set of assumptions or premises to this methodology. These premises have been shaped to some extent by the tools of each discipline, but the premises also go beyond the limitations imposed by their tools. Consider, for example, the issue of a task orientation.

When an AI worker goes about investigating an idea (say, an idea about the organization of human memory), the first step is to find a suitable task orientation. What input will the system process? What output will be produced? If the problem is human memory organization, what I/O (input–output) behavior will provide a suitable environment for the investigation? All computer programs must be characterized in terms of desired I/O behavior, so a commitment to specific I/O behavior constitutes a critical first step. The term "task orientation" is used to describe decisions about targeted I/O behavior, and all AI workers treat the commitment to a task orientation as a necessary premise of their research effort.

Many potential task orientations are rejected for one of two reasons:

1. Input cannot be adequately specified.
2. Output cannot be adequately specified.

For example, suppose we wanted to investigate all the ways that people are reminded of painful memories. We could say that the output of the task orientation is a pointer to a painful memory, but what is the input? The potential range of input for this task is so vast (visual images, auditory stimuli, memories of related events, experiences resulting in related emotional reactions, etc., etc.) that we are immediately forced to narrow the question to a more tractable input set, if such a restriction is possible.

Alternatively, suppose we want to investigate memory representations for narratives. Now the input is not troublesome (narrative texts are well specified), but the output is problematic. How do we evaluate the validity or adequacy of an internal memory representation? What criteria can be invoked to argue for or against the myriad possibilites? In this case, we must look for a second task orientation to provide us with suitable output. For example, question answering behavior or paraphrase behavior takes the internal memory representation as input and gives us specifiable output. If we can chain two task orientations together to produce paraphrases of an input text or answers to questions about an input text, then we can demonstrate the adequacy or limitations of the underlying memory representations, a hidden component of human information-processing. By feeding output from the understanding phase as input to the paraphrase or question answering phase, internal memory representations become explicit components in a chain of task orientations.

Hypothesis Testing

Suitable task orientations doubtlessly influence the types of questions that AI researchers ask. Questions cannot be too broad or too fuzzy for a clear I/O specification, and task orientations must often be chained together to build systems that can be evaluated in terms of human capabilities. Although the notion of a task orientation has been readily embraced by cognitive psychologist who work on information-processing problems and formulate their questions in terms of input and output, it is important to understand that psychology experiments can be conducted very nicely without any reference to task orientations at all.

For example, suppose a psychologist has reason to believe that people with blue eyes have slower reaction times than people with brown eyes. To confirm this hypothesis, a reaction time experiment must successfully show that people with blue eyes respond to a stimulus more slowly than their brown-eyed counterparts. If it works for one stimulus, it would be nice to see it work for a few additional stimuli as well, but the choice of stimulus items is not critical to the general hypothesis. This hypothesis is not dependent on a specific task orientation, it is only dependent on the speed with which stimuli are processed. Experimental subjects will unquestionably be performing some task (the task of responding to a stimulus), but this task does not constitute a task orientation because the hypothesis being tested could be tested just as effectively using different subject tasks.

If one tried to force this hypothesis into a task orientation framework, eye color becomes an input variable and the speed of the response becomes targeted output. But this I/O specification does not describe an information-processing task in the sense that AI workers speak of information-processing. Eye color and reaction times simply do not qualify as I/O specifications that can be manipulated or produced by an information-processor. They are more appropriately classified as biological factors that influence and characterize biological systems that happen to engage in information-processing. So while the hypothesis about eye color may be loosely described in terms of information-processing (people with blue eyes process information more slowly than people with brown eyes), an AI worker would not consider this relevant to symbolic information-processing. Confusion about what does and does not constitute "information-processing" may seem to be subtle or of little consequence, but it contributes to a number of heated disagreements, including the knee-jerk reaction that many AI people exhibit when they are confronted with a reaction time experiment.

To understand what an AI person means by "information-processing," one must understand that this expression is an abbreviation for "symbolic information-processing" which in turn refers to the notion of a computer as a general symbol manipulator (Simon, 1969). The hypotheses that can be tested in AI are hypotheses about rigorously-defined processes of symbolic manipulation: hypotheses have to be about algorithms. And of course, the best way to test an algorithm is to implement it. But keep in mind that all computer programs implement algorithms, and this does not make all computer programs AI programs. In addition to implementing an algorithm, an AI program should be "testing" its implementation in some non-obvious manner—usually in terms of a human simulation. It would be trivial to write a table look-up program that: (a) inputs an eye color and a subject task, in order to; (b) output a reaction time that is consistent with the results of various reaction time experiments. This is a computer program, but it is not an AI program because it does not implement a theory of symbolic information-processing for the subject task that was input to the system. It does not in any sense simulate the information-processing that a human subject executes while performing the experimental task.

Because AI workers are preoccupied with simulations of intelligent behavior in terms of symbolic information-processing, they are primarily interested in programs that try to simulate some aspect of an information-processing model. They are also interested in psychology experiments that shed light on such issues, but the light must be directed toward an algorithmic issue to be of value. The eye color of a biological system is not an algorithmic issue. Nor is the speed with which a biological system operates *unless* those reaction times are presented as evidence for or against some specific algorithm that is under investigation.

Whereas psychology can be conducted without recourse to task orientations and algorithmic hypotheses, experimental results in cognitive psychology are often dismissed by AI workers as being either totally irrelevant (there is no task orientation) or interesting but non-constructive (there is a task orientation but there is no algorithm to discuss). The best reception a psychologist can expect is one with the latter flavor unless special attention is paid to the algorithmic angle—a slant that is usually very difficult for anyone who has never written a computer program. Knowing this, many AI people are both receptive to the interesting experiments and furthermore willing to build whatever algorithmic bridges might be needed to bring the results home. Still, it is much harder for AI people to compensate for a lack of task orientation in an experimental hypothesis. This is similar to asking a fish to ignore the fact that he's been pulled out of the water. So the lack of a task orientation operates to filter out a number of otherwise constructive communications that psychologists try to direct toward AI workers. Ironically enough, the exact same issue also serves to impede communication in the other direction, although the obstruction in moving from AI to psychology arises from a class of task orientations that psychologists are generally trained to reject a priori.

Local and Global Task Orientations

Everyone in AI works with task orientations, but psychologists can design experiments that do not assume task orientations. Task orientations are not crucial for experimental psychology. Psychologists have other premises that constrain the questions they can ask. Most importantly, they must worry about the variables that enter into an experiment, and how to control for these variables. If an experiment has too many uncontrolled variables, its results cannot be interpreted in terms of a well-defined function.

Psychologists are trained to break problems into functions and variables. It is valuable to know which variables contribute to which functions in what ways. If reading comprehension (a function of many variables) is affected by the color of ink used in the experimental materials (one particular variable), this effect can be confirmed by controlling all the other variables and exhibiting significantly different comprehension levels across various experimental groups (using black ink, red ink, green ink, etc.). In a "clean" experiment all possible variables are controlled with the exception of one experimental variable to be manipulated by the experimental design. In a "messy" experiment a number of different variables can contribute to the experimental results.[2] Messy experiments occasionally occur because the experimenter didn't think of everything, but they can also occur when there is just no way to avoid the mess. In either case, the experimenter is liable to be chastised by his colleagues and have difficulty publishing his results. Reputable psychologists tend to steer clear of messy experimental designs unless they are obviously unavoidable and absolutely necessary. Many psychologists unconditionally avoid such experiments in any case as a matter of principle, and this extreme (but common) attitude obstructs many contributions that the AI community offers to psychology.

The difficulty can be traced to the relationship between task orientations and controllable variables. Some task orientations are clean in the sense that an experiment can be designed to control all the variables of that task. Associate-pair learning tasks are extremely clean task orientations, but other memory-related task orientations can be exceedingly messy. Sentence comprehension is messy because of the many variables involved in comprehension. If two subjects say that they understand a sentence, how can we be sure they have the same understanding? Are causal relationships perceived? Are deviant descriptions recognized as such? Do people understand, "Colorless green ideas sleep furiously," in the same way that they understand, "Exhausted young children sleep soundly?" How can the experimenter control for all the possible variations of semantic content? In addition to problems with experimental materials, how can an experimenter control for variations across subjects in terms of active vocabulary, passive vocabulary, idiosyncratic connotations, and personal associations? These are the variables of unavoidably messy experiments.

The psychologists' concern for cleanliness has led to what AI researchers consider to be an extremely restricted investigation of natural language processing phenomena. For example, the *Journal of Verbal Learning and Verbal Behavior* was dominated for years by experiments on list learning behavior. All the messy semantic variables had been factored out of these experiments by restricting experimental tasks to manipulations of nonsense syllables. Real words (along with real morphemes) were avoided in the name of clean psychology. The only thing linking these task orientations to human verbal behavior is the presence of phonetically feasible but otherwise disembodied syllables. Eventually, psycholinguists moved up to real sentences, but only for the sake of syntactic manipulations. All questions about semantic manipulation have been largely ignored until very recently. Within the last decade a few notable psychologists have been paving the way for more ambitious undertakings in language language research, and many have been careful to explain these departures from cleanliness in acknowledgments of disciplinary trends and conventions.

---

[2]"Messy" experiments are not to be confused with "sloppy" experiments where the experimental design might be fine, but the execution of instructions or data collection is loosely conducted.

I do not want to foster the belief that AI task orientations always foster messy experiments. Feigenbaum's implementation of discrimination nets in EPAM was inspired by paired-associate experiments, and presented completely within that psychological tradition (Feigenbaum, 1963). EPAM successfully simulated the human memory phenomena of oscillation and retroactive inhibition, and thereby achieved strong status in terms of psychological validity. More generally, EPAM illustrated the relevance that computational information-processing models can have for psychology, and the feasibility of psychological simulations in the AI tradition. But it is important to understand that EPAM's happy impact on psychology was enhanced by its clean task orientation, a feature that EPAM does not share with most AI programs.

From the AI perspective, a task orientation is roughly either local (fragmented) or global (self-contained). For example, free word association is local, whereas answering a question is global. Recognition for nonsense syllables is local, but word recognition is global. Verbatim recall for a sentence is local, but paraphrase recall for a story is global. Very short-term memory recall is local, whereas long-term recall is global. To distinguish local tasks from global tasks it is useful to try to answer the question, "How often do normal people engage in this task in their normal daily routines? Is the task subsumed by other tasks, or does it stand alone as a self-contained activity?" In general, local tasks are only encountered in psychology experiments, while global tasks are things people do every day. It follows that local task orientations are often described as being "artificial," "unnatural," or "odd," while global task orientations are "realistic," "natural," and "normal." Many AI workers are extremely uncomfortable with local task orientations, for reasons that are frequently misunderstood by psychologists.

¹ The field of artificial intelligence is ultimately concerned with the design of machines that perform intelligent tasks for people. Intelligent tasks that are useful to people tend to be global tasks, so AI is primarily concerned with global task orientations. It is also the case that computer scientists are typically schooled in the "divide-and-conquer" approach to problem solving: every large problem can be broken into smaller problems whose solutions contribute to the solution of the large problem. This is the only way to write a large computer program, and AI programs tend to be very large by anyone's standards. So AI people tend to be fairly experienced at this business of breaking down large problems—so much that they develop their own techniques for examining problems and finding decompositions that are advantageous from a programming perspective.

If life were simple, all the local task orientations would contribute to the solution of some global task orientation. Then it would make sense to work on all local task orientations because each advance at the local level would eventually be valuable at some global level. Unfortunately, some problem decompositions are not as useful or workable as others. AI people know this in general, although they often disagree among themselves about which breakdowns are preferable for a given problem. For example, there has been a long-standing controversy among natural language researchers about the autonomy of syntax and semantics. For years, many people concentrated on syntax problems in the belief that this would eventually hook up with semantic processing in some useful way. Now the trend is to integrate syntax and semantics: it may be possible to study syntax without semantics, but it is not so easy to reintegrate a purely syntactic analysis within a larger system. Similar difficulties arise when linguistic functions are excised from higher memory processes, and problems in inference are factored out of sentence comprehension.

Because AI people see these difficulties every day, they are very nervous about wasting time with local task orientations that have dubious value in the long run. Psychologists are not trained to think about global and local task orientations, so criticisms of this sort from the AI community are often misinterpreted as hostile and irrelevant attacks. Psychologists do not have to worry about the long-run in the same way that AI people do. So we have a potential impasse between AI and psychology that is further exacerbated by the fact that local tasks encourage clean experiments, whereas global tasks tend to breed messy experiments.

When a psychologist tries to talk to an AI person about a local task orientation, he will typically be met with a diatribe about the importance of some global task orientation that will never lend itself to clean experimentation. It is natural for the psychologist to try to redirect the conversation back to some nice clean (and local) task orientations. But of course this only feeds the fire under the increasingly perplexed and frustrated AI proponent. An interaction of this sort can only circle and escalate along counterproductive lines, creating bad feelings all around.

The depth of this particular impasse is considerable and deserves special sensitivity. For example, the introduction to a highly respected monograph in cognitive psychology [see Anderson and Bower (1973)] contains the following quote:

> It is commonplace that the Zeitgeist in current psychology opposes global theories such as the one to be presented. It is said, instead, that one ought to work on limited hypotheses for small, manageable problems—categorization effects in free recall, verification latencies for negative sentences, search of items in short-term memory, and so on ad infinitum. Indeed, we have been told by many respected colleagues in psychology that we will surely fail because we "are trying to explain everything." . . . In rejecting the earlier global theories, modern research on human memory has overreacted to the opposite extreme; it has become far too narrow, particulate, constricted, and limited. There is no overall conception of what the field is about or even what it should be about. There is no set of overarching theoretical beliefs generally agreed upon which provide a framework within which to fit new data and by which to measure progress [p. 1].

An AI person who reads these words might conclude that the author understands all about local–global tensions in task orientation decisions. It would be difficult to read these words in any other way. The reference to "earlier global theories" sounds like a reference to Bartlett's famous book on memory (Bartlett, 1932), a work that exemplifies bad psychology, yet is nevertheless a perennial favorite in AI circles. But the local–global issue can be discussed at many levels, and the level that is operating here becomes more clear in a subsequent passage (Anderson & Bower, 1973) about the task orientation of question answering:

> When we observe psychologists struggling with the full complexities of question answering, we cannot help but be struck by the analogy to a physicist walking through a forest and trying to explain all the physical events that are occurring about him. No doubt he might be able to provide a enlightening qualitative description of principle forces that govern the descent of a leaf to the ground, but surely he would never set himself the task of simulating in the laboratory the descent of that leaf and surely would not consider simulating the entire forest. Similarly, we find it mistaken to attempt to mimic the question-answering behavior of the human [p. 414].

In AI question answering is a most attractive task orientation. It would be very useful to have computers that could respond to people in natural language, and this is a problem in question answering behavior. Because AI people would tend to agree that their science cannot be profitably conducted with methodologies that are appropriate to physics, it is difficult to understand why a psychologist (especially one who flies against the Zeitgeist) would worry about physicists simulating forests. But we have a paradigmatic trade-off at work here, and any psychologist who advocates broader task orientations can be easily squeezed into a tricky position.

Exploring and Kludging

One final observation about task orientations can be made concerning the ways that AI programmers use their task orientations. Just as a psychologist who tries to account for everything will surely fail, so will an AI researcher. Because global task orientations tend to involve a large number of theoretical problem areas, it is especially crucial for AI projects to be very clear about which areas are targeted for research and which areas are destined to be "kludged." A kludge is a piece of code that finesses a serious problem—it essentially saves the programmer from having to solve everything at once. Whenever an AI program claims to implement a theory of X (where X is some ambitiously global undertaking), one can be sure that the program is kludging around some aspect of the problem. Reputable AI people are quick to recognize and acknowledge programming kludges as areas of theoretical weakness or incompleteness, and many serious discussions among AI workers revolve around the question of when it is wise or unwise to kludge something. A questionable kludge can point to a serious

152

B6√8b

---

616 .

153.12

C√63N

Churchland

152
B988N

130 — Visa
120 — RENT
50 — LOANS
30 — Phone

330
321

130 - x
120
80

330

difficulty with the way that a program is decomposing its original problem, and students in AI quickly learn that a poorly placed kludge can betray a lack of theoretical understanding.

People outside of AI sometimes detect a programming kludge, and often interpret their discovery as evidence of cheating. Some kludges are easier to defend than others, and it is standard for AI people to point out their more glaring kludges to discourage misplaced accusations. When an audience pounces on a questionable kludge (as opposed to the merely obvious or reasonable kludges), the speaker knows he is addressing a knowledgeable group.

Even a local task orientation is liable to involve some kludging. For example, when Feigenbaum tackled the task of associate-pair learning in EPAM, he realized that he could not simulate the visual pattern recognition processes that people use to recognize letters. The state-of-the-art in vision processing just wasn't up to such problems in 1963. But this part of his targeted task orientation could be reasonably kludged without jeopardizing the remaining implementation. The theoretically significant part of EPAM (the discrimination net) would not be affected by a kludge in the initial stages of input processing, so Feigenbaum wisely finessed the problem of visual pattern recognition. Even if the necessary vision capabilities had been available, there would have been no theoretical advantage in utilizing them—a television camera at the front end of EPAM might have been very impressive to a lay audience, but no one in a position to appreciate the theoretical significance of EPAM would have been influenced by its presence.

Issues of kludgery can be very tricky and are sometimes not resolved without first making a few mistakes. In fact, AI students are often encouraged to try to implement something (anything) for the sake of seeing how one can learn from an implementation mistake. It is usually impossible to anticipate all of the difficulties one will encounter in a global computer implementation. Indeed, vast problem areas can emerge in the course of writing an AI program that would otherwise go unnoticed. This is because it is very easy to kludge over subtle theoretical issues without even realizing they are there. Realizations surface as soon as the program behaves inappropriately on some piece of input, and the error is traced back to its point of origin. Unintentional kludges are at the heart of the AI discovery process—which brings us to our last point about task orientations in AI.

At the onset of a particular investigation, an AI researcher will often find it useful to implement an exploratory system, just to get a feel for the critical problems. To see how unintentional kludges surface in these systems, consider the TALE-SPIN program designed by James Meehan (Meehan, 1976). TALE-SPIN was an exploratory system designed to investigate specific knowledge representation techniques in the task orientation of story generation. When TALE-SPIN encountered harmful kludges in its design, the result was sometimes a bizarre or ''mis-spun'' tale:

MIS-SPUN TALE #1:

> One day Joe Bear was hungry. He asked his friend Irving Bird where some honey was. Irving told him there was a beehive in the oak tree. Joe threatened to hit Irving if he didn't tell him where some some honey was.

Here the program makes a mistake because it does not interpret Irving's answer appropriately. To understand how a beehive is relevant to Joe, TALE-SPIN needs to use knowledge about beehives that was not available to the part of the program that processed Irving's answer. In short, TALE-SPIN was kludging around some inference problems. A serious repair for this mistake would require some theory of conversation along with a general inference capability based on associations between physical objects. Meehan did not choose to expand TALE-SPIN in these directions, opting instead to gloss over character interactions with intentional kludgery.

MIS-SPUN TALE # 2:

> Once upon a time there was a dishonest fox and a vain crow. One day the crow was sitting in his tree, holding a piece of cheese in his mouth. He noticed that he was holding the piece of cheese. He became hungry, and swallowed the cheese. The fox walked over to the crow. The end.

TALE-SPIN was set up to produce "The Fox and the Crow," when this was generated. The fox was supposed to see the cheese and trick the crow into dropping it. Unfortunately, the crow got hungry before the fox arrived. To prevent this from happening in a serious way, TALE-SPIN would have to have had some notion of a plot guiding its goal states to produce reasonable problems and problem resolutions. But the representational techniques of TALE-SPIN did not attempt to tackle the notion of a plot, so Meehan suppressed this particular glitch in the program by making sure the crow had eaten recently. The bug was fixed, but with an intentional kludge designed to get TALE-SPIN around the problem of what constitutes a plot.

In the task orientation of story generation, kludges can manifest themselves as amusing errors—silly things a person would never do. In other task orientations, a kludge that is not anticipated or intended is liable to bring down the whole system. Most programs will not run properly when they encounter thoughtless kludges. So unintended kludges demand immediate attention, and the successful use of an intended kludge requires a thorough understanding of how a process model can or cannot be broken down into separable processing components.[3] When an exploratory system is implemented to investigate decomposition issues,

---

[3]The concept of a "kludge" in AI is very similar in spirit to the philosophical notion of a "homunculus," as it is described by Dennett (1978).

it is necessary to concentrate on a global task orientation to get at the heart of the matter.

The AI concern for problem decomposition has led to some interesting departures from our standard academic boundary lines. Consider, for example, the boundary between linguistics and psychology. In 1965, Chomsky's formulation of transformational grammar divided language phenomena into "competence" and "performance" issues, and thereby encouraged a separation of syntactic phenomena from semantic phenomena. If a linguist is interested in competence issues, she is usually studying formalisms that describe the syntactic regularities of a language. Questions about how people actually use language to communicate with each other fall under the category of performance, which is really a problem for psychology. So the study of language was sharply constrained by transformationalists to exclude semantic phenomena, leaving semantic issues for psychologists. But the experimental paradigm of psychology can not easily embrace questions about how people use language to communicate with each other, because this points to a very global task orientation: it cannot be handled with clean experiments. As a result, questions about natural language semantics were never pursued by the traditional scientific disciplines. Semantics was then left for the philosophers, and more recently for AI researchers.

Semantic issues are quite central to AI research in language, where performance takes precedence over competence, and global task orientations do not violate standards of clean experimentation. From the perspective of the AI paradigm, problems in language use are primarily problems of a semantic nature— semantics that cannot be naturally separated from syntactic structures or human memory organization. We try to illustrate this perspective on language by taking a detailed look at a representative problem area. In the next section we consider a specific memory phenomenon documented by psychologists, and explore an AI explanation for this phenomenon from the perspective of natural language processing.

MAPPING THE COGNITIVE TERRAIN

My own research in recent years has been involved with the design of a computer program named BORIS (Dyer, 1981a,b; Dyer & Lehnert, 1980; Lehnert, 1979, 1980, 1981; Lehnert, Black, & Reiser, 1981; Lehnert, Dyer, Johnson, Yang, and Harley, 1983). BORIS processes prototype narratives that are designed to take us beyond our current understanding of inference and memory representation. As such, the system incorporates a range of processing strategies first implemented in earlier systems along with processing techniques unique to BORIS.

The stories BORIS reads are much more sophisticated than the texts handled by its predecessors because BORIS integrates the inference capacities of multiple knowledge structures (see Appendix A for some sample BORIS I/O). Although SAM processed stories using only scripts (Cullingford, 1978), and PAM concentrated on plans and goals (Wilensky, 1978), BORIS uses a number of different knowledge structures including scripts, plans, goals, themes (Kolodner, 1980), mops (Schank, 1982), thematic affect units (Dyer, 1981a), and plot units (Lehnert, 1981).

Our initial design for question answering in BORIS borrowed a computer metaphor about memory that is intuitively pleasing and easy to overlook because of its naturalness. Quite simply, most of us are willing to assume that the data structure we call memory is only altered by storage processes. In particular, memory structures are not altered by retrieval processes. This seemingly reasonable assumption leads to likewise natural divisions between understanding and remembering, as well as between learning and using. But experimental evidence in psychology has brought this assumption into question, at least insofar as question answering is thought to be a process of pure information retrieval.

Elizabeth Loftus has established a ''rewriting'' effect during question answering (Loftus, 1975, 1980) which indicates that established memory representations can be altered to incorporate false presuppositions present in questions. If a subject sees a slide containing a stop sign, and is subsequently asked about the activities near the yield sign, a later query about the type of sign present will reveal memory of a yield sign. (We will see more about this later). But these phenomena cannot be simulated in a system that treats question answering as a purely passive retrieval process. Memory alterations can only occur if the retrieval process somehow acts on memory, to refine or alter its previous contents.

The notion of an ''active'' (vs. passive) retrieval strategy is also consistent with ideas about ''reconstructive memory retrieval'' (Linton, 1979; Norman & Rumelhart, 1975). Although little has been done with reconstructive memory techniques in terms of computer simulations, the recent CYRUS system (Kolodner, 1980) suggests that such techniques will prove to be critical for large data bases encoding mundane knowledge domains.

While we did not set out to simulate the Loftus effects in BORIS, we did try to unify the search processes that BORIS used at the time of story understanding, with the search processes BORIS used at the time of question answering. It seemed both wasteful and wrong to maintain separate sets of search routines for both task orientations since the same memory structures were being located and examined in either case. Interestingly, our attempts to integrate these search processes endowed BORIS with a potential for making memory alterations during question answering in the same ways that memory was altered during story understanding. The increasingly artificial distinction between memory creation and memory access in BORIS began to look like an arbitrary division that was imposed on the system by our intuitions alone—with little else to recommend it in terms of efficiency or ease of design.

With this turn of events we began to look more closely at the Loftus results. Exactly when is memory altered by the act of answering a question? To control potential alterations in BORIS we need to know what constraints are operating to suppress potential alterations. Some false presuppositions can be recognized by human subjects and rejected accordingly. While the specifics of these constraints are not yet understood, the ramifications of unified search techniques are significant. If search processes used at the time of story understanding are identical to search processes used at the time of question answering, then we should expect to see some mirror images across the two task orientations. More specifically, problems in text comprehension may be connected to analogous problems in question answering behavior. If this is the case, it will be crucial for us to consider text comprehension and question answering within a single theoretical framework.

Memory Interactions during Question Answering

When a story is read, symbolic structures are generated in memory to represent the conceptual content of that story. Any subsequent questions about the story are then interpreted in terms of that underlying context. In particular, potentially ambiguous questions are unconsciously interpreted in whatever manner makes it possible to answer the question. For example, suppose a subject reads the following story about two executives:

> John and Bill went to a restaurant to discuss a business deal. When the meal was over, John left a very large tip, and apologized to Bill for wasting his time.

Now consider the question, "Was the dinner a success?" This question can be reasonably interpreted two ways. From the perspective of a restaurant visit, the answer should be "yes." An inference about their dinner can be made on the basis of the large tip: restaurant patrons do not normally leave generous tips unless the meal meets with their expectations. But at another level of comprehension "the dinner" is merely a social setting for the higher level goal of a business deal. For the dinner to be successful at this level, the business deal had to be negotiated satisfactorily. Because John apologized to Bill, we should infer that no agreement was reached, and the higher goal failed to be realized. The dinner did not go well at this level.

Question answering behavior in this case must be analyzed qualitatively. A positive response indicates that "the dinner" was interpreted very literally, and an inference about restaurants was successfully made. A negative response indicates that "the dinner" was interpreted as an instrumental setting, and a probable inference about apologetic behavior was made. A response of "I don't know," suggests that neither inference was made. Possibly the best response would be "yes and no," indicating that the ambiguity of the question was perceived, and all the relevant inferences were made. In any case, we are in a position to assess a subject's ability to make inferences and interpret events appropriately.

In our simple example question, the ambiguous reference to "the dinner" had to be resolved by information in memory. We could alter the story context to manipulate the question's reference by eliminating either the higher level goal (not mentioning the business deal or John's apology) or the lower level experience (not mentioning the tip). If either of these concepts were absent from the story text, the question about "the dinner" would have to be answered in terms of the remaining information. Words used in questions are conceptually interpreted by whatever constraints a relevant memory representation imposes.

Potential ambiguities in questions can occur at many different levels. In addition to the potential ambiguities of single words, the entire focus or conceptual emphasis of a question may be ambiguous. For example, consider another story about John:

John had just bought a new car. He was so happy with it that he drove it at every possible opportunity. So last night when he decided to go out for dinner, he drove to Leone's. When he got there he had to wait for a table . . .

If asked, "Why did John drive to Leone's?" subjects will naturally explain something about John liking his new car. The question has been interpreted to be asking about the act of driving. John could have walked to the restaurant or taken a cab—why did he drive? Now consider a slightly different context:

John had a crush on Mary. But he was so shy that he was happy to just be in her proximity. So he was in the habit of following her around a lot. He knew that she ate at Leone's very often. So last night when he decided to go out for dinner, he drove to Leone's. When he got there he had to wait for a table . . .

And now reconsider the same question, "Why did John drive to Leone's?" Suddenly the focus of the question has shifted to the destination of John's trip. Why Leone's? Why not the corner diner?

The question is lexically invariant, but its conceptual interpretation changes as the context around it changes. We interpret the question in whatever way allows us to answer it. This means that our conceptual memory representation for the story text is being accessed and searched as we read the question. In this case a potential ambiguity of focus must be resolved by the internal memory representation (Lehnert, 1978).

Another type of question ambiguity involves partial event specifications. For example, consider the following text:

John closed up his office and went to an office party after work on Friday. While he was there, he overheard a conversation between two executives concerning a special account that John had been working on. From what they were saying, it became clear that John was not receiving credit for the time and energy he had devoted to the project. After hearing this, John felt extremely frustrated and in no mood to socialize. In an effort to get his mind off his troubles, John excused himself and went out to a movie. After the movie, he went out and got drunk.

Twenty five subjects in an informal classroom experiment heard this text and then answered the question, "Why did John leave?" All 25 subjects answered the question by explaining that John was upset by a conversation he had overheard, or he was preoccupied. Although there were many conceptual distinctions across the answers, all answers addressed the cause of John's early departure from the party. When asked if the question was unclear or ambiguous, none of the subjects indicated any conscious confusion. Yet there is a glaring ambiguity in this question. Why did John leave *where?* The question could have been interpreted to be asking why John left his office, or why John left the movie. Yet everyone immediately interprets this question to be asking about why John left the party. Intuitively, this seems natural because the bulk of the story describes why John left the party. But this means that the memory representation for the story is being examined to arrive at a preferred interpretation of the question.

All these examples illustrate how an internal memory representation can affect the conceptual interpretation of questions in a "top-down" or predictive manner. In each of these cases, we appear to have a memory interaction that carries information from the story representation to the conceptual interpretation of a question. This predictive processing can be used to "disambiguate" multiple word senses, potentially ambiguous focus assignments, and conceptual ellipses (omissions) in questions. We call these top-down influences *predictive residues* because they incorporate top-down predictions from some story context (Lehnert, 1982).

Interference Effects from Question Answering

At the same time that predictive residues aid in the interpretation of questions, other influences are operating in the opposite direction. In a series of experiments designed to examine the reliability of eyewitness testimony, Loftus has determined that "leading questions" can be used to alter a subject's memory of an event or physical description (Loftus, 1975, 1980). In one experiment, subjects were shown a series of slides depicting successive stages of an automobile accident. In one slide a car approaches an intersection with a stop sign. After the slides were shown, subjects were asked some questions about what they had seen. Subject A was asked, "Did another car pass the red car while it was stopped at the stop sign?" Group B was asked, "Did another car pass the red car while it was stopped at the yield sign?" Half of the subjects (in both groups A and B) had been shown a slide with a stop sign, while the other half had seen a slide with a yield sign. A week later the subjects were tested for their memory of the slides. As many as 80% of the subjects given the false-presupposition question identified the sign in question incorrectly. This error rate is significantly higher than the error rate for subjects who received consistent questions, and it is also significant when compared to the error rate for subjects who received no questions after the initial viewing. There was no significant bias favoring memory for stop (yield) signs over memory for yield (stop) signs.

In the Loftus experiments, questions affect internal memory representations. Contradictory information in a question can alter memory with the incorporation of new (and incorrect) information. We will call these effects retroactive residues, because residual information from a question is retroactively altering the content of a memory representation. Additional experiments conducted at Yale (Lehnert, Robertson, & Black, 1981) have shown that retroactive residues will operate when memories are constructed from reading text as well as from viewing slides. At first glance it might seem that alterations occur only when contradictory questions are asked. Questions that are consistent with the original stimuli should act only to reinforce existing memories. But in fact, it is possible to alter memory with information that is not contradictory. For example, consider once more our original story:

> John and Bill went to a restaurant to discuss a business deal. When the meal was over, John left a very large tip, and apologized to Bill for wasting his time.

Q: Was the dinner a success?

Here we have an example of a question that will alter memory by providing a further specification for "the meal." The original text does not specify if John and Bill were meeting over breakfast, lunch, or dinner. But the question assumes they met over dinner. We would expect to find a retroactive residue operating here to refine the original concept of a meal into the more specific concept of a dinner.

Predictive and Retroactive Processing Residues

It appears that cognitive processes are operating between questions and internal memory representations in two directions. First, predictive residues in memory affect the conceptual interpretation of questions. Secondly, retroactive residues from questions operate to alter memory.

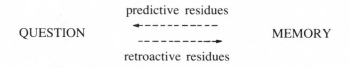

predictive residues

QUESTION ← - - - - - - - - - MEMORY
- - - - - - - - - →

retroactive residues

Predictive residues have been the subject of a few AI investigations and are most naturally studied within the AI paradigm. A computer program can very easily indicate when predictive residues are present by failing to process sentences in context correctly. Whereas people find the contextual interpretation of sentences to be an effortless and trivial task, it is an extraordinarily difficult problem for computers. On the other hand, the effects of retroactive residues that are central to the Loftus experiments are best left to experimental psychology. Only an experiment with human subjects can indicate when retroactive residues are acting to alter memory representations. Human intuitions about this phenomenon are not reliable and must be tested with human subjects.

We are then confronted with two distinct phenomena: (1) predictive processing strategies for natural language understanding; and (2) memory interference effects that result from answering questions. The first falls into the province of AI, and the second is a problem for psychology. The interference effects of Loftus are furthermore problematic for AI models in the sense that these phenomena are not obviously motivated by computational considerations. What will compel computer models of human memory to make these retrieval errors? Is it possible to simulate these effects at all? Is there a larger framework (possibly a larger task orientation) that naturally forces such imperfections on the system? What computational strategies could lead us to Loftus-type retrieval errors as a natural trade-off for some larger advantage? Following these questions out to their logical conclusion, is it possible that natural language computer systems are doomed to simulate human information-processing phenomena all the way down to undesirable levels of human imperfection and fallibility? Can computers do no better than people at tasks involving natural language?

For years I wondered about the significance of the Loftus results for AI models of language, watching for some computational evidence that might relate this memory phenomenon to our computational models. When the connection finally surfaced, it seemed embarrassingly obvious. Predictive residues and retroactive residues are the effects of identical processes, separated only by the task orientations of text comprehension and question answering. Yet this realization became obvious only when we tried to consolidate the two separate sentence analyzers that BORIS had been using for story understanding and question answering (Dyer, 1981b).

The unified and integrated memory techniques now operating in BORIS represent a significant departure from Q/A techniques of similar systems (Carbonell, 1978; Kolodner, 1980; Lehnert, 1977). All memory interactions in BORIS that operate at the time of story understanding also operate at the time of question understanding. Memory-based predictions at the time of story understanding are extended to memory-based predictions at the time of question answering to give us the predictive residues needed to interpret questions appropriately. But these predictions are likewise integrated with routines designed to build a memory representation. So the routines used for memory construction at the time of story understanding become retroactive residues at the time of question answering. In this way, memory alterations at the time of question answering are a necessary side-effect of predictive text processing.

To see how these two ideas naturally dovetail with each other, suppose our original example story continued on:

> John and Bill went to a restaurant to discuss a business deal. When the meal was over, John left a very large tip, and apologized to Bill for wasting his time. Bill thanked John for the dinner, and they went their separate ways.

Here we have mentioned "the dinner," which must be understood to be a reference to the meal John and Bill just had in the restaurant. Predictive inference mechanisms at the time of story understanding are able to merge the dinner concept with its proper memory token as a routine requirement of effective text processing. This *very same mechanism* will also operate when "the dinner" appears in a question about the story. In the above text, "the dinner" is interpreted to mean the meal (as opposed to the business meeting). We know John left the tip, so we would infer that John paid the bill. This gives Bill good reason to feel grateful for the meal. It makes less sense to feel grateful about an unsuccessful business deal, so a sophisticated token merger must not interpret "the dinner" in the wider sense of a business meeting in this case. Had the story gone on to explain that "the dinner" was destined to change Bill's life, a smart token merger would have to interpret this as a reference to the unsuccessful business meeting that happened to be conducted over a dinner meal.

Whereas token merging can become a complex problem in its own right, we are only claiming that the same general mechanism used to arrive at an interpretation for "the dinner" in the story text can also work to interpret "the dinner" when it appears within a question about the story. Any residue that served to merge memory tokens for our extended story will therefore produce the same effect during question answering. It doesn't really matter if the reference to "the dinner" is encountered while we are reading the story or afterwards. If it is accepted as a further specification for the meal at the time of story understanding, then it will be accepted in an identical fashion during question answering. We would furthermore expect that a human processor can be no more conscious of such things during question answering than during story understanding. Indeed, recent experiments by Loftus suggest that people are incapable of controlling these mechanisms, even when the experimenter explicitly warns subjects to watch out for false presuppositions.[4]

While the unification of predictive and retroactive residues suggests a compel-

---

[4]Department of Psychology Seminar at Stanford University (3/6/81).

ling computational reason for memory interference during question answering, a great deal of work must still be done with respect to human experimentation. What conceptual conditions of question content enhance the Loftus effect? Preliminary experiments at Yale suggest that it may be easier to mislead people with false states than false actions (Lehnert, Robertson, & Black, 1981), but this is just a beginning. Can retroactive residues cause a "ripple effect" in memory that alters information that is causally connected to the direct manipulation? Our preliminary data says yes, but we need to find out exactly how conceptual factors affect these memory ripples. Can we separate ripple effects generated at the time of a misleading question from reconstructive processing at a later time? When can subjects reliably detect false presuppositions, thereby suppressing normal memory integration? Answering these questions will be very difficult. Yet all of these questions must be at least partially answered by studying human subjects. Some very fundamental questions about memory representation might be resolved by conducting refinements of the Loftus experiments in conjunction with computer simulations.

While the issue of predictive and retroactive residues may very well raise more questions than it answers, it nevertheless shows how an AI perspective can integrate radically different phenomena concerning language and memory. Psychological simulations in AI impose an essentially holistic perspective on problems of language and memory, a perspective actively discouraged by other experimental traditions. In particular, the fragmentary studies encouraged by psychology cannot integrate seemingly disparate phenomena into large cohesive frameworks. The Loftus experiments are typically described as memory experiments: there was no reason to suspect that these phenomena were necessary side effects of natural language processing techniques. In the AI paradigm, processes of language are inseparable from memory phenomena, and this integrated view made it easy to interpret the Loftus results as a critical feature in a larger cognitive terrain.

COGNITIVE MOUNTAIN CLIMBING

One of the questions frequently put to AI workers by psychologists involves the issue of theory validation: "When will your model be ready for quantitative evaluation?" The answer is usually, "Not yet," although a few AI workers prefer to sabotage the question by answering, "Never—by your methods." Of course many ideas about language processing have already inspired successful psychology experiments. For example, the notion of a script has received a lot of scrutiny from cognitive and developmental psychologists (Abelson, 1980). But this does not mean that the original implementation of scripts in the SAM system has been tested (or should be).

The trick is to separate the serious kernel of a programming implementation from the unavoidable kludges that surround it. SAM was not designed to investigate script management (how specific scripts are triggered in long-term memory and activated as needed), so SAM cannot be viewed as a comprehensive model of script application. But SAM did serve to illustrate how scriptal knowledge structures can account for a class of inferences people generate when they read narratives. More importantly, it caused a lot of people to think about explicit structures in memory for organizing large bodies of information. Psychologists are not inclined to worry about memory in terms of explicit processes that access and search for complex structures. Ask a psychology student about memory access and you'll most likely be reassured that it's all done with intersection searches. Systems like SAM can be instrumental in forcing people to confront problem areas that are not receiving the attention they deserve. The notion of a script lacks rigor insofar as no one has been able to explain exactly where scripts cut off and other things take over, but ill-defined scripts are at least one order of magnitude closer to a rigorous explanation than ill-defined intersection searches.

I suspect that this last sentence has had a negative effect on some of my readers. Whenever I get into a conversation about "relatively correct" explanations, I sense a rising impatience in the psychologists. This leads me to suspect that AI people are comfortable with the idea of successive approximations toward a theory, whereas psychologists require immediate verification. It is almost as though psychologists partition the world of ideas into two classes: the right ones and the wrong ones.

I would guess that this has to do with task orientation differences. Psychologists assume that any verification of an idea at a local level will contribute to the ultimate verification of some larger idea at a global level. They can then feel justified in working up from their local investigations, certain that each local idea is a step in the right direction, bringing them closer to their ultimate answers. This strategy is highly reminiscent of hill-climbing algorithms: if you want to reach the top of the highest mountain, follow the steepest gradient over some local radius. Unfortunately, hill-climbing techniques fail more often than not. It is very easy to get stranded on local maxima when your vision is confined to nearby topography. Worse yet, there is no way to detect failure once a local maximum is reached; a nearsighted hill-climber can easily assume that he has scaled Mount Everest when he is really only on top of Bald Mountain.

Knowing this, AI workers are much more skeptical about the value of local verification in the long run, preferring to work their way down from imperfect global ideas (which can't possibly account for everything correctly). Some people feel comfortable with these global approximations and other people can't tolerate them.[5] People who thrive in AI generally harbor a bit of an engineer as

---

[5]This theme is described very entertainingly in (Abelson, 1981) as a conflict between "neat" and "scruffy" ideologies.

an alter ego: they find it gratifying to build things that do something. The thing may not work perfectly, but next year's model will be better.

So the question of ultimate verification is not a natural one for AI workers, who typically argue that they must build adequate systems first. From an engineering viewpoint, it seems premature to worry about how we're going to compare and evaluate two competing explanations (systems) when we don't even have one explanation (system). When we actually have some theories to compare, we will know a lot more than we know now, and it just might be obvious how to proceed. Was it possible to think about Environmental Protection Agency standards for cars before the car was invented? It might have been possible to anticipate the current EPA concerns in a very vague way, but specific ideas can only evolve in a cumulative fashion.

I realize that many psychologists and philosophers are having difficultly evaluating AI models, and that this is a source of some concern. It would be nice to have some systematic methodology to determine whether a model is failing to perform a task because of "temporary ignorance" that will be corrected as research progresses, or critically erroneous assumptions that render the model inherently inadequate. Unfortunately, the only methodology that seems wholly reliable involves a lot of waiting. Highly respected computer scientists have argued that it is impossible to develop viable techniques to verify the intended behavior of even simple computer programs—let alone complex ones (DeMillo, Lipton, & Perlis, 1979). Because effective extensibility of computer programs is an even murkier area, it seems quite possible that rigorous evaluation techniques for AI systems cannot be devised. Perhaps the fairest argument for AI (and cognitive science) relies on common sense. Philosophy has been around for a few thousand years, psychology and linguistics at least a hundred. AI, which could not exist before the advent of computers, is roughly 20-years-old, and cognitive science has been around for all of 5 years. One can debate ad infinitum the relative contributions that each area has made to our understanding of human cognition. But whatever the final score, 20 years isn't much time by multidisciplinary standards. Five years isn't even worth discussing. For the moment, the best vantage point for cognitive scientists may be a simple time line. From that perspective it seems that a little patience is in order.

ACKNOWLEDGMENTS

The author is indebted to Chris Riesbeck, Steve Shwartz, Peter Polson, Walter Kintsch, George Mandler, John Haugland, and Bill Clancy for many helpful suggestions that guided me in the completion of this paper. The views expressed here are nevertheless solely my own, for which I assume all responsibility. The research described in section three was supported in part by the Advanced Research Projects Agency under contract N00014-75-C-1111 and in part by the National Science Foundation under contract IST7918463.

APPENDIX A

Below is an execution trace of the BORIS program running on a divorce story:
(Input to the program is in lower case; output in uppercase.)
TOPS-20 Command Processor 4(560)-1
@BORIS-DEMO.EXE.5

Top-Level of LISP
*(EXPFS 50000)
*(BORIS' DIVORCE-STORY)

PROCESSING PARAGRAPH 1:

> Richard hadn't heard from his college roommate Paul for years. Richard had borrowed money from Paul which was never paid back, but now he had no idea where to find his old friend. When a letter finally arrived from San Francisco, Richard was anxious to find out how Paul was.

READY TO ACCEPT Q/A:

> What happened to Richard at home? <
RICHARD GOT A LETTER FROM PAUL.

> Who is Paul? <
RICHARD's FRIEND.

> Did Richard want to see Paul? <
YES, RICHARD WANTED TO KNOW HOW PAUL WAS.

> Had Paul helped Richard? <
YES, PAUL LENT MONEY TO RICHARD.

> Why didn't Richard pay Paul back? <
RICHARD DID NOT KNOW WHERE PAUL WAS.

> How did Richard feel when the letter appeared? <
RICHARD FELT GLAD BECAUSE PAUL AND HE WERE FRIENDS.

> go <

PROCESSING PARAGRAPH 2:

> Unfortunately, the news was not good. Paul's wife Sarah wanted a divorce. She also wanted the car, the house, the children, and alimony. Paul wanted the divorce,

but he didn't want to see Sarah walk off with everything he had. His salary from the state school system was very small. Not knowing who to turn to, he was hoping for a favor from the only lawyer he knew. Paul gave his home phone number in case Richard felt he could help.

READY TO ACCEPT Q/A:

> What was the letter about? <
PAUL AND SARAH WERE GETTING A DIVORCE.

> Did Paul write Sarah? <
NO, IT WASN'T SARAH, IT WAS RICHARD WHO GOT THE LETTER.

> Why was Paul upset about the divorce? <
PAUL AND SARAH WERE FIGHTING OVER THE FAMILY POSSESSIONS.

> What did Paul do for a living? <
PAUL WAS A TEACHER.

> What did Sarah want? <
SARAH WANTED TO HAVE THE CAR AND THE HOUSE AND THE KIDS
AND THE ALIMONY.

> Why did Paul write to Richard? <
PAUL WANTED RICHARD TO BE HIS LAWYER.

> go <

PROCESSING PARAGRAPH 3:

Richard eagerly picked up the phone and dialed. After a brief conversation, Paul agreed to have lunch with him the next day. He sounded extremely relieved and grateful.

READY TO ACCEPT Q/A:

> Why did Richard call Paul? <
RICHARD WANTED TO ARRANGE A MEETING WITH PAUL.

> Did Richard contact Paul? <
YES. RICHARD CALLED PAUL.

> How did Richard contact Paul? <
BY PHONE.

> How did Paul feel when Richard called? <
PAUL WAS HAPPY BECAUSE RICHARD AGREED TO BE PAUL'S LAWYER.

REFERENCES

Abelson, R. P. The psychological status of the script concept. *American Psychologist,* 1980.

Abelson, R. P. Constraint, construal, and cognitive science. *Proceedings of the Third Annual Conference of the Cognitive Science Society.* Berkeley, Calif., 1981.

Anderson, J. R., & Bower, G. H. *Human associative memory.* New York: Wiley, 1973.

Bartlett, F. C. *Remembering: A study in experimental and social psychology.* Cambridge University Press, London: 1932.

Carbonell, J. G. POLITICS: Automated ideological reasoning. *Cognitive Science,* 1978, *2,* 27–52.

Chomsky, N. *Aspects of the theory of syntax.* Cambridge, Mass.: MIT Press, 1965.

Cullingford, R. *Script application: Computer understanding of newspaper stories.* Department of Computer Science Research Report 116. Yale University, 1978.

Demillo, R., Lipton, R., & Perlis, A. Social processes and proofs of theorems and programs. *Communications of the Association for Computing Machinery,* 1979, 22, pp. 271–280.

Dennett, D. C. *Brainstorms.* Cambridge, Mass.: MIT Press, 1978.

Dyer, M. The role of TAU's in narratives. *Proceedings of the Third Annual cognitive Science Conference.* Berkeley, CA, 1981. (a)

Dyer, M. Integration, unification, reconstruction, modification: an eternal parsing braid. *Proceedings of the Seventh International Joint Conference on Artificial Intelligence.* Vancouver, British Columbia, 1981. (b)

Dyer, M., & Lehnert, W. *Memory organization and search processes for narratives.* Department of Computer Science Research Report 175, Yale University, 1980.

Feigenbaum, E. A. The simulation of verbal learning behavior. In E. A. Feigenbaum & J. A. Feldman (Eds.), *Computers and thought.* New York: McGraw-Hill, 1963.

Kolodner, J. *Retrieval and organizational strategies in conceptual memory: A computer model.* Department of Computer Science Research Report 187, Yale University, 1980.

Lehnert, W. Question answering in a story understanding system. *Cognitive Science,* 1977, *1,* 47–73.

Lehnert, W. *The process of question answering.* Hillsdale, N.J.: Lawrence Erlbaum Associates, 1978.

Lehnert, W. Text processing effects and recall memory. *Cognition and Brain Theory.* 1982, 5(1), 3–28.

Lehnert, W. Narrative text summarization. *Proceedings of the First Annual National Conference on Artificial Intelligence.* Stanford University, 1980.

Lehnert, W. Plot units and narrative summarization. *Cognitive Science,* 1981, *4,* 293–331.

Lehnert, W., Black, J., & Reiser, B. Summarizing narratives. *Proceedings of the Seventh International Joint Conference on Artificial Intelligence.* Vancouver, British Columbia, 1981.

Lehnert, W., Dyer, M., Johnson, P., Yang, C. J., & Harley, S. BORIS—An experiment in in-depth understanding of narratives. *Artificial Intelligence,* 1983, *20,* 15–62.

Lehnert, W., Robertson, S., & Black, J. *Memory interactions during question answering.* Text Comprehension Symposium at the Deutsches Institut fur Fernstudien an der Universitat Tubingen. University of Tubingen, W. Germany, 1981.

Linton, M. I remember it well. *Psychology Today,* July, 1979, 81–86.

Loftus, E. F. Leading questions and the eyewitness report. *Cognitive Psychology,* 1975, *7,* 560–572.

Loftus, E. F. *Memory.* Reading, Mass.: Addison-Wesley, 1980.

Meehan, J. *The metanovel: Writing stories by computer.* Department of Computer Science Research Report 74, Yale University, 1976.

Norman, D. A., & Rumelhart, D. E. *Explorations in cognition.* San Francisco: W. H. Freeman and Company, 1975.

Schank, R. C. Reminding and memory organization: An introduction to MOPs. In W. Lehnert & M. Ringle (Eds.), *Strategies for natural language processing.* Hillsdale, N.J.: Lawrence Erlbaum Associates, 1982.

Simon, H. A. *The sciences of the artificial.* M.I.T. Press, Cambridge, Mass.: MIT Press, 1969.

Wilensky, R. *Understanding goal-based stories.* Department of Computer Science Research Report 140, Yale University, 1978.

3 Methodology for Building an Intelligent Tutoring System

William J. Clancey
Stanford University

Introduction

Over the past 5 years my colleagues and I have been developing a computer program to teach medical diagnosis. Our research synthesizes and extends results in artificial intelligence (AI), medicine, and cognitive psychology. This chapter describes the progression of the research, and explains how theories from these fields are combined in a computational model. The general problem has been to develop an "intelligent tutoring system" by adapting the MYCIN "expert system."[1] This conversion requires a deeper understanding of the nature of expertise and explanation than originally required for developing MYCIN, and a concomitant shift in perspective from simple performance goals to attaining psychological validity in the program's reasoning process.

Others have written extensively about the relation of AI to cognitive science (e.g., Boden, 1977; Lehnert, this volume; Pylyshyn, 1978). Our purpose here is not to repeat those arguments, but to present a case study that will provide a common point for further discusstion. To this end, to help evaluate the state of cognitive science, we outline our methodology and survey what resources and viewpoints have helped our research. We also discuss pitfalls that other AI-oriented cognitive scientists may encounter. Finally, we present some questions coming out of our work that might suggest possible collaboration with other fields of research.

---

[1] A glossary appears at the end of the chapter.

GOALS: INTELLIGENT TUTORING SYSTEMS

An *intelligent tutoring system* is a computer program that uses AI techniques for representing knowledge and carrying on an interaction with a student (Sleeman & Brown, 1981). Among the most well-known systems are WHY (Collins, 1976, uses Socratic principles for teaching causal reasoning in domains like meteorology), SOPHIE (Brown, Burton, & Bell, 1974, provides a "simulated workbench" in which a student can test electronic troubleshooting skills), and WEST (Burton & Brown, 1979, coaches a game-player on methods and strategies for exploiting game rules). This work derives from earlier efforts in computer-aided instruction, but differs in its attempt to use a principled or theoretical approach. First and foremost, this entails separating subject material from teaching method, as opposed to combining them in ad-hoc programs. By stating teaching methods explicitly, one gains the advantages of economical representation (the methods can be applied flexibly in many situations and even multiple problem domains) and the discipline of having to lay out subject material in a systematic, structured way, independently of how it is to be presented to the student. So the primary application of AI to these instructional systems is in the representation of teaching methods and domain knowledge. Ideally, this enterprise involves having a theory of teaching and the nature of the knowledge to be taught.

When we separate domain knowledge from the procedures that will use it, we say that we are representing knowledge "declaratively" (Winograd, 1975) (with respect to those procedures). For example, in a medical domain, we would represent links between data and diagnoses so they could be accessed and used for solving any given problem. A strong advantage of this approach is that the tutoring system can cope with arbitrary student behavior: no matter what order the student chooses to collect data (or troubleshoot a circuit, or make moves in a game), the program can evaluate partial solutions, and use its teaching knowledge to respond. Typically, the declaratively-stated knowledge base of diagnostic rules, causal relations, and the like is used during a tutorial to generate an "expert's solution," which, when compared to the student's behavior, provides a basis for advising the student.[2] The combination of a knowledge base of this kind and an interpreter for applying it to particular problems constitutes an *expert system,* with an intelligent tutoring system having an expert system inside it (Fig. 3.1).

In general, an *expert system* is a kind of AI program that is designed to provide advice about real world problems that require specialized training to

---

[2]In such a "first order" system, the model of the student's knowledge, as built by the program, is a subset of the internal, idealized knowledge base. This kind of model does not take into account student misconceptions or "bugs," an important area of research (see, for example, [Stevens, Collins, & Goldin, 1978] and [Brown & Burton, 1978]). The research described in this chapter focuses on the (as yet unsolved) problem of constructing the expert knowledge base, the material to be taught.

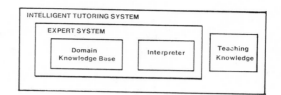

FIG. 3.1 Components of an Intelligent Tutoring System.

master. Some examples are MYCIN (Shortliffe, 1976), which provides advice about antibiotics for infectious diseases; SU/X (Nii & Feigenbaum, 1978), which analyzes sonar signals; and R1 (McDermott, 1980), which configures the components of computer systems. These systems are built by interviewing experts in the given domain, and representing their knowledge in the form of heuristics, or "rules of thumb." For example, in an expert system for field biologists, one might find the rule, "if there are many buttercups and goldfield flowers, then the kind of underlying rock is probably serpentine." We call this kind of conditional statement, consisting of a premise and conclusion, a *production rule*.

In expert systems, there is no attempt to *simulate* how human experts think, for example, to model the order in which they typically attack a problem. Instead, these programs are intended to capture the efficient leaps an expert makes from a problem description to an interpretation. This is what a production rule does. Expert systems differ in the nature of the task they solve (constructive, diagnostic, interpretative, etc.) and in their formalism for representing knowledge ("frames," "semantic nets"), but they all use rule-like associations.

The interpreter of an expert system is a program that controls the order in which the rules are considered. Common control strategies are *backward chaining* (working backwards from a goal) and *forward inferencing* (applying those rules whose conditions are satisfied by the problem description). These strategies correspond to two common ways of structuring the rule base, namely by the goal mentioned in the conclusion and by the problem description mentioned in the premise. By this structuring, the interpreter can index the rules and apply them. By the same token, the structure of a given rule base constrains how it can be used, the possible kinds of strategies the interpreter can use to access it.

The particular tutoring system we will be considering is built upon the knowledge of the MYCIN expert system. MYCIN's rules have to be restructured in order to be applied to teaching; the new system is called NEOMYCIN (Clancey & Letsinger, 1981). Our methodology for building NEOMYCIN is the subject of this chapter. The key idea is that using an expert system for teaching requires a shift in orientation from simply trying to output good solutions, to simulating in some degree of detail the reasoning process itself. The production rules that are used by MYCIN to provide good advice are inadequate for use as teaching material because certain kinds of reasoning steps, whose rationale needs to be conveyed to a student, are implicit in the rules. We need a more explicit, psychologically valid model of problem solving—one that can be understood and remembered by a student and incorporated in his behavior.

FROM MYCIN TO GUIDON (AN AI ENTERPRISE)

MYCIN is an expert system that was developed by a team of physicians and AI specialists. The program was designed to advise non-experts in the selection of antibiotic therapy for infectious diseases. The domain knowledge base (refer to Fig. 3.1) contains approximately 450 rules, which deal with diagnosis of bacteremia, meningitis, and cystitis infections. The interpreter uses backward chaining, working from high order goals, such as "determine whether the patient requires treatment" down to more specific subgoals, such as "determine whether the patient has high risk for Tuberculosis." A typical rule is (roughly stated) "if the patient has been receiving steroids, then his risk for Tuberculosis meningitis is increased." Most rules are modified by a "certainty factor" indicating the rule author's degree of belief, on a scale from −10 to 10, that the conclusion holds when the premise is known to be true. Figure 3.2 shows excerpts from the diagnostic portion of a MYCIN consultation. Rules are chained together, working downwards from the high order goals; the program asks a question when it needs data to apply a rule. After the diagnosis is complete, a therapy program selects the most optimal therapy for the organisms most likely to be causing the infection. Additional tests might also be ordered.

The success of MYCIN as a problem-solver, as measured in several formal evaluations (Yu, Buchanan, Shortliffe, Wraith, Davis, Scott, & Cohen, 1979; Yu, Fagan, Wraith, Clancey, Scott, Hannigan, Blum, Buchanan, & Cohen, 1979), encouraged us to explore its application for teaching. The program's good performance, coupled with an ability to explain its line of reasoning, made it seem particularly suitable as teaching material. The rules had been acquired from physicians over many hours of discussion, comparing the program's behavior to their judgment, modifying rules to improve the program and testing the program on new problems. The rules pertaining to infectious meningitis were especially carefully constructed from experience with over 100 cases from local hospitals and medical journal articles. Therefore, we decided to focus on using the meningitis rules for teaching.

In order to understand what is good about MYCIN's rules and how they fall short for use in teaching, one must understand something about their construction, and what kind of explanation a tutorial program can provide by using them. Rules are not written independently of the whole rule base: a rule author must think about how a given rule will fit. Any given rule must make a conclusion about some goal that appears in at least one other rule premise, otherwise the rule would never be used (recall the mechanism of backward chaining). Moreover, some means must be provided to evaluate the subgoals mentioned in the premise,

----PATIENT-1----
1) Patient's name:
**John Smith
2) Age:
**10 YEARS
3) Sex:
**MALE
• • •

6) Are there any pending cultures for John Smith?
**YES

----CULTURE-1----
7) From what site was the specimen for CULTURE-1 taken?
**CSF
8) Please give the date and time when the pending csf culture
(CULTURE-1) was obtained. (mo/da/yr time)
**8-Nov-88
• • •

15) Has John Smith recently had symptoms of persistent headache or
other abnormal neurologic symptoms (dizziness, lethargy, etc.)?
**YES
• • •

22) Does John Smith have evidence of ocular nerve dysfunction?
**NO
23) Is John Smith a compromised host?
**NO

FIG. 3.2   Excerpt from MYCIN consultation.

by writing other rules to make the appropriate conclusions and/or by making it possible for the system to gain the information from the user. So, in effect, a rule author is writing a kind of program in which goals are chained together by rules.

The author's choice of goals in the program constitutes a decomposition of the problem into reasoning steps. Figure 3.3 shows part of this internal goal structure in MYCIN. One method for determining the type of the infection brings into consideration whether the infection is meningitis and whether the patient has leukemia. To determine if the patient has leukemia, the program checks to see if the patient is immunosuppressed, and so on.

The explanation capability of MYCIN (Scott, Clancey, Davis, & Shortliffe, 1977) is based upon the assumption that these steps, provided by a human expert, will make sense to the consultation program user. Figure 3.4 shows how the steps factor into an explanation of why MYCIN is requesting a datum. After the rule currently being considered is printed, repeated "Why" questions are answered by moving up the inference stack. Thus, the goals of Fig. 3.3 are mentioned in the context of the particular rules now being considered that tie them together. This constitutes the program's "line of reasoning."

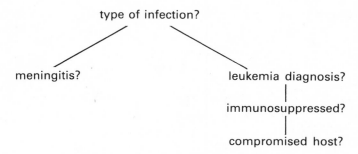

FIG. 3.3   Portion of MYCIN's goal structure.

In the "question-answer" interaction after MYCIN has given its advice, the user can inquire in more detail about how data was used and how conclusions were made see (Fig. 3.5). Again, all explanations are based on the rules and goals of the system.[3]

There are two important kinds of explanations that MYCIN cannot give: *it cannot explain why a particular rule is correct, and it cannot explain the strategy behind the design of its goal structure.* These deficiencies only became important to us in the course of developing GUIDON. In effect, we were forced to reconsider our conception of "transfer the expertise," the model by which we viewed the process of representing expertise and using it in an AI program (see Fig. 3.6).

Instead of thinking of expertise transfer as a process of moving knowledge down a pipeline from an expert (via Davis' Teiresias system [Davis, 1979]) to a client or student, we were forced to be more precise about the different kinds and purposes of knowledge. At a certain level, MYCIN is aphasic—able to perform, but unable to talk about what it knows. Teaching and explanation, we came to recognize, place different demands on an expert than simply solving problems. A teacher can provide analogies, multiple views, and levels of explanation which are unknown to MYCIN. In building MYCIN, we did not make explicit *how an expert organizes his knowledge, how he remembers it, and strategies he uses for approaching problems.*

These shortcomings became clear in the course of developing GUIDON, a project undertaken in a typical AI way as a "constructive experiment." GUIDON was a test of the assumption that the rule base could be used for purposes other than problem solving, so in designing the system we held the rules fixed. Moreover, to test the idea that teaching knowledge could be represented wholly separate from domain expertise, the system was designed to work with any rule base written in MYCIN's language (see van Melle, 1980). Therefore, the rules are treated syntactically: GUIDON's teaching procedures are

---

[3]The AI technology that makes this possible—giving the program knowledge about its representation so that it can dissect its rules—is not of prime concern to us here. The interested reader can find details in Davis (1976).

23) Is John Smith a compromised host?
**WHY

[i.e. WHY is it important to determine whether PATIENT-1 is a compromised host?]

[3.0] This will aid in determining whether PATIENT-1 is immunosuppressed.
   If
      [3.1] *John Smith is not a compromised host*
   then
      It is definite (1.0) that John Smith is not immunosuppressed
   [RULE343]

[back to question 23 . . .]
**WHY

[i.e. WHY is it important to determine whether PATIENT-1 is immunosuppressed?]

[4.0] . . . in order to determine the diagnoses of PATIENT-1.
   If
      [4.1] *John Smith is not immunosuppressed*
   then
      it is definite (1.0) that leukemia is not one of the diagnoses of John Smith
      it is definite (1.0) that lymphoma is not one of the diagnoses of John Smith
   [RULE395]

[back to question 23 . . .]
**WHY

[i.e. WHY is it important to determine the diagnoses of PATIENT-1?]

[5.0] . . . in order to determine the type of the infection.
   it has already been established that
      [5.1] *the infection which requires therapy is meningitis, and*
      [5.2] *only circumstantial evidence is available for this case, and*
      [5.3] *John Smith does not have a tb risk factor*
   Therefore, if
      [5.4] *1) Leukemia is one of the diagnoses of John Smith, or*
      2) Lymphoma is one of the diagnoses of John Smith
   then
      there is weakly suggestive evidence (.3) that the type of the infection is not tb
[RULE056]

[back to question 23 . . .]
**NO

FIG. 3.4   Excerpt of MYCIN consultation showing explanation capability. (User input follows**.)

**DID YOU CONSIDER WHETHER THE PATIENT WAS AN ALCOHOLIC?
RULE535 could have used:
> whether John Smith is an alcoholic

Clause 5 of RULE535 {the patient is an alcoholic} was already known to be
. false for CULTURE-1.

**HOW DID YOU KNOW THAT THE PATIENT WAS NOT AN ALCOHOLIC?
RULE343 was used to conclude that John Smith is not an alcoholic. The
last question asked before the conclusion was made was 23.

**WHAT WAS QUESTION 23?
Question 23 was asked in order to find out whether John Smith is a
compromised host in an effort to execute RULE343.

FIG. 3.5   Excerpt of question/answer interaction after a consultation.

designed to discuss how a goal is achieved in terms of the data appearing in rule
premises and to evaluate a student's hypotheses in terms of conclusions made by
the rules. Figure 3.7 shows an excerpt from a dialogue with GUIDON.[4]

GUIDON was improved over time by annotating the rules to separate the
"key factors" of the rule from contextual and "screening" clauses that determine applicability of the rule (so complex rules with 5 clauses and a table of
conclusions can be summarized by saying "the CSF protein is evidence for viral
infection"). Clause distinctions of this kind are part of the implicit design knowledge that is unknown to MYCIN because it lies outside of the rule syntax.

But simple annotations were insufficient; the knowledge base also lacked
medical knowledge necessary for teaching. We found that students were unable
to remember the rules, even after discussing a single problem with GUIDON
many times. Students who apparently knew what data to collect were unsure of
the order in which to collect it, and consequently had no confidence that their
investigations were complete. This experience suggested that the program
needed to teach a problem solving strategy that a student could follow, as well as
some underlying mnemonic structure for understanding and remembering the
rules. No formal experimentation was necessary, the program plainly lacked the
necessary medical knowledge.

FROM GUIDON TO NEOMYCIN (A COGNITIVE
SCIENCE ENTERPRISE)

In the course of studying the teaching problem, we learned that the expertise and
explanations of MYCIN are narrowly conceived. On the one hand, we have not
captured all that an expert knows, for example, his causal models of disease

---

[4]The teaching procedures are not our main concern here. (See Clancey, 1979a,b for details.)

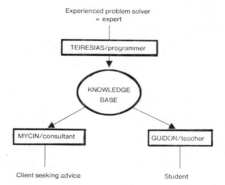

FIG. 3.6 Transfer of expertise: Learning, Advising, and Teaching.

processes by which he understands rules and can use as a basis for violating them. On the other hand, some of what we have captured is *implicit* in the rules, namely the taxonomic structure of diseases and the search strategy (top-down refinement). This knowledge is procedurally embedded in the choice of subgoals and their ordering in a rule. This is illustrated by the alcoholic rule (see Fig. 3.8).

Clauses 1–3 specify that this rule about particular bacteria will only be applied after bacterial meningitis infection is established (three levels of the taxonomic hierarchy). Clause 4 is based on the fact that children are usually not alcoholics, illustrating that the rules are based on implicit knowledge about the world, too. Finally, the rationale for associating alcoholics with the listed bacteria is not represented. Figure 3.9 illustrates the different kinds of knowledge that the human expert relied upon to construct this rule, which we did not represent explicitly in the program. The kinds of knowledge are labeled as strategic, structural, and support knowledge.

The MYCIN program shows us clearly that the task orientation to develop a program with a high level of performance alone does not lead to a process model of human problem solving. MYCIN does subgoaling, as people sometimes do, but it doesn't do diagnosis like people. For one reason, subgoaling is not the key element of diagnostic rule application; focused forward-inferencing is. For teaching purposes, we need to model how an expert uses and remembers his knowledge—not just capturing the associations he makes, but capturing also why these associations come to mind. It is the task orientation of tutoring that makes these considerations relevant and that will be the measure of adequacy for the models we construct.

To recap, in building an intelligent tutoring system, we are forced to move beyond the constraints of performance and consider the psychological constraints of teaching. We need to be able to articulate how the rules fit together, how they are constructed. We have studied MYCIN's rules and developed an epistemology of the kinds of knowledge that relate to the teaching of heuristics (see Fig. 3.9 and Clancey, 1983). Following the theory, a new representation was developed in which the original MYCIN rule set is reconfigured to make these kinds of

**FACTORS

The following factors will be useful to determine the type of the infection:

    3a. whether the meningitis is partially treated

    3b. whether a lumbar puncture has been performed on John Smith

    3c. the duration of the neurological signs

    3d. the CSF protein

(*) 3e. whether John Smith has symptoms of Herpes Zoster virus

{Section omitted}

Factors marked with '*' have already been stated. Based on what we have already mentioned, you should be able to conclude about: 3b.

**USE 3C

The duration of the neurological signs is 7.2 hours.

Using other information available to you now, the duration of the neurological signs enables you to conclude that the type of the infection is bacterial (.2) viral (.2) [RULE524].

**IKNOW 3B

Good. MYCIN has also made a decision.

The site of CULTURE-1 is evidence that a lumbar puncture has been performed on John Smith [RULE112].

Back to our discussion of the type of the infection. . . .

**3D

The CSF protein is 40.

**HYPOTHESIS VIRAL

You're right, there's strong evidence that the type of the infection is viral, given the CSF protein, the duration of neurological signs, and the fact that the patient has symptoms of Herpes Zoster virus.

What other facts about this case tell you that the type of the infection is viral?

(Enter one factor per line.)

>>

FIG. 3.7   Excerpt of GUIDON interaction.

RULE535

   If: 1) The infection which requires therapy is meningitis,

       2) Only circumstantial evidence is available for this case,

       3) The type of the infection is bacterial,

       4) The age of the patient is greater than 17 years,

       5) The patient is an alcoholic,

Then: There is evidence that the organisms which might be causing the infection are diplococcus-pneumoniae (.3) and e.coli (.2).

FIG. 3.8   The Alcoholic Rule.

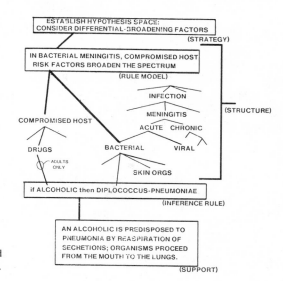

FIG. 3.9 Strategic, structural, and supporting knowledge for a heuristic.

knowledge explicit (Clancey & Letsinger, 1981). Figure 3.10 illustrates the main components of this new system, NEOMYCIN. With its theoretical, epistemological underpinning, NEOMYCIN is designed to represent the subject material that a new version of GUIDON can use to articulate important teaching points.

Figure 3.10 shows that the key feature of NEOMYCIN is separation of domain-specific disease knowledge from general procedures for doing diagnosis. The strategical knowledge "gets a handle on" the disease knowledge by way of alternate "views" or structural organizations of the disease knowledge. It is through these general indexing relationships, such as the hierarchical relationships of "sibling" and "father," that a general procedure can examine and select specific problem-solving knowledge to apply it to a given problem. The causal view indexes the disease hierarchy through causal abstractions. (For example, "double vision" might be caused by "increased pressure in the brain," which might be caused by "a brain tumor," "a brain hemorrhage," and so on.) The process view pertains to general features of any disease, which describe its location, progression of symptoms, degree of spread, and so on—general concepts by which the problem solver can index his knowledge about diseases to compare and contrast competing hypotheses. Figure 3.10 shows the working memory which will be described later.

So far we have been considering how NEOMYCIN, as a representation, adheres to an epistemological theory of knowledge, that is, how it separates out expertise by the divisions suggested in Fig. 3.9. The "content" of NEOMYCIN is a psychological theory for gathering and interpreting new data, in part, the content of the meta-strategy box in Fig. 3.10. NEOMYCIN embodies a psycho-

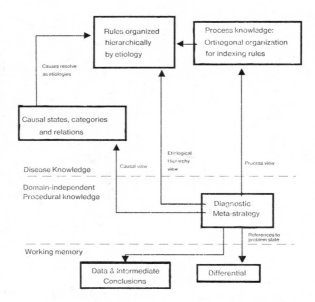

FIG. 3.10   Components of the NEOMYCIN expert system.

logical theory of medical diagnostic reasoning for the purpose of monitoring a student's problem solving and providing assistance that a student can follow. For example, we will be teaching forward-directed inferences—leaps from data to hypotheses—that we represent in NEOMYCIN's *trigger rules*. With this additional knowledge of how people think, GUIDON Version 2 will have leverage for interrupting the student to test his knowledge, as well as having a better basis for understanding a student's partial solutions.

Although this chapter is not primarily about teaching strategy, we hasten to clarify that we do not propose to directly teach students a model of what experts do. Indeed, the epistemological separation of knowledge in NEOMYCIN brings out individual steps of reasoning that we believe are "compiled" in experienced problem solvers, just as in MYCIN's original rules. The point of the decomposition is to provide a rationale for surface expert behavior so a student can understand it. Thus, on the surface NEOMYCIN is designed to behave like an expert in its focusing, data collection, and hypothesis formation. Moreover, the types and organization of knowledge are those of an expert. But the process itself is drawn out here into "diagnostic tasks" (the meta-strategy) that we believe an expert follows when stuck, but generally does not consciously consider, knowing what to do in each situation from years of practice.

Furthermore, we have not specified how this material will be presented to a student. The sequencing of material and various support stories for understanding and memory are part of the theory of teaching which we do not address here.

THE RELATION OF THEORY, AI FORMALISM, AND PROGRAM

NEOMYCIN is more than an ordinary AI system built to simply do some task. It is not an ad hoc system built to get performance—it is an implementation of a theory of diagnosis and certain principles for representing knowledge. Our tutoring goals require that the program combine both: a theoretical model of medical diagnosis, so that the student's problem solving can be interpreted and advice offered; and an epistemological theory of knowledge, so that this model of diagnosis can be articulated to the student. These theories are instantiated in a program by way of AI formalisms for representing and controlling knowledge, some of which are novel and grew out of the theoretical goals. Figure 3.11 shows how theory and model are related in NEOMYCIN. This section describes in more detail how the theories factor into the AI formalisms and the actual code of the system.

The Psychological Theory of Medical Diagnosis

The questions addressed by the theory of medical diagnosis we are developing are: how does a physician use problem data and disease knowledge to formulate hypotheses; to request additional data; and to reach a diagnosis? Issues pertaining to the processing of new information, the structure of disease knowledge, the nature of procedural knowledge, and its relation to disease knowledge, among others, are appropriate. The theory is general, both in its application to multiple problems in a given domain and its potential applicability to other domains, thus the problem of arbitrariness in process models (VanLehn, Brown, & Greeno, this volume) is partially ameliorated. Underlying regularities become manifest through this constant consideration of multiple tasks and multiple domains (Kosslyn, 1980).[5]

The theoretical features described here do not literally appear anywhere in the program. These are descriptions of behavior that were written down clearly and explicitly before any coding began. They were not extracted from the program; they were designed into it.

In writing down the principles of the theory, we were almost always thinking about their implementation in the program, often requiring that we return to be more precise about the theory. For example, we could not simply write down that, "data are used immediately in a forward-directed way." Should every rule that uses new information be allowed to fire? This did not fit our observations.

---

[5]This is on top of the principled character of representation deriving from our epistemological framework. Generality stemmed, first of all, from our need for teaching general principles to students. Ultimately, the enterprise has engineering value: we can lift the representation framework as well as the domain-independent diagnostic strategy into another problem domain and develop a new consultation system with this as a starting point.

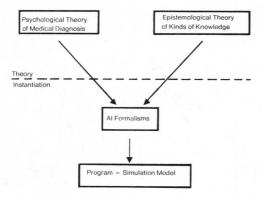

FIG. 3.11   Relation of Theory to AI Formalisms and the Model.

For example, when thinking about "steroids" in the context of a possible meningitis case, inferences are obviously focused by the problem at hand; tuberculosis might come to mind, not the possibility of a law suit in college athletics. In turn, our evolving knowledge representation (also on paper) suggested that this focusing might be modeled by only firing rules that appear in subtrees of the etiological hierarchy below hypotheses currently being entertained (so "tuberculosis" would come to mind because it is below "meningitis").[6]

The following is a brief presentation of the key theoretical features of NEOMYCIN, fairly similar to how they appeared before we wrote any code:

A.   Incoming data are immediately applied by forward-directed reasoning leading to more abstract descriptions of the problem and support for specific diagnostic hypotheses:

1)   trigger rules place hypotheses on the differential (working memory of hypotheses) directly as data is received. The differential is maintained so more specific causes replace general hypotheses;

2)   data are abstracted immediately, for example, "diplopia" is thought of as a "abnormal neurological finding";

3)   process-oriented questions are immediately asked, relevant to the domain, but not directed to any particular hypothesis, for example, asking when a symptom began and how it has changed over time;

4)   data suggest causal state-categories, possibly jumping over a chain of causal links to conjecture some generic problem whose subtypes are later considered (as "brain pressure" suggests "space-occupying substance in the brain" rather than the specific causes of "brain pressure);

5)   data/hypothesis associations are applied in the context of the current differential. Only associations that appear in subtrees below the current hypotheses come to mind;

---

[6]This idea bears some obvious relation to frame theory; an elaboration is beyond the purpose of this chapter.

B. The following knowledge sources are represented separately and explicitly, in accord with the epistemological theory:

    1)   a problem-space hierarchy to which data/hypothesis rules are attached, "etiological taxonomy" (previously implicit as the "context clauses" of rules);

    2)   causal rules that ultimately tie into this hierarchy (see Fig. 3.10);

    3)   world relations that constrain the relevance of data (previously implemented as "screening clauses");

    4)   disease process knowledge that cuts across the etiological distinctions, useful for initial program formulation.

C. A hierarchical set of domain-independent meta-rules constitute a diagnostic meta-strategy. These rules examine the knowledge sources listed above and the current differential to select a hypothesis to focus on and the next datum to request.

Turning now to the content of the strategic rules, we determined that the key strategic idea to teach students is that the purpose of collecting circumstantial evidence, in preparation for making physical measurements, is to "establish the hypothesis space," to determine the *range of possibilities* that might be causing the problem. Strategies for achieving this involve looking for evidence that will broaden the space of possibilities by considering common and unusual causes.

There are two orientations when establishing the hypothesis space: 1) "group and differentiate"—upward-looking, initial problem formulation in which one tries to cluster the data under some generic process (cause); and 2) "explore and refine"—attempting to confirm successively more specific causes. The trigger associations mentioned here bring the problem solver "into the middle" of his problem space hierarchy. These strategies together establish a path to a diagnosis.

The initial problem formulation we want to teach goes beyond MYCIN's expertise, requiring both the strategy of "group and differentiate" as well as additional medical knowledge. Essentially, we want to teach a student not just how to confirm that meningitis is present, MYCIN's task, but when one should think about meningitis, and what it might be confused with. One normally associates these questions with the "primary care" physician, as opposed to a consultant like MYCIN. These perspectives, stemming from our tutorial goals, led us to adopt a more theoretical understanding of the task of diagnosis itself.

The AI Formalisms of NEOMYCIN

A tacit principle of AI is that an AI program must be describable in terms of theoretical formalisms of knowledge representation and control. Thus, in a real sense we might move the theory/instantiation" line of Fig. 3.11 to below the "AI formalisms" box. For just as what we write down about trigger rules in our psychological theory is separable from its implementation as code, the mathematical, logical, and AI concepts of "antecedent rule," "hierarchy," and the like are abstractions for entities and processes in our FORTRAN or INTERLISP programs. However, they are apparently "closer" to our code than is the psychological theory, often even designated by procedure and variable names that make the correspondence explicit to the programmer.

A good example of the use of AI technology in NEOMYCIN are the diagnostic strategies, which are represented as meta-rules, an adaption of a pre-existing formalism (Davis, 1976). These rules are applied as a *pure-production system* for each subtask (e.g., "find a new focus" is a subtask). *Abort conditions* are inherited to simulate shifting of focus (and return to higher goals) as data broadens the differential or exploration suggests that a conjecture is unlikely.

We mention these examples of AI formalisms in NEOMYCIN to illustrate the point that a cognitive scientist doesn't simply sit down and write any program whatsoever as a model of his theory. As in mathematics and logic, there are certain notations that have been developed for couching theoretical relations, and the notations evolve as the theories become more complex. The work of writing AI programs is made much easier by previous efforts to abstract representational devices such as "meta-rules." These devices become like a bag of tools for expressing theories. In order to communicate the NEOMYCIN model to other AI programmers, it was essential to adapt whatever tools were already in common use, rather than inventing new terms or arbitrarily combining old formalisms. So in describing NEOMYCIN, indeed in *thinking about it,* we say that the meta-rules are applied as in a pure production system; the disease process knowledge is represented as a frame associated with each disease; and so on. Furthermore, AI's bag of tools provided a ready-at-hand, suggestive set of organizational and processing concepts for expressing the psychological theory. Finally, in this special case AI provided the data (MYCIN's rules) that enabled us to study human knowledge in a new way.

When is a Program Ad Hoc?

The scheme shown in Fig. 3.11 provides an interesting handle on the question of ad hocness in computational models. It shows that there are multiple perspectives from which the model can be said to be ad hoc. From the AI perspective, code is ad hoc if it is loosely put together without regard for unified, simple, and elegant formalisms. If NEOMYCIN's diagnostic meta-strategy had been implemented in INTERLISP procedures directly, instead of a hierarchy of meta-rules applied cyclically with abort conditions, etc., the implementation would be said to be ad hoc. Here ad hocness would have interfered with our teaching goals as well as program maintenance.

Moving up a level, if we had used MYCIN's rule language, an AI formalism, instead of the NEOMYCIN scheme of an etiological hierarchy combined with meta-rules, our implementation would have been said to be atheoretic from the epistemological perspective. That is, we would have represented different kinds of knowledge in a uniform way, losing distinctions—in some sense the essence of an ad hoc implementation.[7] Indeed, it was the ad hoc representation of strategies and taxonomic concepts in MYCIN rules that limited its usefulness for teaching.

Finally, looking at the left theoretical arm of Fig. 3.11, an implementation can be ad hoc from the psychological perspective. If we had persisted in using exhaustive, top-down refinement, as in MYCIN, and several other medical AI systems, we would have constructed a program that does medical reasoning, but in an ad hoc way, limiting the usefulness of the program for interpreting student behavior. Note that exhaustive top-down refinement is not an ad hoc implementation from the AI formalism perspective, but it is a psychologically implausible model of search.

Observe that from all three perspectives, it is the task orientation that determined what aspects of the implementation were relevant, those which should not be done in an ad hoc way. In general, the question of where we should "draw the line between implementation detail and relevant model content" (Miller, Polson, & Kintsch, this volume) depends on what we want to model, what we want the program to do. The attempt to apply the model to a real world task will provide the empirical feedback that reveals what was ad hoc and now needs to be implemented in a theoretical way. But note again, we do not *extract* the theoretical principles from the program (contrast with Pylyshyn, 1978) we write them down, then build them in. By default all other coding decisions will be ad hoc, and we won't know whether that matters until we do more testing.

Summary of NEOMYCIN as a Model

To summarize, NEOMYCIN is an information-processing model that uses AI formalisms to instantiate psychological and epistemological theories of knowledge and processing:

1. The epistemological theory specifies how different kinds of knowledge interact, specifically how organizational knowledge interacts with strategies.

2. What the expert does is not simply listed: the strategies are domain-independent, they specify how different kinds of knowledge sources are called into play to massage a guess that is being constructed and refined (domain-independence makes the process model more psychologically plausible and extensible, [van Lehn, Brown, & Greeno, this volume]).

3. Associations of data with hypotheses are described in terms of the working

---

[7]Notice the tension between the epistemological and AI formalism levels: without uniformity there is no formalism, but the uniformity chosen may not allow important distinctions to be expressed.

memory and a structured representation of the problem space (following from the diagnostic theory).

4. The model of strategies specifies hierarchical organization of knowledge in the form of rules for achieving tasks; the problem solver is said to be oriented to "what he is trying to do" (diagnostic theory).

5. different kinds of follow-up questions are not simply listed: The model specifies how subgoals can be set up by associations that trigger when data is received, and how immediate follow-up questions are associated with data abstractions (diagnostic theory).

In short, NEOMYCIN specifies organization of different kinds of knowledge and processes by which this knowledge is called into play. It is a model that relates a working memory to the kinds of associations people try to think about and why they remember them at particular times. The overall theory is complex; the computer program provides a practical means of testing the coherency and completeness of the theory.[8]

METHODOLOGY

We began with an extensive review of the medical problem-solving literature (e.g., see Elstein, Shulman, & Sprafka, 1978; Feltovich, Johnson, Moller, & Swanson, 1980; Miller, 1975; Pauker & Szolovits, 1977; Rubin, 1975; Swanson, Feltovich, & Johnson, 1977). We returned to this previous work to augment and refine the NEOMYCIN model of diagnostic strategy, after the study described here was complete.

From the period February 1 through December 1, 1980 we met regularly with a physician consultant with the purpose of revising MYCIN's rules to make the teaching points clear. Protocol analysis (using cases MYCIN had previously solved) was the chief method. We also attended classes taught by this physician and compared them to another physician's handling of the same course. In addition, we presented several cases to our physician's best student to compare his reasoning and explanations to his teacher's.

Our physician's approach was logical and easy to emulate. After listening to several other physicians and sitting in on other classes, we decided that we had found an unusually good teacher, someone who was consistent from case to case, and moreover did what he told students to do. Other teachers we observed were not able to articulate their approach as clearly and seemed to be less sure of what students were thinking. There were common strategical concepts, however, that our experts all used to explain their reasoning ("hit the high points," "consider risk factors"). In our opinion, the reason our physician was a good teacher was

---

[8]See (Kosslyn, 1980) for further discussion of the relation of theories to programs as models.

because his explanations were not as "flat" as other physicians'. Rather than saying, "Well, the patient hasn't traveled, so it isn't Valley fever," he would say "Well, travel would have widened the spectrum of possibilities, so we can rule out things like TB picked up in Mexico." That is, he supplied abstractions that said *what he was trying to do,* how his thinking was oriented.

Our framework of structural, support, and strategic knowledge for organizing, justifying, and controlling the use of heuristic rules served well in knowledge acquisition dialogues. We would always ask ourselves, "What kind of explanation is he giving us? A data/hypothesis rule? Why he believes a rule? Why he thought to consider that association (the indexing, the approach)?" We organized these kinds of knowledge around each rule we discussed (Fig. 3.9), and directed the conversation appropriately. In contrast, several years ago, before deriving this framework, our interviews tended to take a depth-first plunge into pathophysiological details (we always asked, "And what causes that?"), which did not shed much light on the physician's strategies and organization of data/hypothesis relations.

We tape-recorded sessions whenever a case was presented to the physician. A note file was maintained in which we recorded what we learned from each meeting. A summary of the kinds of interactions is given below (in the order they occurred).

A. Informal discussion of a case previously diagnosed by MYCIN.   The experimenter presents data and asks how it is useful. Among the points of discussion: how the expert cuts up the problem (for example, acute vs. chronic), how he remembers data/hypothesis relations (diagnostic values are related to a mnemonic story), the significance of frequently-mentioned problem features ("predisposition," "compromised host"), how urgency and faulty data factor into reasoning. Later comparison of the expert's terms and rules to MYCIN's suggest questions for subsequent meetings.

B. The expert solves a case, while the experimenter actively questions his reasoning throughout.   Initial data is presented, the expert must then request information in any order he desires, and make a diagnosis. Among the items we record: the differential (hypotheses under consideration); strategy (either a domain-specific goal, such as "look for evidence of a focal lesion," or a domain-independent goal, such as "pursue most likely causes first"[9]); rule-like associations ("diplopia suggests increased brain pressure"); and meta-statements about strategy ("think before the lab results, not from the results"; "make reliability checks of data"). NEOMYCIN's strategy rules were first derived from analysis of one of these protocols.

---

[9]These are often stated aphoristically as well: "When you hear hoof beats, think of horses, not zebras."

Designing a general program from a single, typical interaction is a common method in AI. The knowledge base designer idealizes the interaction, specifying knowledge (frames, rules, etc.) and processing (general procedures, strategy rules) that will bring about the desired program behavior in the particular case. GUIDON's tutorial procedures were first sketched out in this way by proceeding from a sample interaction in which we played both the part of the student and teacher (generating realistic student input and then looking in the rule base to find what response would be satisfying). The program is generalized and debugged by testing it on many other cases afterwards. For example, single statements might become separate procedures as the complexity of problem situations becomes better understood. This method presupposes that the general framework of the system (or metastrategy in the case of NEOMYCIN) can be induced, at least in preliminary form, from any particular problem solution.

C. The expert is asked to describe a typical case for each of the main diagnoses. The expert finds this easy to do. This method brings out the diagnostic or invariant associations, as well as what evidence is required to rule-out competing hypotheses. For comparison, the expert is asked to describe atypical presentations of the same disorders. (In these cases, the expert gives the impression that he is telling a joke.) From these analyses, we developed a theory of what makes a case easy or difficult.[10]

D. The expert is asked to present a case to the experimenter, reversing the roles of method B. This helps the experimenter determine whether he has formalized an "executable procedure." This method quickly reveals any gaps in knowledge or approach that have not been extracted from the expert. The expert is asked to present both easy and difficult cases so the evolving model can be more fairly evaluated.

E. The same cases discussed in B and D are presented to different experts. Because we already understand the significance of the data (the data/hypothesis rules) we are especially interested in comparing strategies that bring the data to mind.

F. The developed strategy model is presented to the original expert for his evaluation. What resonates with his thinking? What does he care to elaborate upon? Where do students have problems? (The expert says things like, "Most

---

[10]*Diagnosticity* (sharpness of measurements (classicality); presence of important factors (but not necessary); presence of invariant factors (sufficient); *dissonance* (absence of extraneous, unexplained factors) and inconsistencies (unexpected factors)); *the a priori likelihood of the problem* (expert has less confidence in unlikely diagnoses); and *multiplicity of cause* (before reaching a diagnosis, the expert will struggle to find a simpler, single explanation).

students encounter roadblocks—they're not sure what to do next. They focus too narrowly and specifically on details of the case.'')

G. *The same cases discussed previously are presented to the expert's best student.* We find which phrases have been picked up (''establish a data base'') and how the student carries out the strategies he has learned. For example, a student might verbalize his reasoning more slowly and carefully, providing some details that the expert skips over.

H. *We discuss each rule with the expert, grouping them according to the hypotheses they support* (e.g., rules that conclude ''bacterial meningitis'').
From this analysis, we fill in the structure of data and the hypothesis space (e.g., we find out about different kinds of compromised hosts) and acquire a support story for each rule (why it is believed to be correct).[11] By asking ''when would you think about requesting this datum,'' we are able to cross check our strategic concepts and rules.

In summary, the methodology used to develop NEOMYCIN was task-oriented, namely to acquire the knowledge to place MYCIN's rules in order so they were more useful for teaching. We originally intended to simply ''clean up'' the rules, but decided that a more radical change in MYCIN's control structure was called for (use predominantly forward-directed reasoning instead of backward chaining).

To implement the expert's strategy, we had to translate his task statements (''establish the etiology'') into more procedural terms (''establish a grouping of possibilities by confirming a path upwards in the hierarchy''). The idea that the initial problem formulation takes the expert into the middle of an etiological hierarchy was not stated by the expert. In fact, the concept of ''initial problem formulation'' came from previous work in problem solving.[12]

The general methodology that we are following is summarized by this next list. (NEOMYCIN development is now iterating in Steps 4 to 6):

1. Formulate design guidelines
   (This is the task orientation: What should the system do?
   Who will use it? This conception may change over time.)
2. Model system on paper (hand simulations) (Steps 4–6)
   (This may take several months or more than a year, including the experiments described in this section.)
3. Code/modify program
   (including simplifications for elegance)

---

[11]We discovered that some rules were redundant or simply encoded incorrectly; some problem situations were not considered; some rules were ''folklore'' and not worth teaching.

[12]Significantly, the expert did tell students ''you have to search the tree of possibilities,'' so he knew something about how he organized his knowledge.

4. Experiment with program
    —observe behavior on test cases
5. Analyze program behavior (to determine shortcomings)
    —determine appropriateness (expert perspective)
    —assign credit and blame to code sections, determining if there is a programming error or shortcoming in the general theory or domain specific knowledge
6. Theorize/reformulate model (to eliminate shortcomings)
    —restate theory principles and/or collect domain knowledge through reading and dialogue with expert
    —use, modify, and develop programming technology
7. Go to Step 3.

Testing NEOMYCIN will cover both its performance (comparing it to MYCIN) and use for teaching (incorporating it in GUIDON Version 2): We expect that our experience with students using GUIDON2 will enable us to refine the expert model and to construct, perhaps as a variation of NEOMYCIN, a preliminary model of novice diagnostic thinking.

METHODOLOGICAL PITFALLS

In the course of developing a program like NEOMYCIN, it is possible to lose the way temporarily. The pitfalls of an AI orientation to Cognitive Science include the problems of introspection, non-empiricism, and over-formalization.

Introspection and Representation

In order to understand what the expert was teaching us, we drew diagrams of the hierarchies of data, hypotheses, and rule generalizations. Then, in trying to understand the expert's strategies, we found outselves remembering these diagrams, so we were unable to separate our interpretation of the expert's behavior from our evolving representation of his knowledge. In particular, we came to realize that the structures we had drawn could account for the expert's reasoning in multiple ways, and we had been mistaken to think that we were capturing structures that were isomorphic to something that was "in his brain."

Some examples of this phenomenon might be useful. When the expert learns that the patient has a fever, he frequently will ask for details (severity, periodicity, etc.). This is modeled in NEOMYCIN as "process" questions that are directly associated with the concept of "fever." Yet, one could also say that the expert is thinking about a particular cause of fever, so asks about severity, for example, to see if the fever confirms his guess. This is in fact how Ann D. Rubin (Rubin, 1975) interpreted this kind of question, and it is consistent with her general model of hypothesis formation. However, we found no reason to postulate the intermediate steps of reasoning (setting up an hypothesis), even though the follow-up question is relevant because it is potentially useful.

The point is that in interpreting expert behavior we can easily crank through the reasoning processes and knowledge structures we have already formalized, producing system performance that matches the expert's but which does not simulate his reasoning steps (associations). The cause of this problem is that people's associations can be ad hoc, made efficient through rote, and are not restricted to the principled structures of subtype, causality, process, and so on that we postulate in a system like NEOMYCIN. This is the idea that knowledge can be "procedurally attached" and doesn't need to be stepped through in declarative form (Winograd, 1975). (Anderson's program for modeling learning is based on proceduralization of this form [Anderson, Greeno, Kline, & Neves, 1980].)

In NEOMYCIN, we have attempted to capture the "compiled associations" of the expert, while labeling them to record their principled basis. Thus "acute and chronic," process terms, are placed in what should be a strictly causal network (the etiological taxonomy, Fig. 3.9). Similarly, the expert doesn't always clearly distinguish between the concepts of "subtype" and "cause," so a principled representation that does make this distinction must be interpreted by procedures that blur the difference.

Our investigation indicates that people form associations on any useful basis, and it is not trivial to find principled theories for the basis of these associations. For example, Pople is trying to account for how classificational and causal knowledge are combined. Pople's concept of "bridge concepts" provides a first order theory of how "trigger associations" evolve by combining the two kinds of associations through a form of transitive closure (Pople, 1982). However, this model predicts far more trigger associations than expert behavior demonstrates. We will need to refine this theory by appealing to notions of complexity and usefulness of triggers.

Similarly, we can find "proceduralized associations" which have been learned by rote instead of the kind of composition that Anderson's model describes. For example, an expert considering fungal meningitis tends to ask about travel first; considering virus, he asks about absenteeism in the schools (for a child); considering TB he asks about crowded conditions and previous illnesses. We can explain these questions in terms of the principle "try to confirm the enabling step of a causal process first." Thus, in infectious disease diagnosis one first tries to establish exposure to the causative agent. But this is a rationalization, for neither we nor the expert learned what questions to ask in this way.

In conclusion, one pitfall of modeling using the AI-oriented approach we describe is the tendency to be satisfied with a consistent, coherent model (a knowledge representation and model of reasoning for diagnosis, learning, explanation, etc.) that produces the same behavior as the expert. Because we can learn by rote and we are able to compose factual knowledge with procedure, an expert's associations may be more complex, and not fit the formal elegance of the program. But relying only on introspection, and introspectively observing that we can reach the same results as the expert by reasoning like the program, we can be misled into thinking we have modeled his reasoning. More precise experimentation is necessary if we hanker after psychological validity.

Empiricism and Technology

In developing the first version of GUIDON, we were dangerously close to saying that because we could relate a student's partial solutions to MYCIN's rules, we had an explanation of his reasoning—as if just because a model could be constructed by a program, it was accurate. Similarly, it is easy to suppose that when a program is able to parse a user's English sentence (as in MYCIN's question/answer module), it has determined what the user is trying to say. One never considers that the next question could be a restatement or request for clarification—it is just the "next input." In a variant of the introspective pitfall, the programmer is now thinking like his model of the machine. Rather than thinking in terms of what he can do with his representation (what is suggested by the technology), the AI-oriented cognitive scientist must be oriented to the phenomenon he is trying to emulate. Simulating the program in the problem-solving environment is a valuable approach.

The technological pitfall is exacerbated by those who never get their program working, so they don't get the hard shock of empirical test. In short, a program isn't a "functional model" (Pylyshyn, 1978) if it isn't functional.

Validity and Elegance

As in the hard sciences and mathematics, it is important that a computational model be formally simple and elegant. However, programming provides special opportunities for reframing and reorganization that adds nothing to the theory being programmed, and tends to even obscure its implementation. On the other hand, a theory sometimes profits from reorganization of the code that implements it, in the same way a physicist can find formal clarity by manipulating his equations, looking for symmetry and the like.

One measure of improvement is the perspicuity of the code. If the new rules or frames make it easier for a colleague to understand the theory (to see the theory in the code), the representation (and accompanying interpretation) has probably been improved. For example, a programmer may rerepresent a single rule with multiple steps in its action as a set of ordered rules with identical premises, producing what he takes to be a more elegant representation with only single steps in each rule. But this obscures the simple idea of a procedure being a block of steps. More effort is required to interpret the code to see the procedure within it, just as the problem solver would need to exert more effort to carry out the procedure. Requiring a rule for each step of the procedure therefore violates our understanding of the theory we are implementing, so we say that the representation is not improved.

AREAS FOR COLLABORATION

We now list some research problems that have been suggested by our work. In doing this, we have two purposes: first, to demonstrate that a computational model like NEOMYCIN can suggest new areas for psychological research; and second, to encourage non-AI cognitive scientists to contribute methodological assistance for attacking these problems. The list of research problems follows:

1. The structure of working memory.   Is the differential a simple list? A hierarchy? Does it include a stack of goals? For example, when refining a hypothesis, moving down a hierarchy, how is each child visited in turn? By a strategy that iteratively focuses on siblings, as in NEOMYCIN, or by a separate, "saved" list of waiting hypotheses to consider?

2. Identifying lines of reasoning.   The expert stated a rule generalization (Fig. 3.9) that might be used in multiple ways. One could think in terms of "differential broadening factors," leading to consideration of "compromised host risk factors" (data orientation). Or one could think in terms of "unusual causes," leading to consideration of "gram-negative organisms" (hypothesis orientation). Is it possible to say that the expert is following one line of reasoning and not the other? Could he in some sense be doing one thing that combines the goal and method, namely "trying to broaden the differential by considering compromised host risk factors"? Is it possible to get at the expert's line of reasoning without being misled by his rationalization? Or is it wrong to say that there is some explicit, conscious line of reasoning that we can discover?

3. The effect of problem context.   Our expert supplied details to make the cases presented to him seem more realistic ("I'm at the patient's bedside" or "I'm in the emergency room and this patient comes up to me, accompanied by her mother"). Presenting a case twice, separated by many months, we saw that this story can change the expert's approach, even leading him to explore completely different hypotheses. How does the expert's imagination of the situation affect his reasoning? What variables must be specified to control for this effect?

4. Clustering of hypotheses for manageability.   One diagnostic task is to refine a category by considering what causes it. Thus, the physician considers the types of chronic meningitis. However, a physician does not run through the several dozen organisms that might be causing bacterial meningitis. He thinks in terms of common and unusual causes to make the set more manageable. What happens when there are too many common causes to entertain? What other kinds of groupings are useful?

5. *Experimentally verifying diagnostic strategy.* How can we test NEO-MYCIN's diagnostic strategy? For example, how do we confirm that focusing on a hypothesis and asking a question to confirm it are best described as two separate decisions, made independently? Or that an expert requests details before following-up on the implications of data (process-oriented questions before making associations with hypotheses)? How can we test the control structure of strategies: a pure production system at the task level, tasks arranged hierarchically, and inherited "abort conditions"?

6. *Explanatory theory of strategy.* Can we construct a principled, explanatory theory that could in some sense generate the diagnostic meta-strategy? Viewing the processor ideas as constraints—a differential (working memory), focused activation, hierarchical problem space and problem features, trigger associations, and strategic control—how do we derive a diagnostic procedure? For example, "reviewing the differential" is not motivated by computational needs, but is a reflection of human forgetting. Rather than viewing this as a "forced imperfection" in the system, the review process (and indeed, the structure of the differential) might follow from a deeper model for retrieval of disease knowledge, along the lines of Lehnert's model of question answering (Lehnert, this volume).

7. *Modeling belief.* What makes an expert believe that a hypothesis is confirmed or unlikely? Are there general principles for dealing with missing data, for knowing when to drop a losing line of inquiry, or to return to a previously discarded hypothesis?

8. *Shifts of attention and noticing subproblems.* When the problem solver gets more data, he may be receiving information that supports a hypothesis he is not currently considering. What determines whether he does/can shift attention temporarily? The NEOMYCIN model allows for focused associations to other hypothesis, but does not allow for "filing a reminder" to take something up later or noticing that a hypothesis is ruled out, so it is not considered later. What does the problem solver notice about other parts of the problem as he moves along and what kinds of notes does/can he make to himself to affect his performance later? What kinds of errors might shifts of attention cause? How does the problem solver avoid retracing his steps? If the current differential is poorly grouped, circumstantial evidence might support widely different hypotheses. Might this ambiguity be a likely point of error, in which one of the interpretations is missed? Are there meta-cognitive strategies for checking these errors?

9. *Effect of level of abstraction on problem formulation.* In discussing the same case separated by the period of months, the expert stated his initial differential (guess) differently. In one case he said "mass lesion." In the other case he broke this down into subtypes. Very clearly, stating the subtypes brought other

associations to mind, leading to a quite different exploration (using the same strategies). How can we account for this choice in level of abstraction? There is a clear trade-off, for the expert forgot to consider a traumatic problem when he was so busy reciting and considering the subtypes of mass lesion. What reasoning strategies do people use to maintain a manageable level of abstraction in working memory? What errors occur?

10. Observation strategies. We need to deal with the richness of the data collection procedure: partial stories are corrected later, making backtracking necessary; data must be verified; questions must be asked so they are understandable to the layman; therapeutic benefit, urgency, and availability of medical equipment must be factored in. Expertise surely requires a good deal of common sense. Just how the two are cross-related and build upon one another are difficult questions.

THE PROSPECTS FOR COLLABORATION

In carrying out the NEOMYCIN research, we have not had as many collaborative discussions as might have been useful. Few computer science graduate students, the most likely collaborators, have the necessary LISP programming experience, a background in AI techniques, a willingness to learn medical technology, and an inclination to do psychological research. Therefore, the most immediate methodological problem we face is superficial: a lack of trained people to share in the research. But what kind of collaboration is possible? Should we think of cognitive scientists as hybrids, or as specialists sharing in a common project?

Looking at the fields of cognitive psychology and AI today we find a wide spectrum of interests and methods, particularly along the dimensions of experimentation and programming. In cognitive psychology, we find, for example, Bower at one end, doing traditional psychology experiments and no programming, but making some use of AI concepts (Bower, 1981). In the middle, we find someone like Feltovich, doing traditional experiments, but whose analyses and questions tend to be based in information-processing terminology. At the other end, John Anderson is experimenting and writing programs, to the extent that people in computer science might think of him as being in AI.

In AI we find the same kind of spectrum. On the one hand, we find researchers with a psychological bent whose main goal is to build a working program, but who periodically say "It would be interesting to find out if people work this way" (e.g., Fahlman, 1980; Friedland, 1979). This group includes the "knowledge engineers" (Feigenbaum, 1977): who have practical objectives; have fears about "listening to experts too closely" ("experts can't really explain how they reason"); and avoid the "paper modeling" of the psychologists. They want to build useful tools, therefore they are concerned with difficult, realistic problems (and never toy blocks). They want programs to be better than people, involving formalization of computational methods that perhaps people don't and can't use. Experimentation, to determine "what anybody's grandmother could have said," (as Gordon Bower puts it) is unnecessary. Talk to an expert and incorporate his heuristics. Test the program by asking the expert to point out shortcomings.

Finally, we find AI researchers using the behavioral studies of the cognitive psychologists to build a complex system for doing some real task (e.g., NEO-MYCIN, Lehnert, 1980). These researchers are output-oriented like the first group, but their task involves human interaction in such a way that the program's reasoning should model human performance. This group also includes researchers who believe that the performance of AI programs can be enhanced if we better understand how people solve problems. When they listen to an expert, they are oriented to understanding how he is reasoning, not simply filling in their representation of slots, rules, and so on. Potentially, this group could include any researcher in AI; work in learning, natural language understanding, and intelligent tutoring systems seems especially likely to benefit from cognitive studies.

In considering collaboration between AI and other fields in cognitive science, we should consider that people differ along these dimensions of interest and methodology. It is not at all clear that only people doing both experimentation and programming should be called cognitive scientists. It seems more likely that cognitive science will be made of people using interdisciplinary analogies and sharing research results.

The easiest form of collaboration is by evolution of common interests. We may not talk to each other directly very often, but we will communicate in the literature, translating ideas to our own application. For example, this is the way in which GUIDON research benefits from the work of Tversky concerning biases in human judgment (Tversky & Kahneman, 1977).

A second possibility is "mission-based" collaboration, in which we work together on a single project, sharing tasks according to our expertise. We might work in parallel—we might work with someone to precisely define a problem and months later he would return with experimental results.

It is important to remember the dialectic power of a program. The strength of cognitive science is surely in the way theories are changed and suggested by the very process of building computational models. Besides worrying that perhaps not enough formal experimentation is being done, we should be concerned that not enough cognitive scientists are writing programs, or helping to write programs. Too often experimental analysis seems to fall short by not being precise enough to be programmable. Or the simplifications to make an experiment tenable eliminate the very points that we need to build a working system (as fixed-order experiments in medical diagnosis eliminate focusing and data selection strategies).

Within the GUICON/NEOMYCIN project, the experimentation that we do in the future, outside of continuing to interview experts, will consist of having students use GUIDON2. In many respects, these trials will resemble the experiments carried on by Feltovich, and others. (As his experiments have prepared us for the kinds of diagnostic errors students make.) Our theory of knowledge representation and strategies, and our lower-level concepts of the working memory and control structure will evolve as we change the program to meet the needs of the task. It is an open question just how detailed an "explanatory theory" is needed to build a reasonably effective intelligent tutoring system. In our collective work on diagnostic tasks, "bugs" and epistemology, we are already going beyond what the average teacher knows about reasoning. As the knowledge engineers, we reach for computational methods that surpass human expertise. However, in building an intelligent tutoring system, it is not sufficient to seek improvements in formal efficiency and elegance alone; we must also ask why people fail.

ACKNOWLEDGMENTS

The members of this project at the time of this writing (1981) include the author, Research Associate in Computer Science at Stanford University; Bruce Buchanan, Adjunct Professor of Computer Science; two Research Assistants, Reed Letsinger (MS candidate in AI program) and Bob London (PhD candidate in the Department of Education); and Dr. Timothy Beckett, Research Fellow at Stanford University Medical Center. Funding has been provided in part by ARPA and ONR Contract N00014-79C-0302. Computing resources are provided by the SUMEX-AIM facility (NIH Grant RR00785).

GLOSSARY

An attempt is made here to generalize terminology beyond the medical application, though the reader should realize that some definitions are peculiar to our research project and others have a slightly different meaning in other areas of AI.

causal rules—productions of the form, "if A then B" with the interpretation that "A is caused by B."

compiled association—composition of a chain of productions into a single production, e.g., "if A then B" and "if B & C then D" might be compiled to "if A & C then D."

compromised host—in medicine, a patient in a weakened condition that increases susceptibility to disease.

data—facts about a problem in the form of direct measurements or circumstantial evidence.

differential—a list of hypotheses that the problem solver is considering as possible solutions to the diagnostic problem.

diagnostic problem—a situation, entity, or event that the problem solver attempts to explain (characterize its nature) by observing its appearance and behavior over time.

disease—in general, some underlying condition or process in a system that has an undesirable effect on the system.

disease process knowledge— discriptive facts about diseases that have been previously observed in a system along the lines of how the disease is caused and how it affects the system over time.

domain-dependent knowledge—with respect to a given kind of diagnostic problem (e.g., electronic troubleshooting) and a given problem being diagnosed (e.g., a Zenith computer terminal), those facts about the design of the system and its functionality, as well of scientific theories pertaining to its operation, that are useful for explaining how the system operates.

domain-independent knowledge—facts and reasoning procedures brought to bear in problem solving that are not domain-dependent.

etiological taxonomy—a hierarchy of diseases or possible causes of a diagnostic problem, in which the leaf nodes of the hierarchy are well-defined specific causes and intervening nodes are abstract categories of diseases.

expert—a problem solver with sufficient knowledge to make correct diagnoses a high percentage of the time and to know when a problem cannot be confidently solved using the knowledge available to him.

expert system—an AI computer program that is designed to solve problems at the expert level in some scientific, mathematical, or medical domain.

forward-directed inferences—associations between data and hypotheses that are made by the problem solver at the time new data comes to his attention.

group and differentiate—a diagnostic strategy that attempts to compact the differential so the hypotheses under consideration fall under a single node in the etiological taxonomy, generally by ruling out alternatives through discriminating data collection.

hypothesis—a disease or more general causal category that the problem solver is considering as a solution of the diagnostic problem.

intelligent tutoring system—an expert system whose domain of expertise is teaching, containing an expert system within it relevant to the area the tutoring system is teaching about.

interpreter—a program that generally follows a simple control policy for applying knowledge to the problem at hand. The interpreter for disease knowledge determines how new problem data leads to inferences being made to augment working memory. The interpreter for strategical knowledge determines how planning knowledge is used for collecting new data or changing the phase of problem solving.

knowledge base—domain-dependent knowledge represented in various AI formalisms.

knowledge engineering—the art of building expert systems by working with experts to codify their knowledge.

meta-strategy—a hierarchy of general tasks related by meta-rules, by which a problem solver directs his attention during diagnosis.

problem formulation—the task of characterizing a diagnostic problem so that the correct etiological category is brought into the differential.

procedurally embedded—knowledge that is implicit in the design of a program; for example, the rationale for ordering a sequence of steps in a particular way. A procedure is represented *declaratively* if the knowledge behind its design is explicitly represented in the system so that an interpreter can be applied to the design and domain knowledge to execute the procedure.

production rule—an association of the form, "if A then B," whose interpretation is such that when A is considered, believed, or accomplished by the problem solver, it is valid (according to some unspecified justification) to consider, believe, or achieve B.

screening relation—an association between data of the form, "A screens (for) B," with the interpretation that A should be considered before B with the justification that B might be derived from knowledge of A. For example, the sex of a patient screens for whether or not the patient is pregnant.

structural knowledge—any organizational constructs based on domain-independent relations ("sibling of," "location," "process question follow-up," "screening question"), used by a meta-strategy to index domain-dependent knowledge.

subgoal—in MYCIN, a reasoning step that appears as a clause in the premise of some production rule; for example in the rule "if A & B then C," A and B are subgoals.

subtype—a relation between disease categories, synonymous with "kind of."

top-down refinement—the diagnostic strategy of searching the etiological taxonomy in breadth-first manner starting at some node of the tree; called "refinement" because each level of the tree specifies a finer or more precise diagnosis.

triggers—production rules of the form, "if A then B" where A is a conjunction mentioning problem data which are said to "trigger" or "suggest directly" the hypothesis B, that appears in the etiological taxonomy.

REFERENCES

Anderson, J. R., Greeno, J. G., Kline, P. J., & Neves, D. M. Acquisition of problem-solving skill. In J. R. Anderson (Ed.), *Cognitive skills and their acquisition*. Hillsdale, N.J.: Lawrence Erlbaum Associates, 1981.

Boden, M. A. *Artificial intelligence and natural man*. New York: Basic Books, 1977.

Bower, G. H. Mood and memory. *American Psychologist*, 1981, *36*, 129–148.

Brown, J. S., Burton, R. R., & Bell, A. G. *Sophie: A sophisticated instructional environment for teaching electronic troubleshooting*. BBN Report No. 2790, 1974.

Brown, J. S., & Burton, R. R. Diagnostic models for procedural bugs in basic mathematical skills. *Cognitive Science*, 1978, *2*, 155–192.

Burton, R. R., & Brown, J. S. An investigation of computer coaching for informal learning activities. *The International Journal of Man-Machine Studies*, 1979, *11*, 5–24.

Clancey, W. J. *Transfer of rule-based expertise through a tutorial dialogue*. Computer Science Doctoral Dissertation, Stanford University, STAN-CS-769, August 1979. (a)

Clancey, W. J. Tutoring rules for guiding a case method dialogue. *International Journal of Man-Machine Studies*, 1979, *11*, 25–49. (b)

Clancey, W. J., & Letsinger, R. NEOMYCIN: Reconfiguring a rule-based expert system for application to teaching. *Proceedings of the Seventh IJCAI*, 1981.

Clancey, W. J. The epistemology of a rule-based expert system. *Journal of Artificial Intelligence*, 1983, *20*(3), 215–251.

Collins, A. Processes in acquiring knowledge. In R. C. Anderson, R. J. Spiro, & W. E. Montague (Eds.), *Schooling and the acquisition of knowledge*. Hillsdale, N.J.: Lawrence Erlbaum Associates, 1976.

Davis, R. Applications of meta-level knowledge to the construction, maintenance, and use of large knowledge bases. STAN-CS-76-552, HPP-76-7, Stanford University, July 1976.

Davis, R. Interactive transfer of expertise: Acquisition of new inference rules. *Journal of Artificial Intelligence*, 1979, *12*, 121–157.

Elstein, A. S., Shulman, L. S., & Sprafka, S. A. *Medical problem-solving: An analysis of clinical reasoning*. Cambridge, Mass.: Harvard University Press, 1978.

Fahlman, S. E. Design sketch for a million-element NETL machine. *Proceedings of the First Annual National Conference on Artificial Intelligence*, Stanford, Calif., August 1980.

Feigenbaum, E. A. The art of artificial intelligence: I. Themes and case studies of knowledge engineering. *Proceedings of the Fifth IJCAI*, Cambridge, Mass.: MIT, August 1977.

Feltovich, P. J., Johnson, P. E., Moller, J. H., & Swanson, D. B. *The role and development of medical knowledge in diagnostic expertise*. Presented at the Annual meeting of the American Educational Research Association, San Francisco, April 1980.

Friedland, P. Knowledge-based experiment design in molecular genetics. *Proceedings of the Sixth IJCAI*, Tokyo, August 1979.

Kosslyn, S. M. *Image and mind*. Cambridge, Mass.: Harvard University Press, 1980.

Lehnert, W. Narrative text summarization. *Proceedings of the First Annual National Conference on Artificial Intelligence*, Stanford, Calif., August, 1980.

McDermott, J. *R1: A rule-based configurer of computer systems*. Department of Computer Science, Carnegie-Mellon University, CMU-CS-80-119, April 1980.

Miller, P. B. *Strategy selection in medical diagnosis*. Project MAC, Massachusetts Institute of Technology, MAC TR-153, September 1975.

Nii, H. P., & Feigenbaum, E. A. Rule-based understanding of signals. In D. A. Waterman & F. Hayes-Roth (Eds.), *Pattern-directed inference systems*, New York: Academic Press, 1978.

Pauker, S. G., & Szolovits, P. Analyzing and simulating taking the history of the present illness: Context Formation. In Schneider/Segvall Hein (Eds.), *Computational linguistics in medicine*. North-Holland Publishing Company, 1977.

Pople, H. E. Heuristic methods for imposing structure on ill-structured problems: The structuring of medical diagnostics. In Szolovits (Ed.), *Artificial intelligence in medicine*. Boulder, Colorado: Westview Press, 1982.

Pylyshyn, Z. W. Computational models and empirical constraints. *The Behavioral and Brain Sciences, 1978, 1*, 93–127.

Rubin, A. D. *Hypothesis formation an evaluation in medical diagnosis.* Artificial Intelligence Laboratory, MIT, Technical Report A1-TR-316, January 1975.

Scott, A. C., Clancey, W. J., Davis, R., & Shortliffe, E. H. Explanation capabilities of production-based consultation systems. *American Journal of Computational Linguistics,* 1977, Microfiche 62.

Shortliffe, E. H. *Computer-based medical consultations: MYCIN.* New York: Elsevier, 1976.

Sleeman, D., & Brown, J. S. *Intelligent tutoring systems.* London: Academic Press, 1981.

Stevens, A. L., Collins, A., & Goldin, S. *Diagnosing student's misconceptions in causal models.* BBN Report No. 3786. 1978.

Swanson, D. B., Feltovich, P. J., & Johnson, P. E. Psychological analysis of physician expertise: Implications for design of decision support systems. MEDINFO77, 1977, 161–164.

Tversky, A., & Kahneman, D. Judgment under uncertainty: Heuristics and biases. In P. N. Johnson-Laird & P. C. Wason (Eds.), *Thinking: Readings in cognitive science,* Cambridge University Press, 1977.

van Melle, W. *A domain-independent production-rule system for consultation programs.* Computer Science Doctoral Dissertation, Stanford University, 1980.

Winograd, T. Frame representations and the declarative/procedural controversy. In D. G. Bobrow & A. Collins (Eds.), *Representation and understanding.* New York: Academic Press, 1975.

Yu, V. L., Buchanan, B. G., Shortliffe, E. H., Wraith, S. M., Davis, R., Scott, A. C., & Cohen, S. N. Evaluating the performance of a computer-based consultant. *Computer Programs in Biomedicine,* 1979, *9,* 95–102.

Yu, V. L., Fagan, L. M., Wraith, S. M., Clancey, W. J., Scott, A. C., Hannigan, J. F., Blum, R. L., Buchanan, B. G., & Cohen, S. N. Antimicrobial selection by a computer—a blinded evaluation by infectious disease experts. *Journal of the American Medical Association,* 1979, *242,* 1279–1282.

4

First Among Equals

John Haugeland
University of Pittsburgh

Cognitive science of AI-driven. By this I mean two things. First, pre-AI linguistics (e.g., Chomsky), psychology (e.g., Sternberg), and epistemology (e.g., Quine) can do perfectly well without one another—save the occasional boundary dispute or interdisciplinary platitude. Only AI makes it seem that these fields are closely related, not just in the millenium, but *now* (or soon, anyway). Second, the older disciplines stand to be transformed, both in practice and in theoretical superstructure, by the proposed merger, whereas AI stands mostly to be broadened and enriched, without much changing its basic character. My principle question is: Why is AI "first among equals" in cognitive science?

More specifically, I focus on the relation between AI and psychology, with particular reference to work on language behavior, and even more particular reference to the preceding articles by Bill Clancey and Wendy Lehnert. Insofar as it would be nice to have a unified cognitive science, it would be nice to say that AI and psychology study the same thing (viz., intelligence), but in different manifestations (viz., artificial and natural). In fact, it would be *so* nice to say this that many people just do. But the two chapters just mentioned, as well as countless others bearing on the issue, show that there is much more that can and should be said.

I begin by summarizing a number of prima facie contrasts between AI and psychology, trying, as I go, to group them into four main categories. Needless to say, these lines are drawn sharper and cleaner than the divisions really are; seldom does a given specimen fall neatly on one side throughout all the distinctions, and a few cases simply defy consistent classification in these terms. Furthermore, it is becoming fairly common for authors versed in both traditions to try consciously, in their own research, to bridge the gap; naturally, we shall look elsewhere for examples of the gap they're trying to bridge.

The first of the four categories I dub the *contrasts of modality*. Here we find the classic distinction: psychology studies intelligence in one specific form, the form in which humans[1] happen to manifest it, whereas AI studies intelligence in general, intelligence by whatever means possible. So, research on short-term memory limitations, child development, narcotic or fatigue effects, and so on, would all count as psychology, because the results (presumably) would be ideosyncratic to human intelligence. On the other hand, designing knowledge-based expert systems counts as AI, because, as Clancey puts it, "'knowledge engineers' . . . want programs to be better than people [p. 78]." Knowledge engineering is not concerned with the foibles and peculiarities of natural expertise; the goal is the maximum intelligence achievable, consistent with such broad-spectrum design constraints as reliability, flexibility, and efficiency. And, given that perspective, research on the merits of procedural vs. declarative representation, or forward vs. backward chaining, can seem bound only by the fundamental nature of intelligence itself.

This suggests a second contrast in the modality group. Empirical psychology, it is said, is bottom-up or data-driven, whereas AI, by comparison, is more top-down or theory-driven. The point is that psychology starts from known facts about given instances of intelligence (human, naturally), and tries to build up from these to as general a theory as it can manage. The available facts are not restricted to ones obtained from psychological experiments, but may include any enlightening tidbits garnered from neurophysiology, clinical practice, or even evolutionary biology. AI, however, in its relative top-down-ness, will be concerned with the most abstract, essential features of intelligence as such. Insofar as can be, it will work down to a detailed theory, starting from definitions, first principles, and intuitive ideas about what's crucial and what's incidental to the nature of intelligence; thus, outside help is more likely to be found in mathematics, systems engineering, or even philosophy.[2]

In a similar spirit, Dan Dennett (1978) has proposed a variant up/down scheme, according to which traditional epistemology and philosophy (he men-

---

[1]That is, people. It's a matter of some dispute just how much cognitive science and science fiction overlap, but they at least share the dubious distinction of referring to people as "humans."

[2]The sense of the top-down/botton-up metaphor is fluid and unstable. It is amusing to note, for example, that van Lehn, Brown, and Greeno (this volume) characterize what they call "competence models" as top-down, on grounds not unlike the above (focusing on abstract, fundamental principles, with strong entailments); but the examples they cite are mainly psychological research and their contrasting description of a bottom-up approach applies most readily to typical AI systems. Swinney, on the other hand (this volume), uses "top-down" for systems that are "maximally interactive," as opposed to collections of nearly autonomous modules (compare my discussion below of "molarity," and Lehnert's use of "global"). Finally, Clancey and Lehnert use it in two further ways yet.

tions Kant) form the opposite pole to psychology, with AI falling somewhere in between. Kant and I call this (though Dennett doesn't) the transcendental/empirical distinction. The idea is that an empirical inquiry concerns the way things (of a certain sort) *actually* are, whereas a transcendental inquiry concerns the minimal prerequisites for there being anything of that sort at all—preconditions on the very *possibility* of such phenomena. So AI is more transcendental than psychology, because it investigates and works from the general principles of all intelligence; but it's more empirical than philosophy, because it demands that its theories be physically workable in the real world, and demands a demonstration thereof in an actual implementation.

It should now be clear why these contrasts are grouped under the rubric "modality." The underlying theme is that psychology addresses intelligence as it actually is (already, in people), whereas AI pursues intelligence in all its possibilities.

The second of the four main groups I term *contrasts of molarity*. This is at least part of what Lehnert means by "local" versus "global" [p. 30]; for, despite some difficulties of interpretation,[3] both her choice of terms and her examples show that one concern is how much of the cognitive system is involved at once—and that's molarity. On this line, then, psychologists prefer, so far as possible, to isolate one or a few distinct psychological processes and characterize them separately (c.f., "local"); hence their preoccupation with "controlling for" any other variables that might complicate or muddy the results. AI programs, by contrast, tend to incorporate simultaneously a larger number of interacting procedures, that can then be studied together as an integrated system (c.f., "global").

Essentially the same distinction can be seen, in a different form, in the kinds of data typically sought and cited. Psychologists characteristically want precise measurements that are repeatable in controlled experimentation (and, thus, in one sense, "clean"). AI research, on the other hand, is likely to rest on coarser, more molar data—like what responses would be appropriate at all, in a certain context. Clancey (following Bower) calls such data "what anybody's grandmother could have said [p. 78]."

The two ways of looking at the point can be connected if we think about the consequences of having many factors interacting at a time. Suppose just one element were omitted, or misunderstood, or incorrectly implemented. In a highly interdependent system, such an error can propagate, and under certain circumstances, be amplified. That is, the initial error can throw off course whatever

---

[3] I have a hard time telling, for instance, which tasks "stand alone as self-contained" activities, or which ones are engaged in most often in my "normal daily routines"; these sound like descriptions of small scale, independent primitives, yet Lehnert uses them to explicate "global." I have related problems with "realistic," "natural," and "normal."

connected functions depend on it; they in turn throw many others even further off; and so on.[4] This wouldn't always happen, of course; but in systems where it can happen, there is an especially sensitive test for even rather "small" errors: under the relevant conditions, the observed results are periodically outlandish— not just a little off, but wildly wrong. So all one has to do is try a variety of conditions, and check for consequences in the right ballpark; hence the molarity of AI.

The third main group, *contrasts of criteria,* is illustrated indirectly by the terminology in Lehnert's discussion of "task orientations." When she talks about AI, she uses expressions like "input/output specification," "algorithm," and "program"; but when she turns (in the same context) to psychology, these give way to "variables" related by "well-defined functions." Though Lehnert does not explicitly make anything of it, the shift in wording is clearly no accident; and the implicit distinction is quite important.[5]

The difference lies, I think, in the research objectives of the two traditions, as effectively determined by their standards or criteria for having met those objectives. Thus, the functional relationships among psychologists' variables are preferably quantitative, preferably simple, indeed preferably linear. It is highly desirable that small changes in the independent variables give small changes in the dependent variables, and that these independent variables also operate independently of one another. And, above all, it is essential that the data supporting the various claims be repeatable, convincingly analyzable, and statistically significant. It all leads to one outcome: each relationship (more or less independently of the others) must be empirically established. Verifiability is the touchstone of experimental psychology.

Algorithms and programs, on the other hand, aren't so much measurable relations among variables as they are structured sequences of primitive operations, whose input/output behavior is expressed in terms of (complexes of) the corresponding operands. It is important to realize that Lehnert's use of "algorithm" is broader that some others' (including mine); for she means all effectively specifiable procedures, regardless of how good they are at achieving the desired result. (That is, she would count programmable chess heuristics as algorithms, right along with standard routines for multiplying or alphabetical sorting, despite the fact that chess heuristics don't always give very good answers.)

---

[4]In the real world (and especially in "analog" systems), a lot of design effort goes into preventing this kind of amplification—mainly through restriction of component interactions, redundancy, and negative feedback. Such techniques, however, are most effective for systems (and subsystems) with narrowly specialized and clearly defined operational desiderata—which, lets out much of AI.

[5]I am unpersuaded by what Lehnert seems to be saying about task orientations not being required in psychological experiments, basically because I can find no clear and independently motivated sense of "task" that makes the right discrimination. To add, for instance, that it must be "an information-processing task in the sense that AI workers speak of information-processing [p. 27]" would close in on the desired distinction, but at the expense of begging the question and obscuring the issue.

Be that as it may, the characteristic research objective of AI is an explicit and complete specification of the procedures (whether algorithmic in a strong sense, or merely heuristic) by which some actual or plausible mechanism can realize the intended input/output behavior. Of course, whether any given project is interesting or important will depend on the particulars of the input/output behavior in question. But the more basic issue is whether (and how) the procedures proposed could be implemented on some actual or plausible machine; without that, there's no real program or algorithm to discuss. Thus, mechanical (or computational) realizability is the bottom line; working systems are the touchstone of AI.

I think this contrast of research criteria must be what Clancey has in mind when he complains that:

> Too often experimental analysis seems to fall short by not being precise enough to be programmable. Or the simplifications to make an experiment tenable eliminate the very points that we need to build a working system [p. 78].

Lehnert likewise characterizes AI hypotheses as "about rigorously-defined processes [p. 27]," as if to say that psychology lacks comparably rigorous definitions. But I can just hear some renowned psychologist leaning back in his or her endowed chair, muttering: "Tell me about precision and rigor, you wet-nosed whiz-kids—you've never performed a properly controlled experiment in your lives!" That would be an exaggeration, obviously; but still, a point gets made.

As Zenon Pylyshyn notes (1981), there is more than one way to be rigorous in the pursuit of scientific understanding. He is thinking of rigorous adherence to observations and operational definitions, versus rigorously formal expression of a comprehensive theory, with far-reaching (and telling) consequences (see the top-down/bottom-up distinction in our first group of contrasts). But his point extends naturally to include the kind of rigor endorsed by Clancey and Lehnert: rigorously complete specification of a working mechanism. The lesson, clearly, is that little stands to be gained through protestations about who is more precise and rigorous. Different research criteria support and demand rigor of different sorts; it is the criteria themselves that merit comparison.

Later in that same paper, in fact, Pylyshyn uses essentially the contrast of criteria to explicate the following remark of Allen Newell's (1972):

> I will, on balance, prefer to start with a grossly imperfect but complete model, hoping to improve it eventually, rather than start with an abstract but experimentally verified characterization, hoping to specify it further eventually [p. 375].

Pylyshyn points out the vagueness in several key phrases, and proposes (rightly, I think) to construe "imperfect but complete" as meaning fully implementable systems that imperfectly simulate observed behavior, and "abstract but experimentally verified" as meaning functional relations that capture the data, but don't spell out the mechanisms. On this reading, Newell comes out (not surprisingly) as prototypical AI.

The interpretation is doubly relevant, since Newell made his remark in the course of reporting simulations of some famous work by Saul Sternberg—a prototypical experimental psychologist. In this work, Sternberg demonstrated a number of unobvious properties of short-term recall, with compelling implications regarding search strategies, encoding and decoding procedures, and so on. And it might be supposed that he was as interested as anyone in specifying mechanisms. What is lacking, however, is any nitty-gritty concern with an actual, buildable (virtual) machine, on which the described procedures would really run (and exhibit the observed properties). That's the question Newell was addressing in his article; and it's what I've called the touchstone of AI.

My fourth and final group of contrasts is also the trickiest to catch in an appropriate (and unbiased) term. The best I can do is: *contrasts of cognitivity*. As Lehnert points out [p. 29], traditional psychological investigation of "verbal behavior" has been dominated for decades by work with nonsense syllables, or purely syntactical manipulations of "real" sentences. Within this general tradition, effects attributable to the meanings or connotations of the expressions were systematically eliminated or cleverly finessed. Even those more recent studies (like the ones discussed by Swinney, this volume), which ostensibly address such semantic issues as context and lexical ambiguity, tend to do so only very gingerly and obliquely. Thus, there is no suggestion in Swinney's chapter about *how* coherent interpretations are produced for ordinary, ambiguous texts; he is concerned only to show that lexical access of multiple meanings runs to completion, whatever the context. In other words, the details and complexities of real semantics and pragmatics make no difference to the work at all.

Just the opposite would be the case for your typical AI project on language processing. That is, the particular convoluted skein of cues, implications, and presuppositions that picks out the single "obviously right" interpretation of each given text would be the very focus of research. Finding and dealing with these is where the bulk of the effort would be devoted.

I think the same distinction can be drawn, on a slightly subtler level, between Lehnert's approach and the work she reports by Elizabeth Loftus. The latter was concerned, I gather, mainly to show that the processes invoked in answering questions from memory can systematically (and unconsciously) alter what is remembered—"rewrite" the contents of memory. More specifically, in certain cases a presupposition of the question will be remembered, in place of some contrary fact that would have been remembered if the question had not been asked. Now this does involve semantics, insofar as the presupposition of the question must be apprehended (if not noticed), and that must then alter only the relevant portion of memory. But the semantic involvement does not seem crucial to the basic point. As I understand it, if the relevant parts of memory could be keyed *syntactically,* and if the probe patterns didn't have to be exactly identical to the targets to get a "match," then the fundamental Loftus result could be obtained by showing that deviant probes sometimes replace their targets. In other words, the same discovery about memory could, in principle, have been made with nonsense syllables. This, of course, is not to criticize Loftus's procedure, but only to put the semantic element in it in proper perspective.

How different (and more "cognitive") it all looks in Lehnert's hands! Her central professional interest is in story-understanding systems, and that (if it ever works) will have to involve semantics with a vengeance—not to mention common sense reasoning, pragmatics, hermeneutics, personality, and everything else (compare Haugeland, 1979)—and she knows it. In particular, building a unified (coherent, explicit, unambiguous) memory representation from an ordinary lan- ·guage story is a complex, multi-leveled, semantic juggling act. Much the same act must be performed in question answering; and Lehnert's essential suggestion is: if most of the same apparatus is used both times (as would be efficient anyway), then the Loftus effect is spun off as a side-effect (see Lehnert).

An alternative version of this cognitivity contrast can be put in terms of background knowledge, instead of meanings. In a traditional psychological ex- periment, for example, what the subject already knows (about the real world, etc.) counts as "bias," and gets factored out as much as possible. Thus, even Loftus had to show that there is no antecedent bias (common sense presupposi- tion) favoring cars stopped either stop signs or yields signs (see Lehnert, p. 39). But the effects of prior knowledge—how to bring them about, not how to avoid them—are a primary concern in AI; this is true above all for expert systems, of course, but also quite conspicuously for story understanding, ordinary problem solving, and so on.

Perhaps in part this just reflects the extent to which everything we do is conditioned by what we know: the pervasiveness of common sense, as it were.[6] Thus, if you try to home-in experimentally on anything else, you'll find that getting prior knowledge out of the way is one of your uppermost headaches; but if you try to build up an intelligent system from scratch, you'll find you're always having to cram more background knowledge into it, to get it to work right. If you want to study anything else in the ocean, the water will always be getting in the way; but when you want to make one, you can't get enough.

I suspect that the ubiquitous relevance of common sense will come to haunt Clancey, as he carries on his transformation of an expert system into a more realistic model of human expertise—whether for application in a tutoring system, or otherwise. Perhaps he already senses this, in his brief allusions to "initial

---

[6]Dreyfus (1982; see his introduction) and Searle (1978) have recently raised an important issue about how much of the prerequisite background for everyday understanding (of language, situations, etc.) can properly be termed "knowledge," even in a broad sense. I won't pursue the issue here; but I think it fair to say that it could not even have been formulated (let alone resolved) in terms accessible to the traditional experimental psychologist.

problem formulation," "circumstantial evidence," establishing an "hypothesis space," and the "effect of problem context" (Clancey, pp. 65 & 75). But I wish he had pulled these threads together in an explicit discussion of the relation between expertise and common sense, especially for systems that go beyond the level of electric encyclopedia (that is, systems that don't rely on their operators for common sense).

That completes my survey of four groups of contrasts between psychology and AI. So we can return now to the original question: *Why* does AI have its peculiar first-among-equals status in cognitive science? More specifically, why is it AI which first promises imminent unity among fields which otherwise could have kept a polite distance for generations; and why do the other disciplines stand to be transformed by the union, while AI just gets broader and deeper? In pursuing these larger questions, I will still concentrate on the psychology/AI relation (with emphasis on language behavior), so as to take advantage of the contrasts just delineated; the case is, I think, sufficiently typical to be useful in general.

Many commentators find it attractive to characterize AI with as little reference to computers as possible—as if the intellectual essence of the discipline should rise above the accidents of its birth among the keypunches, its weaning on teletypewriters, and its social debut in resplendent color video. And it might seem that the foregoing contrasts, with all their variety and lofty methodological tone, would support this abstract view. But, I think the computer is the heart of the matter, in two quite distinct ways, and that only by appreciating this can we find any coherence in the diversity of the contrasts, and have any hope of answering our original questions.

Computers are fantastically cheap, consistent, and reliable. That makes it feasible to build, maintain, and intellectually manage systems of unprecedented complexity. Perhaps the single most important reason that traditional psychology, linguistics, and philosophy of mind can get along so well in mutual ignorance is that their common problems are so overwhelming. The standard tools of intellectual mastery—taxonomic classifications, schematic diagrams, formal calculi, causal histories, and so on—are simply inadequate to accommodate the range and subtlety of, for instance, a teenager's telephone conversation. So each (sub)discipline has found some little tag-end of the phenomena that it can get a relatively firm grip on, and then stuck with that. Coming to terms with cognition and meaning as a whole, in any terms other than speculative generalizations, has just been out of the question.

The new mastery of complexity made possible by computers promises to change all that. In their potential to extend the reach of science, computers are often compared to slide rules or telescopes—only more so. But that misses the point. Scientists could already multiply and peer into the distance; they could just do it better with their new gadgets. I think pen and ink make a more appropriate comparison (though maybe just a tiny bit exaggerated). The introduction of quick durable writing made new *kinds* of achievement possible, such as: accumulating a large body of data (record keeping), manipulation of abstract symbols, publication, and so on. Moreover, these are all general purpose capabilities, not particularly associated with any special disciplines or professions. Mastery of complexity is a comparable increment: new in kind, and general purpose.

The complexity that's at stake is not just being composite of many parts, all intricately interrelated. That much is true of snowflakes and television sets, and they are intellectually manageable without computers. Nor is the issue one of scale, or sheer number of distinct "variables." It is rather what I call *dynamic* complexity, which only really ramifies in a succession of states, each *qualitatively* contingent on a number of simultaneous features of its predecessor. Even in an ordinary amplifier or engine, of course, later states will depend in manifold ways on earlier states. But generally, if one or two earlier states are purturbed modestly (e.g., some static or friction is introduced, or a part changes a little as it warms up), then the later states will be purturbed only modestly as well, and often only briefly.

Qualitative dependence, on the other hand, is like a fork in the road; a small purturbation at the crucial juncture makes all the difference in the world to successive states. The most familiar example is the way chess moves depend on their predecessors: frequently, if just one piece had been displaced just one square, the entire subsequent game would have been dramatically different. In dynamically complex systems, such qualitative contingencies are the rule rather than the exception. During normal operation of the machine, there are (at nearly every step) numerous dependencies, each of which effectively constitutes a separate "fork in the road"; and this is the case step after step. In principle, these separate alternatives can be combined without restriction, and they need not even be independent—leading to an enormous branching factor, and a truly staggering combinatorial explosion.

Dynamic complexity makes the output (or final state) of a system astronomically difficult to predict by analytic means; this is, in fact, the principle of standard pseudo-random number generators. The surprising thing really is that such systems are ever anything other than chaotic, except of course, for net statistical balances. Thus, if you consider a gas sample at the molecular level, it exhibits high dynamic complexity (a small variation in the angle at which A and B collide determines whether A collides with C next or with D); and, the system is paradigmatically chaotic. But at the macroscopic (statistical) level, all the differences cancel, and the system is quite orderly and predictable. Likewise, there might be some nice statistical regularities in the character frequencies in the output of a large time-sharing installation (something a printer manufacturer might care about, e.g.).

However, there is a crucial difference: seen relatively locally, the computer output is (usually) not at all random or chaotic. While this seems fairly clear intuitively, saying exactly what it means is more difficult. I think the essential point is whether there are intermediate levels of description at which generalizations can be expressed, and predictions made. Thus, you can predict a gas's future at the molecular level, by tracing each trajectory and collision (which quickly gets out of hand, of course), or at the macroscopic level, by citing the statistical regularities; but there is no other level "between" these, at which anything useful could be said.

Now there are analogues of both these levels for the computer output: tracing individual causal histories of character tokens, versus predicting distributions on the basis of known frequencies. But, in this case, there are also many intermediate levels, that are often much more powerful and useful for prediction. For instance, if the machine is playing chess, you can guess its output on the basis of what moves would be legal and plausible in the position; if it is preparing an inventory for a hardware store, you can guess many of the items listed, and ballpark figures on the quantities and prices; and so on. I am unable to characterize these intermediate levels in any general way. But notice one respect in which they differ in kind from the macro-statistical level common to thermodynamics and printer design: they do not consist merely of an averaging process, which finds a coarse-grained smoothness by simply ignoring, blurring, and cancelling out all fine-grained distinctions. Regularities in chess players or inventory controllers depend crucially on a sharply etched structure, right down to single character discimations (P-K4 versus P-Q4; 200 nails versus 200 pails).

I think such sharply structured intermediate descriptions (predictions and generalizations) are the basis of our intuition that computer outputs are not chaotic, in the way that gas molecule motions are. Thus, what big computer installations have and gas samples lack is not just dynamic complexity (they both have that, the gas sample more so), but rather *structured dynamic complexity*. This is the possibility that is metaphysically surprising; but, of course, it's commonplace—computer systems being only one conspicuous example. But computers enter in a deeper way: special cases aside, the only known way to study structured dynamic complexity is by tinkering with it, preferably by designing, building, and modifying the relevant systems. The cheap, reliable consistency of computers makes that possible as a research technique; and (special cases aside again) nothing else does.

This is not to suggest that theoretical understanding of such systems is impossible or unnecessary. Quite the contrary: intellectual mastery (= understanding) is the main goal; and in the meantime, random or unmotivated "tinkering" is hardly likely to yield much insight. Rather, my point is that the required conceptual tools can only be developed hand in hand with efforts to build actual systems, and get them working. Only in the course of such work do the problems emerge that require solutions; and only there can the solutions be tested. Even very basic notions like subroutines, data types and structures, global vs. local variables, procedure calls, and error recovery, are difficult to grasp apart from machines (though students pick them up readily with a little hands-on experience). More abstract intellectual inventions—like the very idea of a control structure, or taxonomies of search heuristics and the sorts of problems to which they're suited, or the notion of levels of "languages" (= machines) with their intervening compilers and interpreters—these, I think, would have been well nigh impossible without the ability to build (and tinker with) concrete instances of the problems to which they were eventual solutions.[7]

Once the inventions are made, however, they enter into general scientific currency; it becomes possible, as it was not before, to wonder about the functional architecture of the human virtual machine, or to ask whether common sense is encoded in productions or declarative schemata, and so on . This should make it apparent why AI first makes a unified cognitive science seem feasible. If one only supposes (as sounds eminently plausible in any case) that our "cognitive system" is an instance of structured dynamic complexity, then one can see why the general problems were hitherto so overwhelming, and can also see AI (the acknowledged vanguard of hairy computer science) as thundering to the rescue. But more than that, we also get easy rationales for our second and third groups of contrasts between psychology and AI. The contrasts of molarity fall directly into place, for the study of structured dynamic complexity is nothing if not the study of integrated systems, with many factors operative simultaneously. And the contrasts of criteria are equally obvious: fueling the growth of understanding with the results and problems of tinkering (in the absence of adequate analytical methods) would be nonsense apart from real running systems to tinker with.

Facilitating intellectual mastery of structured dynamic complexity is the first of two respects in which computers are central to the first-among-equals status of AI in cognitive science. The second respect is, I think, less visible and more fundamental, for it rests on a general prejudice about cognition as such. It is often remarked that computer simulations of intelligence are unique among computer simulations at large in that one is supposed to get, not some symbolic surrogate, but the real thing (in whatever measure). Thus, meteorologists can simulate rainstorms without fear of getting wet; or, biologists could simulate lactation without expecting any (real, drinkable) milk (compare Searle, 1980). But when Clancey and Lehnert ask their systems questions, they expect answers—real, intelligible answers, and not mere symbolic descriptions of what real answers would be like. More generally, they hope and expect their systems to exhibit genuine expertise, or genuine story understanding. Indeed, if we were to follow the terminological precedents of the food, drug, and gem trades, artificial intelligence should instead be called "*synthetic* intelligence," for the implication is not meant to be "fake substitute," but rather "the real thing, manmade."

This is truly a remarkable difference; and it reflects a deep presupposition

---

[7]This paragraph is partly a response to some perceptive remarks made by John Seely Brown at the conference from which the present volume derives. My larger point about complexity and the contribution of computer science is inspired in general, though not in detail, by Pylyshyn (1981).

about the nature of intelligence: intelligence *just is* "semantic" or "symbolic" information-processing. This, in turn, is a remarkable doctrine; but it has a history that lights it up. The same history also helps to clarify the distinction between "information-processing" in the relevant sense, and the sense in which hurricane simulations, after all, process information, too.

Seventeenth–century philosophy invented and bequeathed the modern problem of *intentionality:* How can mental states (or any other meaningful items) mean what they do? More specifically, how comes it that they denote some particular object or objects in the outside world? There have been, of course, many approaches to these questions through the years; but we shall have to jump over to a relatively recent variation, that came into its own in 20th–century mathematics. The idea of a purely formal (syntactical) system, separate from, but coupled with an intended interpretation, put semantics on an entirely new footing. In this picture, the individual terms acquire their semantic properties from their formal positions within the larger system, which in turn "earns" an overall interpretation holistically, in virtue of its structure. Or, putting it the other way around, the global formal structure vindicates a global interpretation, which then devolves on the various syntactic constituents.

The ins and outs of this earning/vindicating step are a long story, both philosophical and technical (various aspects of which are, of course, by now outdated). But the basic point can be got by thinking fairly casually about notations for arithmetic. Any conveniently manipulable tokens could be conscripted into service as numerals and signs (that is, as symbols for numbers and operations). Because nothing about the tokens themselves can render them unsuitable for this role, nothing of that sort renders them particularly well suited either; something else must determine whether they really stand for numbers, operations, and so on—in short, whether they are really symbols.[8]

This "something else" must be found (if we're interpreting a purely formal system) in the syntactical transformations that are allowed by the system's rules. These are not mainly the well-formed-ness rules (though the latter are surely relevant in an ancillary way), but rather the procedural rules, that govern how you "get the answers." For instance, in Arabic notation, '5+7=13' is a syntactically well-formed equation, but it is not allowed by the rules of addition. Or (as I think it more perspicuous to say) if your current position is '5+7= ', then writing '13' to the right of the '=' is not a legal move. And it is *because* '12' is the only legal move, that it makes sense to say that the various tokens are actually symbols for addition, equality, and the numbers five, seven, and twelve.

Or rather, we should say, the overall interpretation is vindicated because, in the system as a whole, only equations like '5+7=12' are ever allowed, and never ones like '5+7=13'. What does "like" mean here? The answer is almost

---

[8]Any closet metaphysicians who find "really" too strong in this context are free to substitute something more mealy-mouthed, like "can legitimately be characterized as," or whatever.

too obvious, but the question is deep: five plus seven equals twelve, not thirteen. That is, if the tokens are interpreted in the standard way, then (the numeral) '12' is the right answer—the equation is true. If, given some proposed interpretation, the rules always permit only right answers (true equations), then that interpretation is vindicated overall (at least for arithmetic systems).

Giving the right (true) answers may be called the *cogency condition* on interpreting a formal system as an arithmetic system; for other interpretations of other systems there will be other cogency conditions. Thus, intuitively and to a first approximation, the cogency condition on interpreting a system as one which, say, diagnoses a certain class of infectious diseases, or answers questions about short stories, would be that the diagnoses or answers be consistently plausible, given the symptoms or stories, respectively. Unfortunately, there is a large gap between such intuitive approximations to cogency conditions, on the one hand, and full formal specifications of systems which consistently meet them, on the other. In a way, mathematicians could blaze the formalist trail precisely because they had it easy, with strictly correct answers as the only relevant cogency condition.

Three further points, each too large and difficult to treat here, must at least be mentioned. First, most mathematical systems are purely permissive (non-deterministic); that is, they specify only the moves that would be allowed at any point, leaving entirely aside the question of which move ought to be chosen, consonant with some goal. But, in many cases, the latter question is the more important, especially for implementing actual machines; thus, an automated mathematical system that chose allowable steps randomly would mainly "prove" a lot of valid trivialities (just as a chess player that moved legally but randomly would play a rotten game). In real AI systems, however, the situation is even worse: the distinction between rules restricting what's allowable and heuristics guiding optimal choices, is itself difficult or impossible to make out. Hence, the possibility of specifying independent cogency conditions on successful implementations is problematic in principle.

Second, the cogency of sophisticated systems (mathematical or otherwise) depends not only on having the right answers pop out at the end. Rather, in any multi-step procedure (such as a proof) each step must be individually cogent in the manner appropriate to its status (lemma, hypothesis, etc.), and there must be meta-systematic reasons for believing that any steps allowed by the rules would be cogent in this way. In more general terms, there must be (meta-systematic) rationales for the cogency of larger procedures in terms of the cogency of their constituents—this, indeed, is the principle tenet of the cognitivist credo.

Third, and finally, mathematics holds out a further promise that has tantalized philosophers since Plato, and was especially inspiring to the 17th–century rationalists; namely, the promise of perfect clarity and distinctness—or, to put it another way, perfect explicitness. The idea is that everything that is relevant to the validity of a proof is, or at least can be, articulated (that is, *said,* "in so many words''), either as a premise or as a rule of inference. Nothing is left to the imagination (or inspiration, hunch, savvy, intuition, faith, etc.); everything, save the mechanics of reading and writing themselves, is spelled out right there on the page.[9] The generalization of this ideal to all of thought, first broached by the rationalists, is now essential to cognitive science; for it is nothing other than the suggestion that cognition can be mechanized (by a well-defined, finite, formal mechanism). Whence it is that explicit representation of common sense, conversational goals, mutual knowledge, intellectual ''skills,'' and the like is of such importance in AI.

But what does all this have to do with our original topic: the premier status of AI within cognitive science? Well, subject to all the foregoing hedges and uncertainties, formal systems with earned interpretations offer an enticing (potential) solution to the traditional problem of intentionality. Thoughts (and cognitive processes in general) are regarded as syntactical processes on formal tokens, in an overall system that entitles them to the ''intended'' interpretation. The former explains how the existence of minds can be compatible with materialism, while the latter explains how they can still be mental (i.e., have states with meanings). The primitive rules of the system, needless to add, are understood as ''built in'' to the cognitive mechanism, tantamount to the primitive operations of a running implementation; and hence, the system can be an independent entity, thinking for itself, and so on, without need for any operator, external manipulator, or traditional homunculus.

Now, however, AI ends up holding all the aces. Studying (and tinkering with) arbitrary formal systems, with their rules implemented as procedures, is precisely what computers are best for; in other words, they are the ideal medium in which to conduct the search for a system ''entitled'' to be interpreted as intelligent. In short, AI emerges as official scientific sponsor for the interpreted formalism solution to the problem of intentionality. And this is why AI stands only to be enlarged by the proposed unification of cognitive science, whereas the older but junior partners (insofar as they have not yet embraced the interpreted formalism solution) stand to be transformed. It is also why AI simulations count not as substitutes but as the real thing—the ''simulation'' is as much a genuine symbol manipulator as any old fashioned ''neural implementation,'' and just as entitled to serious interpretation. And, last but not least, we can understand in these terms our first and fourth groups of contrasts between psychology and AI. If interpreted formalism is the key to intentionality (and thus intelligence) in general, then AI is automatically the study of any possible intelligence (due to the well-known generality of computers); people are just a certain venerable, special-purpose design. So much for modality; and cognitivity is even easier. The main claim to fame of the entire enterprise is to have solved the mystery of

---

[9]Just what the ''mechanics of reading and writing'' are turned out, of course, to be a nontrivial problem, not solved till the invention of the Turing machine.

semantics; and the burden of defending this claim is exactly the problem of building systems that actually meet realistic cogency conditions.

Assuming, therefore, a successful outcome, computer-based AI is first among equals in cognitive science.

REFERENCES

Dennett, D. C. *Brainstorms.* Cambridge, Mass.: MIT Press, 1978.

Dreyfus, H. L. (Ed.) *Husserl, intentionality and cognitive science.* Cambridge, Mass.: MIT Press, 1982.

Haugeland, J. Understanding natural language. *Journal of Philosophy,* 1979, *76,* 619–632.

Haugeland, J. *Mind design.* Cambridge, Mass.: MIT Press, 1981.

Newell, A. A theoretical exploration of mechanisms for coding the stimulus. In A. W. Melton & E. Martin (Eds.), *Coding processes in human memory.* New York: Winston, 1972.

Pylyshyn, Z. Complexity and the study of artificial and human intelligence. In J. Haugeland (Ed.), *Mind design.* Cambridge, Mass.: MIT Press, 1981.

Searle, J. Literal Meaning. *Erkenntnis,* 1978, *13,* 207–224.

Searle, J. Minds, brains, and programs. Reprinted in *Mind Design,* Cambridge, Mass.: MIT, 1981.

▌▌ LINGUISTICS

In this section, two linguists describe their research. They have something important in common: both perceive their research as relevant to cognitive science in the double sense that they see their theories as potentially important for other work within cognitive science that is concerned with language, and in the sense that research in psychology, artificial intelligence, or anthropology is in principle, important for their own work, too. Thus, they would claim that an AI program that employs a parser that contradicts or neglects the principles upon which their own work is based is on the wrong track. Conversely, they would be willing to change their own theory if confronted with a well-established, contradictory result from another discipline, such as a computational limit from AI, or some experimental result from psychology. That is why we are concerned with their work here.

At the same time, the chapters by Joan Bresnan and Ronald Kaplan and by Talmy Givón illustrate quite different approaches to the study of language within linguistics. Their methods are the same—involving the use of "starred examples," that is ungrammatical or otherwise unacceptable sentences—but their goals are quite different. While Bresnan and Kaplan aspire to a formal theory, Givón has little use for formalisms. Most importantly, however, they differ in the range of data that they consider significant. For

Bresnan and Kaplan it is the standard linguistic corpus of grammatical sentences in the language, while Givón insists on studying language in the context of actual communication situations.

The third chapter in this section is a discussion of the two linguistics chapters by psychologists. Given that everyone agrees that psychology (as well as other disciplines within the cognitive sciences) is in principle relevant to the work of these researchers, what does psychology actually have to contribute? The answers to that question that the authors give are, of course implicit in the chapters they wrote, but as Herbert Clark and Barbara Malt will show, they are not always the same as the ones a psychologist studying language would give.

These three chapters elaborate several of the themes we have introduced in our discussion of methodological issues in Chapter 1. Central here are questions about ecological validity; what are the proper observations upon which to build linguistic theory: natural discourse, embedded in its rich context, both linguistic and non-linguistic, or judgments about the well-formedness of isolated sentences? Closely interrelated with that issue are questions concerning the role that formalisms should play in the linguistic theory and the nature and source of constraints for that theory.

5 Grammars as Mental Representations of Language*

Joan Bresnan
Stanford University and
Xerox Corporation Palo Alto Research Center

Ronald M. Kaplan
Xerox Corporation Palo Alto Research Center

THE PROBLEM OF THE PSYCHOLOGICAL REALITY OF GRAMMARS

A longstanding hope of research in theoretical linguistics has been that linguistic characterizations of formal grammar would shed light on the speaker's mental representation of language. One of the best-known expressions of this hope is Chomsky's (1965) *competence* hypothesis: "a reasonable model of language use will incorporate, as a basic component, the generative grammar that expresses the speaker-hearer's knowledge of the language . . . [p. 9]. Despite many similar expressions of hope by linguists, and despite intensive efforts by psycholinguists, it remains true that generative-transformational grammars have not yet been successfully incorporated in psychologically realistic models of language use (Fodor, Bever, & Garrett, 1974).

In discussing grammars as mental representations, it is essential to understand that the terms 'grammar' and 'theory' are both used on two entirely different levels of description. On the lower level of description, we speak of the grammar of a particular language such as Navajo. At this level, a grammar is a set of rules within a formal system. The grammar generates the language that it is a grammar of, and on analogy with the technical usage in other formal systems, the grammar is sometimes called a "theory" of the language that it generates. On the higher

*This chapter is reprinted with permission from J. Bresnan, Editor, "The mental representation of grammatical relations", MIT Press, 1982; copyright by the Massachusetts Institute of Technology.

level of description, we speak of a theory of grammars. This is a set of primitives, axioms, and rules of inference (often unformalized) that characterizes the class of possible grammars of particular languages. A theory of grammar is sometimes referred to as a "Universal Grammar."

These dual concepts of 'grammar' and 'theory' play very different roles in discussions of the mental representation of language. To learn to speak Navajo, one must acquire specific, if tacit, knowledge of the sentence patterns and pronunciation of Navajo. A grammar of Navajo, in that it provides specific rules for the construction of Navajo sentences, represents the kind of knowledge of that language that one must have to speak it. It is such grammars, grammars on the lower level, that we assume will represent the stored knowledge in competence-based models of linguistic performance. Grammar on the higher level, the "Universal Grammar" that is a theory of grammars, is not necessarily represented in such models in the same way. For example, principles of Universal Grammar might characterize aspects of the structure of the language-using device.

In response to the fact that generative-transformational grammars (on the lower level) have not been successfully incorporated in realistic models of language acquisition, comprehension, or production, many psycholinguists have come to the view that it is a mistake to adopt Chomsky's competence hypothesis: there need not be *any* transparent mapping between linguistically motivated formal grammars and psychological models of language use; for example, the language-user may employ agrammatical heuristic strategies as knowledge representations (Fodor, Bever, & Garrett, 1974). If this is so, the knowledge representations actually used in language acquisition, production, and comprehension will not satisfy the postulates of Universal Grammar, which makes quite specific claims about the form, organization, and interpretation of rules of (lower level) grammars. But then what mental structures is Universal Grammar a theory of?

In response, Chomsky (1980) has taken the view that it is a mistake to regard 'psychological reality" as anything other than whatever linguistic theory is about: "Challenged to show that the constructions postulated in that theory have 'psychological reality', we can do no more than repeat the evidence and the proposed explanations that involve these constructions [p. 191]." Comparing the linguist to an astronomer studying thermonuclear reactions within the sun, Chomsky (1980) argues, "[I]n essence . . . the question of psychological reality is no more and no less sensible in principle than the question of the physical reality of the physicist's theoretical constructions [p. 192]." But neither of these responses—neither that of Chomsky nor that of those who have abandoned the competence hypothesis—is satisfactory.

Consider first the view that rejects the competence hypothesis. In rejecting the hypothesis, proponents of this view do not also reject the notion that some form of stored linguistic knowledge is employed in all forms of language behavior. For example, it is generally acknowledged that models of linguistic comprehension must include a set of previously learned phonological and syntactic patterns for the language (English, Chinese, or Navajo) being comprehended. Similarly, models of linguistic production require stored information about the syntactic and phonological structure of the language being spoken, in order to convert the input (the speaker's message) into the form of speech. Likewise, models of language acquisition assume that the learner constructs and stores linguistic knowledge structures, forming an 'internal grammar' on the basis of the primary data in the linguistic environment and a set of built-in constraints on the induction process. Each state of the learning process can be represented by a distinct internal grammar; but the sequence of these internal grammars must converge, in the final state of the learning process, on a 'target' grammar that represents mature knowledge of linguistic structure. Finally, the formal grammars of linguistics are themselves abstract models of mature knowledge of language, as reflected in linguistic judgments and the other adult verbal behavior studied by linguists.

If it is uncontroversial that stored knowledge structures underlie all forms of verbal behavior,[1] the question arises how these different components of linguistic knowledge are related. To reject the competence hypothesis is to adopt the theoretical alternative that a different body of knowledge of one's language is required for every type of verbal behavior. Although this alternative might represent the true state of affairs, it is the weakest hypothesis that one could entertain, because it postulates multiple stores of linguistic knowledge that have no necessary connection. In contrast, the competence hypothesis postulates an isomorphic relationship between the different knowledge components and is thus the strongest and simplest hypothesis that one could adopt. On methodological grounds, it should be given priority over weaker alternatives: it enables us to unify our theories of the mental representation of language, to construct our processing models on the basis of a theoretical understanding of the structure of the knowledge domain, and to bring mutually constraining sources of evidence to bear on studies of process and of structure.

Granting the competence hypothesis is desirable in principle, though, is it tenable in fact? In particular, if we do maintain the competence hypothesis, how can we then explain the conflict between psycholinguistic studies and linguistic theories of the mental representation of language? This is the scientific challenge posed by work on the "psychological reality" of grammars, as presented by Fodor, Bever, and Garrett (1974), Levelt (1974), and others. In response to this challenge, Bresnan (1978) pointed out that these psycholinguistic studies presupposed a transformational characterization of linguistic knowledge that could

---

[1]Some philosophers consider this controversial. Matthews (1982), for eample, argues against this view as a form of "computational reductionism." However, he overlooks the fact that the attribution of an internal representation to a computational device may itself involve intensional descriptions. To take an example due to Brian Smith, the description of a machine as running LISP cannot be reduced to purely extensional terms, because what counts as an instance of the LISP language depends on the functioning of the program and not on any particular electronic or mechanical configuration that realizes it.

simply be wrong, a possibility that had also been suggested in the ATN-based work of Wanner, Kaplan, and others (Wanner & Maratsos 1978; Kaplan, 1972). Wanner and Kaplan's studies showed that a psycholinguistic model of syntactic processing based on a competence grammar can be computationally implemented and used to generate detailed and experimentally testable predictions. Bresnan argued that a more radical decomposition of competence grammars into an expanded lexical and contracted syntactic component promises to have far greater explanatory power than the current versions of transformational grammar, permitting a unification of linguistic and psycholinguistic research. Subsequent collaborative research by Bresnan, Kaplan, Ford, Grimshaw, Halvorsen, Pinker, and others has begun to bear this out, as the studies in Bresnan (1982b) demonstrate. This work shows that the competence hypothesis is indeed tenable.

Consider next Chomsky's view, that psychological reality is whatever linguistic theory is about. Recalling the quotation cited earlier, we see that Chomsky (1980) construes the problem of psychological reality as an *ontological* problem: "[I]n essence . . . the question of psychological reality is no more and no less sensible in principle than the question of the physical reality of the physicist's theoretical constructions [p. 192]." In other words, he takes the question to be whether or not the rules and other constructs of linguistic theory 'have reality,' whether they describe real mental entities and processes. Whereas Chomsky's answer to this question is surely a reasonable one, this is the wrong *question*. The cognitive psychologists, computer scientists, and linguists who have questioned the psychological reality of grammars have not doubted that a speaker's knowledge of language is mentally represented in the form of stored knowledge structures of some kind. All theories of the mental representation of language presuppose this. What has been doubted is that these internal knowledge structures are adequately characterized by transformational theory—or indeed, by any grammatical theory that is motivated solely by intuitions about the well-formedness of sentences. The challenge to Chomsky's theory is not the philosophical question that he addresses (whether theoretical constructs correspond to real mental entities and processes), but the scientific question (whether these theoretical constructs can unify the results of linguistic and psycholinguistic research on mental representation and processing). To the latter question, Chomsky's (1980) response is plainly adequate: "Challenged to show that the constructions postulated in that theory have 'psychological reality,' we can do no more than repeat the evidence and the proposed explanations that involve these constructions [p. 191]."

On Chomsky's view, then, a grammar is psychologically real if it contributes to the explanation of linguistic judgments and the other verbal behavior studied by linguists, and nothing more need be said. This, however, is a much weaker conception of psychological reality than we would like. Consider the historical derivations of English words from Indo-European roots. (The following examples are taken from *The American Heritage Dictionary,* and particularly the

Appendix "Indo-European Roots.") For example, the English words *baritone* and *grieve* both derive from an Indo-European root *gwer-* 'heavy'. Historically, the word *baritone* came into English from Italian (ultimately from Greek), whereas *grieve* came into Middle English from Old French (ultimately from Latin). Latin and Greek, of course, emerged historically from distinct ancient dialects of the common Indo-European mother tongue. Now it is possible to construct a formal system of morphophonemic rules and abstract representations that deductively account for these historical relations. By such rules one can formally derive English words from their Indo-European roots. Thus, the labiovelar *gw* is the source of both the initial labial *b* in *baritone* and the initial velar *g* in *grieve*. The same relationship appears in *bar* and *gravid,* as well as many other examples. Would such a formal rule system be psychologically real? With Chomsky's conception of psychological reality, we could answer affirmatively. The rule system might well contribute in some ways to an explanation of English speakers' linguistic judgments of what are well-formed sentences of their language. It could even be argued that the rules and representations of the system do characterize the *competence* of an idealized speaker-hearer, abstracting way from 'performance' limitations such as memory, distraction, and the like.

Given the stronger conception of psychological reality that we would like to maintain, this conclusion is absurd. It is most implausible that the remote historical derivations of the English lexicon are part of contemporary English speakers' mental representations of their language. Although this illustration slightly exaggerates the practice of some linguists, it serves to make two important points. First, linguistically motivated descriptions of a language need not bear any resemblance to the speaker's internal description of the language. Any number of regularities in a body of linguistic data may appear because of remote historical changes or even accidental correlations. Therefore, one cannot justifiably claim "psychological reality" for a rule system (in any interesting sense) solely because it expresses formal regularities in the linguistic data. Second, the concept of *competence* has often been abused; in the above argument, for example, it now appears to mean that a linguistic rule system need not play *any* role in *any* model of performance. But the true import of the competence hypothesis is exactly the opposite: it requires that we take responsibility not only for characterizing the abstract structure of the linguistic knowledge domain, but also for explaining how the formal properties of our proposed linguistic representations are related to the nature of the cognitive processes that derive and interpret them in actual language use and acquisition. Chomsky's current conception of psychological reality represents a retreat from this more ambitious, and scientifically far more interesting goal.

One might think that abuses of the above kind could be ruled out by appealing to theoretical simplicity: perhaps the rules deriving English words from Indo-European roots would not fit in elegantly with the simplest grammar for English. But simplicity is itself a theory-bound notion: as Chomsky (1970) has argued, the choice of a simplicity metric is made on the same empirical grounds as the choice of a theory. Moreover, it is easy to imagine even highly elegant and deductively satisfying rule systems that lack psychological reality in the sense we would like. There is evidence, for example, that the standard mathematical axiomatization of arithmetic differs from the system of conceptual competence that children display in counting (Greeno, Riley, & Gelman, in press). Although the two rule systems may be extensionally equivalent, it appears that they are built up from different sets of basic concepts and procedures. If this is so, we would *not* want to say that the standard mathematical axiomatic system is 'psychologically real'; for while it does describe the conceptual structure of the knowledge domain, it appears to differ in essential ways from our internalized characterization of that knowledge.

Another response to our example would be to deny that there is anything absurd about it. The rules and representations of a linguistic system characterize the competence of an *ideal* speaker-hearer, who interprets his or her knowledge fully and faithfully, who is immune to distraction, has no memory limitations, does not make errors, and so on. The on-line behavior of real native speakers, to which a formal linguistic system may have no discernible relation, is jointly determined by that system *and* these other performance factors, and the latter are the source of any processing discrepancy. For this reason, processing considerations can have no direct bearing on grammatical theory.

As Chomsky (1980 and elsewhere) has often pointed out, idealization has been a crucial ingredient for progress in the so-called mature sciences. Science progresses by abstracting away from the flux of experience to reveal underlying and invariant relationships. Likewise, linguistics and cognitive science will not make scientific progress unless we identify and focus on crucial idealizations, such as the ideal native speaker. There is nevertheless something suspicious about this appeal to idealization. It is used mainly to restrict the kind of evidence that may be brought to bear on representational issues. Thus, it appears to be a way of insulating linguistic theory from the other cognitive sciences.

This is not the role that idealization should play in science. The lawful relations that are constructed around such concepts as ideal gases, frictionless planes, infinitesimal masses, or infinite distances become scientifically interesting only when it can be shown that the behavior of real gases, planes, and particles can be made to come arbitrarily close to the postulated ideal as extraneous sources of variation are identified and controlled. In other words, there is a scientific responsibility to show that the real *does* asymptotically approach the ideal under appropriate circumstances. If this responsibility is taken seriously, there are more, not fewer, ways in which processing results might bear on representational issues, contrary to what Chomsky's appeal to idealization suggests. In particular, we must discover ways of showing that the actual behavior of real native speakers converges on the ideal behavior predicted by our grammatical theory, as interfering performance factors are reduced. Developing methods of reducing such interferences is of course a very hard scientific problem, but analogous problems had to be solved in the other sciences as a precondition for their major advances. Only as we make progress toward this goal will we be justified in speaking of a grammar as a model of the ideal native speaker's knowledge.

In attributing psychological reality to a grammar, then, we require more than that it provide us with a description of the abstract structure of the linguistic knowledge domain: we require evidence that the grammar corresponds to the speaker's *internal* description of that domain. Because we cannot directly observe this 'internal grammar,' we must infer its properties indirectly from the evidence available to us (such as linguistic judgments, performance of verbal tasks in controlled experimental conditions, observation of the linguistic development of children, and the like). The data of linguistics are no more or less privileged for this inquiry than any other data. However, the formal representations of linguistic theory, when joined with the information-processing approach of computer science and with the experimental methods of psycholinguistics, provide us with powerful tools for investigating the nature of this internal grammar and the processes that construct and interpret it. The methods and results of these different approaches can mutually constrain the form of a competence-based model of linguistic performance. In the following we suggest that constraints on knowledge representation derived from considerations in theoretical linguistics may serve to limit the class of compatible information-processing models and, likewise, that computational considerations may have direct implications for the form in which linguistic representations must be cast.

THEORETICAL CONSTRAINTS ON KNOWLEDGE REPRESENTATION

Theoretical constraints on representations of knowledge affect the classes of computations that can construct or interpret those representations. Thus, the choice of grammars can limit the choice of process models. A grammar may prove to be unrealizable in explanatory models of language processing because of the basic representational assumptions that it embodies. To amplify this point, let us compare the representational assumptions of two very different theories of grammar—the transformational theory and the lexical-functional theory. We will focus on the representation of grammatical relations: these are the associations between the semantic predicate argument structures (or thematic role structures) of sentences and their surface constituent structures. The term *grammatical relations* is thus used here in a theory-neutral sense, to be distinguished from *grammatical functions* such as SUBJ(ect) and OJB(ect).

Let us consider first a simple, concrete example illustrating the different representational assumptions of the two theories. The basic motivation for syntactic transformations during the early days of generative grammar was the systematic relationship exhibited by pairs of sentences like *Mary kissed John* and *John was kissed by Mary*. In the first (active) sentence, the noun phrase preceding the verb has the agent or 'kisser' role, and in the second (passive) sentence, it has the theme or 'kissed' role. If there were no transformational rules that moved the noun phrase following the verb to the position in front of the verb (so it was argued), we would have to postulate two different lexical entries for *kiss*—one with an agent subject that appears in the active form, and one with a theme subject in the passive form. The postulation of syntactic transformations removes this redundancy and expresses the regularity that systematically relates active sentences to passive sentences.

Given the original assumptions of transformational grammar (which lacked any mechanisms for representing grammatical relations other than transformations of phrase structures), this argument was certainly persuasive. At a time when the lexicon was considered to be simply a collection of idiosyncratic properties of morphemes, it seemed obvious that a single lexical entry is better than two. When a richer theory of lexical rules and representations becomes available, though, the argument loses its persuasiveness. Lexical entries can represent semantic predicate argument structures independently of phrase structure forms, and lexical rules can capture the redundant relationship between the two lexical entries for *kiss*. For the two sentences given, we need to know only that in one, the subject is the agent and the object is the patient, and in the other, the subject is the patient, and the *by* object is the agent. In the lexical-functional theory, we represent this information in the lexical entries for the active and passive verbs, as in (1) and (2). (In (2) we have designated the English *by* object by the more general function name 'OBL(ique)$_{AG(ent)}$'.)

$$kiss\text{:}kiss<(SUBJ)\ (OBJ)> \tag{1}$$
$$\phantom{kiss\text{:}kiss<}AGENT\ PATIENT$$

$$kissed\text{:}kiss<(OBL_{AG})\ (SUBJ)> \tag{2}$$
$$\phantom{kissed\text{:}kiss<}AGENT\ PATIENT$$

Note that both lexical entries (1) and (2) have the same lexical predicate argument structure: kiss<AGENT PATIENT>. They differ in the grammatical functions that express the agent and patient arguments. The grammatical functions SUBJ, OBJ, and OBL$_{AG}$ are universal, but their phrase structure realizations vary from language to language, just as the surface forms of particular languages vary. Thus in English, the SUBJ is realized as an NP preceding the verb; the OBJ is realized as an NP following the verb; and the OBL$_{AG}$ is realized as a prepositional phrase marked with *by*. Finally, in order to capture the systematic relationship between the two lexical entries, one may propose a lexical rule that changes SUBJ to an optional OBL$_{AG}$ and OBJ to SUBJ:

$$(SUBJ) \rightarrow (OBL_{AG})/\emptyset \qquad\qquad (3)$$
$$(OBJ) \rightarrow (SUBJ)$$

By such a rule, the passive lexical form (2) can be derived from the active form (1). Bresnan (1982d) gives a detailed analysis of passive constructions in these terms. In general, systematic relations between lexical forms can be expressed by means of lexical rules or principles for associating grammatical functions with predicate arguments.

With this example of the active-passive relation in mind, let us now consider the underlying representational assumptions that result in these particular analyses. Despite continually changing theoretical assertions, the basic representational assumptions of transformational grammar have remained surprisingly constant since at least the "standard theory" (Chomsky, 1965). The first assumption is that there is a one-to-one correspondence between the semantic predicate argument structure of a sentence and a set of ('deep') grammatical functions, such as '(logical) subject,' '(logical) object,' and so forth. The second assumption is that these grammatical functions can be reduced to a set of canonical phrase structure configurations: for example, the 'logical subject' becomes the NP immediately dominated by S in the deep phrase structure representation of the sentence, and the 'logical object' becomes the NP immediately dominated by VP. These two assumptions permit semantic predicate argument structures to be represented by canonical, or 'deep,' syntactic phrase structures. Because the surface forms of sentences are also represented as syntactic phrase structures, the mapping between semantic arguments and surface form (which constitutes the grammatical relations of a sentence) must be expressed through operations on phrase structures. This naturally gives rise to the third representational assumption, that there is a set of structure-dependent operations, or syntactic transformations, that map between the deep phrase structure (representing semantic predicate argument structure) and the surface phrase structure. Analogues of these assumptions underlie current versions of transformational theory. For example, in Chomsky's (1981) government and binding theory, the one-to-one correspondence between predicate argument structures (thematic structures) and deep phrase structures (D-structures) appears as the *theta criterion*. The mapping between D-structures and surface phrase structures is accomplished in two segments: first, syntactic transformations (restricted to the form 'Move α') map D-structures onto S-structures (abstract surface structures that contain indexed empty categories); second, a component of "stylistic" transformations, deletion rules, and morphophonological rules map S-structures onto surface forms.

Under these assumptions, the grammatical relations of the sentence *John was kissed by Mary* must be represented in a particular way. First, by the assumption that there is a one-to-one correspondence between semantic predicate argument structure and deep grammatical functions, the surface subject of the passive sentence must have the same deep function as the surface object of the active sentence *Mary kissed John,* for these two NPs have the same thematic role of patient. Second, by the assumption that such deep functions reduce to a canonical set of underlying phrase structure configurations, *John* must be represented in the same deep structure position in both the active and the passive sentence. Third, by the assumption that the mapping between thematic role structure and surface form is effected by syntatic transformations, *John* must be moved from its deep structure position to its surface structure position. Thus, these basic assumptions about the representation of grammatical relations require that one of the constructions in question be formed by the movement of an NP from an underlying phrase structure position (as the logical object, for instance) to a surface phrase-structure position (as the surface subject).

In sum, the guiding representational idea of transformational theories of syntax is that semantic predicate argument structure must be characterized in the vocabulary of constituent structure representation, and must, in fact, be directly reflected in the forms of phrase structures. This is the meaning of Chomsky's *theta criterion.* As we have seen, this immediately leads to a theory in which the grammatical relations of sentences are encoded by phrase-structural computations (such as the transformational derivation).

The theory of lexical-functional grammar (LFG), in contrast, maintains a very different set of representational assumptions. First, on this theory, there is *no* one-to-one correspondence between semantic predicate argument structure and grammatical functions: the 'theta criterion' of transformational theories is rejected. Instead, a single predicate argument structure may have several alternative lexical assignments of grammatical functions, governed by universal principles of function-argument association (see Baker, 1983; Bresnan, 1982a,e; Grimshaw, 1982; Rappaport, 1983). This is illustrated by the active and passive lexical forms of the verb *kiss* in (1) and (2). Second, on the lexical-functional theory, grammatical functions are *not* reducible to canonical phrase structure configurations: on the contrary, the phrase structure categories themselves appear reducible to functional primitives (Jackendoff, 1977; Bresnan, 1982a), and the relation between structural configurations and grammatical functions is clearly many-to-many, varying across language types and even within languages (Bresnan, 1982a; Mohanan, 1982; Simpson & Bresnan, 1982). Because this theory requires no 'normalized' phrase structure representation to express predicate argument relations, the structural component of the grammar can be vastly simplified, the entire transformational derivation being replaced by a single level of phrase structure representing the surface form of a language, the *constituent structure (c-structure)* (see Bresnan, 1982a; Kaplan & Bresnan, 1982). Third, in the lexical-functional theory, the mapping between thematic role structure and surface form is *not* effected by syntactic transformations (or equivalent structural computations). Rather, it is effected by correlating the grammatical functions that are assigned to lexical predicate argument structures with the grammatical functions that are syntatically associated with c-structure forms: *functional structures* (*f-structures*) formally represent these correlations. See Fig. 5.1 for a simple illustration. F-structures represent grammatical relations in an invariant universal format that is independent of language-particular differences in surface form. The f-structures are semantically interpreted (Halvorsen, in press), whereas the c-structures are phonologically interpreted.

Thus, in the lexical-functional theory, unlike the transformational theory, phrase-structure computations play only a very restricted and superficial role in the mapping from predicate argument structure to surface form: the greater part of this mapping is lexically encoded independently of phrase structure form by the assignment of universal grammatical functions to lexical predicate argument structures. The guiding idea of the LFG theory is that *only* lexical rules can alter these function-argument associations. Syntactic rules must therefore *preserve* function-argument correspondences. This is called *the principle of direct syntactic encoding* (Bresnan 1982d, Kaplan & Bresnan, 1982). One consequence of this principle is that active and passive verbs, because they induce different grammatical relations, *must* have different lexical entries. Another consequence is that there can be no NP-movement transformations like the passive transformation.

Because the various function-argument correspondences are already encoded in lexical entries, no phrase structure manipulations are needed to express the grammatical relations of sentences. Instead, active and passive lexical items are

c-structure:

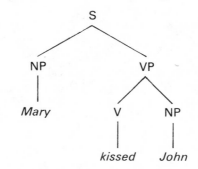

f-structure:

FIG. 5.1   Constituent structure and functional structure for *Mary kissed John*.

$$\begin{bmatrix} \text{SUBJ} & [\text{PRED 'MARY'}] \\ \text{TENSE} & \text{PAST} \\ \text{PRED} & \text{'KISS}\langle(\text{SUBJ})(\text{OBJ})\rangle\text{'} \\ \text{OBJ} & [\text{PRED 'JOHN'}] \end{bmatrix}$$

lexically inserted directly into surface constituent structures (c-structures). The syntactic instantiations of the grammatical functions can be read off from functional annotations to these surface structures, as described in Kaplan and Bresnan, 1982. A phrase structure satisfying the active verb *kiss* must have both an object and a subject; a phrase structure satisfying the passivized form of the verb must have a subject, but no object. How the phrase-structure subject is interpreted—whether as "the kisser" or "the one kissed"—depends on which lexical form of the verb is satisfied by the phrase structure tree.

This theory of grammatical representation explains why passivization in English appears to involve NP movement (and why other grammatical-relation changing rules of English do so as well). The lexical rule of passivization shown in (3) refers only to the grammatical functions SUBJ, OBJ, and OBL$_{AG}$, and not to the phrase structures in which these functions are syntactically expressed. In English, the SUBJ and OBJ functions are structurally expressed by different positions in the phrase structure: the subject NP appears before the verb and is immediately dominated by the S node, while the object NP appears after the verb and is immediately dominated by the VP node. It follows that when an active verb like *kiss* is lexically inserted into a structure, its patient argument, which has been lexically assigned the OBJ function (as in (1)), will be structurally expressed by a postverbal NP; and when a passivized verb like (*be*) *kissed* is lexically inserted into a structure, this same patient argument, which has now been lexically assigned the SUBJ function (as in (2)), will be structurally expressed by a preverbal NP. The lexical change in these function-argument correspondences will therefore induce a change in the phrase-structure positions that express these arguments. Thus, the syntactic effects of passivization in English will appear to involve the 'movement' of a NP from one phrase structure position to another.

But this apparent NP movement is only an illusion, an effect arising from the way that these universal grammatical functions happen to be syntactically encoded in the surface forms of English. In other languages, the SUBJ, OBJ, and OBL$_{AG}$ functions are syntactically expressed through morphological case markings, and not through distinctive positions in phrase structure. And in such languages, predictably, the syntactic effects of passivization involve, not the apparent movement of an NP, but an apparent change in morphological case (Mohanan, 1982). The syntactic mechanisms of LFG successfully generalize across radically different language-types. For detailed analyses of varying language types within LFG, see studies in Bresnan (1982b), as well as, Fassi Fehri (1981), Neidle (1982a,b), Simpson (1983), Simpson and Bresnan (1982), Klavans (1982), Bresnan, Kaplan, Peters, and Zaenen (1982), and Levin, Rappaport, and Zaenen (1983). In contrast, syntactic transformations do not generalize across these language types, and therefore fail to provide a universal mechanism for representing grammatical relations. Recognizing this, transformational theories, including the government and binding theory (Chomsky, 1981), simply make the unsatisfying assumption that language types may differ in their fundamental syntactic mechanisms.

Other explanatory advantages of the LFG theory are discussed in Bresnan (1978, 1982b). They include explanations for the boundedness and structure-preserving properties of relation-changing rules and the fact that rule-interactions follow from the basic organization of the grammar rather than from stipulated conditions on rules or representations.

IMPLICATIONS FOR PROCESS MODELS

Let us now consider the implications of these representational theories for process models. To establish a clear basis of comparison, we will make explicit several definitions and assumptions. First, for the reasons given earlier, we assume that there is a *competence grammar* that represents native speakers' tacit knowledge of their language. Next, suppose that we are given an information-processing model of language use that includes a processor and a component of stored linguistic knowledge K. As a minimum, we assume that K prescribes certain operations that the processor is to perform on linguistic representations, such as manipulating phrases or assigning grammatical functions. K may also include other kinds of information as well. For example, it could contain indexing information that helps the processor quickly determine which representational operations it should perform in a given situation, frequency information to aid in making heuristic decisions, and so forth. We call the subpart of K that prescribes representational operations the *representational basis* of the processing model. (The representational basis is the "internal grammar" of the model.) Because not all components of the internal grammar are necessarily utilized in every linguistic behavior, we do not require all information in the representational basis to be interpreted by every processing model. However, we do require that every rule of the representational basis be interpreted in a model of *some* behavior; thus, the internal grammar cannot contain completely otiose rules. We can now say that a model satisfies the *strong competence hypothesis* if and only if its representational basis is isomorphic to the competence grammar.

We are now in a position to see how the choice of grammatical theories can affect the choice of process models. Natural languages frequently require highly intricate feeding relations among rules. This is as true of phonological and morphological rules as it is of the rules that determine syntactic relations. Examples are easy to construct. Consider (4):

  a Someone is giving too many gifts to politicians. (active)
  b Someone is giving politicians too many gifts. (active-dative)
  c Politicians are being given too many gifts. (active-dative-passive)
  d Too many gifts are being given to politicians. (active-passive)

   e Too many gifts are being given politicians. (active-passive-dative)

   f There is someone giving too many gifts to politicians. (active-*there*-insertion)

   g There is someone giving politicians too many gifts. (active-dative-*there*-insertion or active-*there*-insertion-dative)

   h There are politicians being given too many gifts. (active-dative-passive-*there*-insertion)

   i There are too many gifts being given to politicians. (active-passive-*there*-insertion)

   j There are too many gifts being given politicians. (active-passive-dative-*there*-insertion or active-passive-*there*-insertion-dative)   (4)

In these examples, the mappings between the surface subjects and objects and their semantic arguments depend upon the feeding relations of the rules. If the dative rule feeds the passive rule, for example, the subject will correspond to a different argument of the verb *give* than if it does not or if the reverse is the case, and if any rule feeds *there*-insertion, the subject will correspond to no argument of *give*. Similar intricacies appear elsewhere in English and in other languages.

In transformational theories of syntactic representation, as we have seen, the predicate-argument structure to surface form mapping is performed by phrase structure computations. In such theories, grammatical relations that are encoded by intricate feeding relations must be represented by an ordered sequence of structural computations which is the transformational derivation.[2]

---

[2]This holds for true feeding relations—relations in which the output provided by each operation creates the necessary input of the next operation in a "cascade." If all syntactic rules could be applied simultaneously, there would be no true syntactic feeding relations. Chomsky (1981) has attempted to maintain a new representational principle, *the projection principle,* which, by making the transformational derivation "transparent" in S-structure, would in effect permit simultaneous application of syntactic transformations. However, there are grammatical phenomena that seem to force a structuralist theory into cascades, at one level of representation if not another. For even if the projection principle could be maintained for the mapping between D-structure and S-structure, the mapping from S-structure to surface form will itself fail to satisfy the principle of simultaneous applicability of rules; this problem is particularly obvious in the case of radically nonconfigurational languages (Klavans, 1982; Simpson & Bresnan, 1982), where the mapping from configurational to nonconfigurational form would involve operations of "scrambling," constituent breakup, and deletion, that produce different surface forms depending on their order of application. Thus, the projection principle does not eliminate feeding relations among transformational operations in the derivation of *surface forms.*

It is interesting to note that combining the projection principle with other assumptions in current transformational theory leads to empirically wrong results. First, as Chomsky notes, it is inconsistent with many analyses within his own framework, which remain to be reconciled with it. Second, that framework lacks any well-motivated analysis of a number of constructions that appear inconsistent

In the lexical-functional theory, in contrast, the predicate-argument structure to surface form mapping is performed by lexical operations and surface function annotations. The feeding relations are expressed by the composition of operations that derive lexical entries. Thus, given the LFG representations, no special phrase structure computations are required to decode the grammatical relations which arise from these intricate feeding relations. As pointed out in Bresnan (1978), even the lexical computations are not required in generating sentences, since such lexical rules, as long as they have a finite output, can always be interpreted as *redundancy rules,* and in fact, there is some independent motivation for doing so (Baker, 1979; Jackendoff, 1975). As such, the rules could be applied to enter new lexical forms into the mental lexicon, and the derived lexical forms could subsequently simply be retrieved for lexical insertion rather than being rederived (Miller, 1978). The search space for any lexical form is bounded, because only a finite number of these lexical rules can be defined in the theory (Pinker, 1982).

Consider now how we might model the process of decoding the grammatical relations of natural language sentences. Let us suppose that the model derives a representation of the predicate argument structure of a sentence from a representation of its surface form together with a store of linguistic knowledge structures represented by a grammar—the 'internal grammar' of the model. Let us further suppose that the model satisfies the strong competence hypothesis. If the linguistic knowledge that is required in this process is represented by a transformational grammar, grammatical relations that are encoded by intricate feeding relations must be decoded by an ordered sequence of phrase structure computations (corresponding to the sequence of transformations in a transformational derivation of a sentence). The complexity of the decoding process is then a direct function of the length of the transformational derivation, a hypothesis known to psychologists as the *Derivational Theory of Complexity.*[3] Despite important early work in its support (Miller, 1962; Miller & McKean, 1964), psychologists now appear to be universally agreed that this theory is false (Fodor, Bever, & Garrett, 1974).

However, if the linguistic knowledge that is required in decoding grammatical relations is represented by an LFG, a range of possible models with very different complexity metrics can be obtained. The intricate syntactic feeding relations

with the projection principle, such as dative-alternation constructions and noncompositional idiom constructions (on the latter, see Bresnan, 1982d and Rothstein, 1982). Third, recent work on "raising" constructions (Bresnan, 1982a), derived nominals; (Rappaport, 1983), subcategorization (Grimshaw, 1981) and "small clauses" (Bresnan, 1982c; Neidle, 1982b; Williams, 1982) has brought forth evidence that optimal grammars for these phenomena will violate the projection principle.

[3]There is a slight subtlety here. The statement that such a model will have computational complexity in proportion to the length of the derivation is a fact about that kind of model. The Derivational Theory of Complexity involves another small step, the claim that complexity of the decoding process in the model will correspond to actual experienced cognitive load under some straightforward assumptions about psychological costs.

that can change grammatical relations will now be represented by sequences of *lexical*, not syntactic, rules. If these lexical rules are interpreted as redundancy rules, then all the possible function-argument correspondences will already be expressed in the lexicon by finite sets of lexical forms. Thus, the lexical entry of a verb like *give* will include passive and dative lexical forms, which were derived by the lexical operations when the active form of the verb *give* was first entered in the lexicon. Because the outputs of these lexical redundancy rules already exist in the stored knowledge component of the model, the processor need *not* perform the operations specified by these rules as the model decodes the grammatical relations of a sentence. If we further assume that all lexical forms are accessed in parallel, then in this model, the complexity of syntactic computations will *not* reflect the complexity of the lexically encoded feeding relations, but only the complexity of the analysis of the surface phrase structure tree. In this model, then, the relative complexity of active, active-passive, active-dative-passive, and other sentences of (4) will depend only on the relative complexity of their surface structure analyses. Let us refer to this as Model I.

We can derive an interesting variant of Model I if we further suppose that semantic interpretation is interleaved with the syntactic analysis—an assumption that is compatible with LFG grammars because of their order-free composition property (see the discussion below and Halvorsen, in press). Assume that the arguments of lexical forms can be accessed both by the *functions* they select (SUBJ or OBJ, etc.) and by their *semantic properties*. Then the process of extracting the grammatical relations of a sentence could be facilitated by semantic information that differentiated the arguments of a lexical form. For example, if we are analyzing a sentence like *The kite was admired by the girl*, and we know that *the kite* denotes an inanimate thing, then we can use that semantic information to match *the kite* to the correct (theme) argument of *admire* (since kites can be admired but cannot admire). But if we are analyzing a sentence like *The woman was admired by the girl*, this semantic accessing of the lexical predicate argument structure would not facilitate the analysis (since women can both admire and be admired). In this model, which we will call Model II, the complexity of syntactic analysis could be reduced by asymmetries in the semantic structure of the predicates. Thus, Model II would produce ''nonreversibility'' effects like those reported by Slobin (1966).

A Model III could be designed to accord with the Derivational Theory of Complexity. Such a model can be obtained by using the lexical rules on line to generate forms for lexical insertion. In other words, the processor performs the operations specified by the lexical rules as the model decodes the grammatical relations of a sentence. Thus, in order to analyze an example like (4)c *Politicians are being given too many gifts*, the dative and passive lexical rules would be applied to the active lexical form *give* and the outputs matched with the syntactic analysis. Thus, Model III would make the same complexity predictions as a transformational-grammar based processing model satisfying the strong competence hypothesis.

Could a transformational-grammar based model make the same complexity predictions as the LFG-based Models I and II? In LFG, the lexicon can store the finitely many lexical forms that are the outputs of the lexical redundancy rules. But in transformational grammars, there is no store of the infinitely many phrase structure trees which are the outputs of syntactic transformations. (This is clear if one notes that the passivized object NP itself can be of arbitrary depth, since NPs are recursive phrase structure categories.) Therefore, in order to decode the grammatical relations of sentences in a transformational-grammar based model that satisfies the strong competence hypothesis, the processor must perform the operations on phrase structure trees that the transformational rules specify. Because of the feeding relations among transformations, derivational complexity effects will result.

Is there any way to get around this problem? One way of doing so would be to use transformations, not to transform the infinitely many phrase structure trees that the phrase structure rules generate, but to transform the finitely many subcategorization frames of lexical entries. The dative, passive, and other structure-preserving transformations would be eliminated from the syntactic component, and corresponding lexical rules would be added to the lexical component. This approach substitutes for the transformational grammar knowledge structures an alternative representation of linguistic competence, one whose empirical divergences from the transformational theory were originally explored in Bresnan's (1976, 1978) extended lexical grammars. Thus, to adopt this approach is to drop transformational grammars as the competence theory in favor of a precursor to the LFG theory.[4]

Another way of achieving the complexity predictions of an LFG-based processing model using a transformational grammar knowledge representation has been proposed by Berwick and Weinberg (1983). They argue that if one changes the underlying computational architecture of the transformational-grammar based model, the Derivational Theory of Complexity no longer holds. What they actually show is that by using a finite amount of parallel processing, permitting parsing actions to be executed simultaneously, the analysis of a simple passive structure can take the same amount of time as the recognition of the corresponding active structure. In other words, the elementary operations of passivization— the NP movement and the attachment of the passive verb to the analysis tree—

---

[4]There are certain differences between a lexical-transformational theory and the LFG theory. For example, in the former, the transformations will operate on structural categories—NP, V, PP, etc.— in the strict subcategorization frames, rather than on the functions SUBJ, OBJ, etc. of the LFG theory. This difference is in fact a disadvantage for the former theory; in recent work, Grimshaw (1981) has given linguistic evidence that lexical items subcategorize for *grammatical functions*, not for phrase structure categories.

can be accomplished simultaneously in unit time. However, to successfully mimic the results of the LFG-based model, they would have to demonstrate that the operations specified by *all* of the sequences of standard transformational operations—dative, dative-passive, dative-passive-*there*-insertion, and so on—can be executed in unit time. But because these operations are in true feeding relationships, in which the necessary input of one operation is created by the output of another, it is simply not possible to execute them in parallel.

For every finite feeding sequence of transformations $t_1, \ldots, t_n$, it is possible to construct a new composite operation T_{1-n} that takes structures from the input form specified by t_1 directly to the output form specified by t_n. As long as there are only finitely many of the original transformations, and as long as they are cyclic, bounded rules, this kind of recoding is possible. In this way the complexity effects of the LFG-based models I and II may be simulated. But *the resulting model no longer satisfies the strong competence hypothesis:* that is, it is no longer true that the transformational grammar specifies the only operations that the processor can perform on linguistic representations. For example, the dative transformation, which formerly played a role in the derivations of sentences like (4)b,c,e,h, and j, no longer exists in the model, and the same is true for the passive transformation, along with all of the other cyclic transformations. Instead there is a new set of transformations, $T_1, T_2, \ldots, T_{1-2}, T_{2-1}, \ldots$, and so on, that have been motivated solely on the grounds of reducing computational complexity in this model. The composition procedure loses information in the sense that given the output of composite rules, it is in general not possible to identify what the original rules were.

It might be suggested that the grammar of composite operations, which does reduce comprehension complexity, is in fact the better grammar, the one that more accurately models the native speakers' competence. It is difficult to support this conclusion, however, because this grammar trades simplicity of comprehension for complexity of linguistic descriptions; sentences exhibiting complex grammatical relations can no longer be classified into simple syntactic types solely on the basis of their derivations. For example, the passive morphology of English could formerly be correlated with a characteristic component of the mapping that encodes grammatical relations (namely, movement from object position to subject position). In the new grammar, this correlation disappears, and passive morphology is associated with an arbitrary set of NP movements. For these reasons, the new grammar could actually *increase* the complexity of language acquisition.

We see, then, that a transformational-grammar based process model cannot achieve the results of the LFG-based Models I and II without relinquishing the strong competence hypothesis. This is because the basic representational assumptions of transformational grammar require that complex grammatical relations be encoded by complex transformational derivations. Grammatical relations cannot be pre-stored in any component of the competence grammar, because it is a finite knowledge representation and the syntactic transformation is an operation on infinite sets of syntactic phrase structure representations. Nor can grammatical relations be computed in parallel, because of the *feeding relation-ships* that grammatical-relation mapping rules enter into. Thus, the Derivational Theory of Complexity is implicit in the fundamental representational assumptions of transformational grammar. To the extent that there is evidence against the Derivational Theory of Complexity, this can be taken as evidence against the psychological reality of transformational grammars.

These conclusions do not mean that we must weaken the strong competence hypothesis or give up the goal of a unified theory of the mental representation of language. There are systems based on very different representational assumptions, such as LFG, which are realizable in more explanatory models of language processing, satisfying both the strong competence hypothesis and the substantive constraints of a theory of Universal Grammar.

THEORETICAL CONSTRAINTS ON KNOWLEDGE PROCESSING

Just as theoretical constraints on knowledge representations affect the classes of computations that can construct and interpret those representations, so the converse is true: theoretically motivated constraints on knowledge processing can affect the choice of knowledge representations. In this way, assumptions underlying models of how linguistic knowledge is processed can determine the choice of grammars.

In this section we discuss an abstract computational theory of syntactic processing that provides a conceptual framework for constructing models of various linguistic behaviors. Because all such behaviors (beyond the word-level ones) involve a mapping between strings and grammatical relations (the so-called "syntactic mapping"), we will proceed by postulating various properties that such a mapping might have. These assumptions imply certain conditions that the representational basis of a processing model must satisfy, and thus, if the Strong Competence Hypothesis is to be maintained, they limit the formalisms that are suitable for linguistic theory.

The *syntactic mapping problem* is the problem of computing, for any human language, the grammatical relations of any string of words of that language.[5]

---

[5]Recall that the term *grammatical relations* is used here in a theory-neutral way to refer to the associations between the surface constituents and the semantic predicate argument structure of a sentence. Thus, the grammatical relations of a sentence can be represented by a (semantically interpreted) constituent structure tree in phrase structure grammars, by a pair of deep and surface phrase structure trees in transformational grammar, by a pair of initial and final relational strata in relational grammar, and by an f-structure in LFG.

This is an extremely difficult problem—first, because of the complex, many-to-many relation between the sentences of any natural language and their grammatical relations, and second, because of the radical variations in surface form across languages. Yet we know that the human brain instantiates a general solution to this problem, for despite the exotic variety of the world's languages, any normal child is capable of mastering any language. To solve the problem in a general way, we will therefore introduce theoretical assumptions about the nature of the computation that will serve to constrain the set of admissible solutions. These mapping constraints are motivated by general properties of the computational problem that linguistic knowledge processes must solve, by properties that are intuitively true of the human mind (the one entity that we know is capable of executing those processes), or by properties that we believe a satisfying scientific theory of these phenomena must possess. As is true of the basic assumptions in any scientific theory, the validity of these postulates is not susceptible to direct empirical evaluation. Rather, they stand at the center of a rich deductive system that has testable consequences at its empirical frontier. These central theoretical constraints will be accepted to the extent that the remote empirical predictions of models that embody them are confirmed. For actual processing models that do embody these theoretical assumptions, and whose specific predictions are the subject of ongoing experimental investigations, see Ford (1982), Ford, Bresnan, and Kaplan (1982), Pinker (1982), and Pinker and Lebeaux (in press).

As general conditions on the syntactic mapping problem, each grammatical string must be paired with a phonological interpretation and the representation of its grammatical relations must be paired with a semantic interpretation; a string is syntactically unambiguous if and only if the mapping assigns it a unique representation of grammatical relations; and a string is ungrammatical if and only if the mapping assigns it no well-formed representation of grammatical relations. The essential properties of all generative grammars reflect certain theoretical constraints on the set of possible processes that compute solutions to the syntactic mapping problem. The constraints are *creativity* (the domain and range of the mapping are theoretically infinite), *finite* capacity (there is only a finite capacity for the knowledge representations used in the mapping), and, though not all generative grammars have turned out to satisfy this constraint, *reliability* (the mapping provides an effectively computable characteristic function for each natural language). Let us briefly review these constraints in turn.

Creativity

Prior to generative grammatical theory, most American structuralist linguists considered the aim of linguistics to be to establish procedures for discovering the grammatical structure of a given corpus of linguistic utterances as presented in field work transcriptions. This would be equivalent to taking the domain and range of the syntactic mapping to be finite. However, Chomsky (1957, 1964), emphasizing the creativity of language use, argued that the most revealing way of looking at the problem is to take the domain and range of the mapping to be infinite sets of data and grammatical structures. Although it is true that the entire body of language that any language-user produces or comprehends in a lifetime is finite, the finiteness of the linguistic corpus appears to be an arbitrary restriction from the point of view of linguistic structure. The problem to be explained is how the language-user can construct mental representations of grammatical relations for endless numbers of *novel* sentences. In principle—idealizing away from all limitations on lifespan, memory, and performance—the language-user can identify the infinite set of grammatical sentences of the language. This is the justification for the *creativity constraint,* which requires the domain of the syntactic mapping to include all strings over the lexical vocabulary of the language, the range to include infinitely many representations as well, and the mapping to characterize the infinite set of grammatical sentences of the language; that is, a string is a member of the language if and only if the syntactic mapping assigns it a well-formed representation of grammatical relations.

Finite Capacity

The second constraint, *finite capacity,* was also proposed by Chomsky (1965). Although the possible data and representations in the domain and range of the mapping are infinite, each language has only finite sets of elementary words and relations and there is only a finite capacity for representing and storing knowledge mentally. Given this constraint, the mapping must consist of the recursive composition of finitely many operations that can project a finite store of knowledge of a particular language onto infinite sets of data. Any mapping defined to be the recursive composition of finite elementary operations can be represented as a pair consisting of the specification in some formal language of the composition rules of those elementaries and a procedure which interprets those composition rules. For the syntactic processing case, the rules of composition are called *the grammar,* and we will use the notation m_G to name the procedure which applies to the rules. Thus, we can assume that the syntactic mapping decomposes into a grammar G of the particular language (which includes, for example, a set of syntactic patterns, or rules, of the language) and a procedure m_G that recursively matches the patterns, or applies the rules, of G to construct infinite sets of representations that relate surface strings of words to semantic predicate argument structures.

Reliability

The creativity constraint implies that our computation must characterize the infinitely many sentences of any natural language. The finite capacity constraint implies that this computation must decompose into a finite grammar G and a recursive procedure m_G for projecting G onto the infinitely many sentences of a language. These constraints are based upon the abilities of an ideal speaker of the language, abstracting away from the actual performance limitations of real language-users. The same is true of the third constraint, reliability. Under the same idealization, the speaker of a language is regarded as a reliable arbiter of the sentences of his or her language. It is plausible to suppose that the ideal speaker can decide grammaticality by evaluating whether a candidate string is assigned (well-formed) grammatical relations or not. The syntactic mapping can thus be thought of as reliably computing whether or not any string is a well-formed sentence of a natural language. This motivates the *reliability constraint* that the syntactic mapping must provide an effectively computable characteristic function for each natural language.

The reliability constraint implies that the subset of data in the domain of the mapping for which there are well-formed grammatical relations is a recursive set (for the mapping must effectively compute whether an arbitrary string is grammatically well-formed or not). Arguments for the recursiveness of natural language were first given by Putnam (1961), who concluded:

> [T]he self-containedness of language [i.e. the classifiability of sentences out of context], the usability of nonsense sentences, and the relative universality of grammar intuitions within a dialect group, taken together support the model of the [human] classifier as a Turing machine who is processing each new sentence with which he is provided according to some mechanical program [p. 41].

In other words, independently of knowledge of specific context, even independently of meaningingfulness, speakers can reliably classify sentences as grammatical or ungrammatical, and they do so with a high degree of consistency across individuals within a dialect group. This convergent behavior suggests that classification of strings as grammatical or ungrammatical is based on an automatic procedure. Even though actual speakers may hesitate or conflict in their judgments of grammaticality, it is hypothesized that if extraneous and confusing sources of variation are controlled and if they are given more time and memory, their judgments will approach the behavior of the ideal classifier in the limit.

Putnam also pointed out the consequent need to constrain transformational grammars if they were to be taken as characterizations of the mentally represented linguistic knowledge that is used to classify sentences. Although Chomsky (1965) assumed that constraints on transformational grammars such as the recoverability of deletions restricted the generative capacity of transformational grammars to recursive languages, Peters and Ritchie (1973) disproved this conjecture. Chomsky has increasingly downgraded the importance of constraints on generative capacity, relying on the idea of an evaluation metric to limit the class of available grammars in a learning model (Chomsky, 1977) or on an analogy between language acquisition and biological maturation (Chomsky,

1980). Despite this deemphasizing of grammatical processing in favor of grammatical acquisition, it nevertheless appears to be a precondition for any satisfying computational theory of syntactic processing that it account for the capability of the ideal speaker to classify natural language sentences. Peters and Ritchie (1973) pointed out that the actual generative transformational grammars that had been written for natural languages were in fact recursive, and subsequently both Peters (1973) and Wasow (1978) discovered properties that would constrain standard transformational grammars to generate only recursive sets.

Psychologically based arguments for the recursiveness of natural language—including Putnam's—have been criticized on the grounds of English speakers' reactions to well known "garden path" sentences like the following.

The canoe floated down the river sank. (5)
The editor the authors the newspaper hired like laughed. (6)

For example, Matthews (1979) claims that the classification of examples (5) and (6) as grammatical or ungrammatical can be predictably affected by their position in a list with sentences of form (7) and (8).

The man (that was) thrown down the stairs died. (7)
The editor (whom) the authors the newspaper hired liked laughed. (8)

Sentences (5) and (6), he asserts, will typically be classified as ungrammatical if they precede sentences similar to (7) and (8), but grammatical if they follow them. He claims that this fact is inexplicable in terms of memory and other performance limitations on an idealized effective procedure for classifying sentences. Matthews (1979) argues that the same evidence also counts against the view that real native speakers instantiate an effective procedure when pairing sentences with their underlying structural descriptions during comprehension. Following Fodor, Bever, and Garrett (1974), he assumes that these phenomena are to be explained by postulating a collection of agrammatical heuristic strategies that directly pair certain simple types of sentences with their structural descriptions; but then—he concludes—there is no need to assume that the language-user instantiates an effective procedure in comprehending or classifying the grammatical sentences of the language.

In fact, however, we can draw precisely the opposite conclusion from such examples. Recent work on syntactic perception (reported in Ford, Bresnan, & Kaplan, 1982) has argued that the heuristic strategies approach fails to explain the ability of speakers to *recover* from garden paths. In general, if speakers are given sufficient time and resources (such as pencil and paper), they are able to recover all of the grammatical analyses provided by the competence grammar. The competence-based model of syntactic processing given in Ford, Bresnan, and Kaplan (1982) can explain both the ideal behavior (the recovery of all of the grammatical analyses) and the actual behavior (the experience of conscious garden paths) in terms of resource-limitations on a recursive procedure for constructing grammatical relations.

In particular, note that examples (5) and (6) are locally ambiguous in their initial segments: with respect to the competence grammar of English, (5) permits a local initial analysis either as a simple sentence (*The canoe floated down the river*) or as a reduced relative (*The canoe* [*which was*] *floated* . . .), and (6) permits a local initial analysis either as a sequence of conjoined NPs (*The editor, the authors, the newspaper,* [*and*] . . .) or as a center-embedded relative clause like (8). General principles of local ambiguity resolution which are incorporated into an effective procedure for analyzing all and only the grammatical sentences of natural language can easily explain the initial false analyses assigned to these sentences (see Ford, Bresnan, & Kaplan, 1982). The observed performance difficulties that many speakers have in recovering the true analyses without contextual clues can be explained by limitations on working memory and other computational resources during the grammatical analysis procedure.[6]

We see, then, that speakers' reactions to examples like (5)–(8) would actually *follow* from specific performance limitations on an idealized effective procedure for analyzing or classifying sentences. Although Matthews (1979) repeats the well known arguments due to Fodor, Bever, and Garrett (1974) that a psycholinguistic theory of comprehension cannot plausibly incorporate a transformational grammar as the mental representation of linguistic knowledge, we have already observed that these arguments do not impugn the competence hypothesis, and hence they do not undermine the motivation for imposing the reliability constraint on the syntactic mapping.

A very different source of arguments against the recursiveness of natural language can be found in considerations of semantic interpretation. The most ingenious such argument is due to Hintikka (1974, 1979). Hintikka found that the acceptability of an infinite set of natural language sentences involving the quantifiers *any* and *ever* depends upon their semantic equivalence to other sentences. He argued that there is no effective procedure for determining the acceptability of these sentences, because (assuming that the semantic equivalence of natural language sentences is adequately modeled by logical equivalence) there is no effective procedure for testing the logical equivalence of even first-order quantification sentences. This conclusion implies that the set of acceptable sen-

---

[6]Specifically, according to the perceptual theory of Ford, Bresnan, and Kaplan, (1982), the true analysis of (5) would require morpho-syntactic reanalysis of *floated* from a past tense verb to a passive participle after the grammatical analysis of the initial hypothesized sentence that contains it has been completed; the difficulty of recovering the correct analysis can therefore be explained by limitations on memory for morpho-syntactic categorizations during the syntactic analysis of sentences. Similarly, the true analysis of (6) requires recognition of a center-embedded syntactic structure, which could impose an excessive burden on working memory during the grammatical analysis of a sentence, as Church (1980) argues.

tences is not only nonrecursive, but not even recursively enumerable, and hence well beyond the power of any generative grammar or Turing machine to specify.[7]

Such arguments for the nonrecursiveness of natural language enforce an important point. The automatic mental construction of grammatical relations must *not* require an evaluation of truth in the world, or logical truth conditions. The grammatical relations of the unacceptable sentences cited by Hintikka are in fact as easily perceived as those of other sentences. For example, Hintikka's unacceptable *Louise ever kisses me* can be assigned perfectly well formed grammatical relations, which specify that *Louise* and *me* are the subject and object arguments of a dyadic tensed active verb *kiss*, modified by a temporal quantifying adverb *ever*. This fact is unsurprising, because well-formed grammatical relations can be assigned even to nonsensical sentences like *Louise pilsely grisps a blawn neddle*. If such examples are judged "unacceptable" or "incorrect," this must then be attributed to some property of their nonsyntactic interpretation—whether it be the non-effectively-computable semantic property formulated by Hintikka or another, as yet undiscovered, property. If Hintikka's argument is correct, then semantics must diverge from syntax in a fundamental way, as he observes. The method of recursively characterizing the structure of natural language by means of a generative grammar may in fact be incapable of characterizing the semantics of quantification.

We can conclude that the reliability constraint on the syntactic mapping problem introduces no incoherence: 'grammaticality' can be a recursive concept so long as it is not a function of truth in the world.

The above constraints—creativity, finite-capacity, and reliability—are familiar from early work in generative grammar. Note that none of them is logically necessary for purposes of grammatical description. Their motivation comes rather from an abstract computational theory of syntactic processing. Exactly the same is true of two further constraints that we will now add: *order-free composition*, requiring that the grammatical relations that the mapping derives from an arbitrary segment of a sentence be directly included in the grammatical relations that the mapping derives from the entire sentence, independently of operations on prior or subsequent segments, and *universality*, requiring that the mapping incorporate a universal procedure for constructing representations of grammatical relations.

Order-Free Composition

The constraint of *order-free composition* is motivated by the fact that complete representations of local grammatical relations are effortlessly, fluently, and relia-

---

[7]See Hintikka (1979, 1980) for counterarguments to various objections to his argument that have been raised in the literature, such as Chomsky's (1980) proposal that a constraint on "Logical Form" accounts for Hintikka's generalization in an effectively checkable way.

bly constructed for arbitrary segments of sentences. For example, the sentence fragments in (9) are immediately associated with grammatical representations that provide the same kinds of information as those of complete sentences.

a There seemed to . . .
b . . . not told that . . .
c . . . too difficult to attempt to . . .
d . . . struck him as crazy
e What did he . . .   (9)

In (9) a *there* is a grammatical but not a 'logical' subject of *seemed*, and is grammatically related to an infinitival complement of *seemed*. In (9)b, the unexpressed subject of *told* can be the passivized indirect object of the verb, and *that* can be the complementizer of an unexpressed sentential complement corresponding to the 'logical' object; other grammatical interpretations arise as well. The fragment in (9)c is syntactically ambiguous in the same way as the full sentence *It was too difficult to attempt to understand,* where the subject of *difficult* can be identified either with the clause *to attempt to understand (something)* or with an object within the clause (such as the object of *understand*). In (9)d, *him* is the grammatical and the 'logical' object of *struck,* and *crazy* is the predicative complement of the unexpressed subject of *struck.* The fragment in (9)e is interpreted as part of a question in which *what* is grammatically related to an unexpressed clause of which *he* is the grammatical subject. In each case, the grammatical relations for the fragment are complete in the sense that they express information that can be directly incorporated without change into the grammatical relations of some full sentence.

It is a fact, then, that given a sentence fragment, one can always pick, entirely out of context, a *possible* reading from the finite set of grammatical alternatives, while *impossible* readings are excluded. It is no more possible to interpret the fragment . . . *easy to justify* . . . so that the subject of *easy* is also the subject of *justify* than it is possible to interpret *John's actions* as the subject of both *easy* and *justify* in the ungrammatical sentence **John's actions are easy to justify yours.* In contrast, the fragment . . . *unlikely to justify* . . . allows just this possibility, exactly as in the grammatical sentence *John's actions are unlikely to justify yours.* This fact suggests a very strong natural constraint on the syntactic mapping problem: the composition of operations as performed by m_G in the syntactic mapping must be order-free, in the sense that from an arbitrary sentence segment, the mapping must derive independently of operations on prior or subsequent segments, a set of possible grammatical relations, any of which can be directly included in the grammatical relations that the mapping derives from some sentence containing the segment. In other words, we have the following *order-free composition constraint* (referred to as 'bounded context parsability' in

Bresnan (1979)): *under any valid interpretation of a string xsy consisting of adjacent segments x,s,y, the grammatical relations in $m_G(s)$ must be included in $m_G(xsy)$.*

Note that this constraint allows for the fact that, in general, segments of a sentence—viewed as strings of words—have more possible grammatical relations in isolation than within the sentence. Example (10) illustrates this point:

a . . . her candy killed Mother See.
b A man who hated her candy killed Mother See.   (10)

The string of words in (10)a can be interpreted so that *her candy* is the subject of *killed* and *Mother See* is the object of *killed*. But the subject-verb relation between *her candy* and *killed* is not included in the grammatical relations of the whole sentence (10)b. However, (10)a can also be interpreted as a sequence of two unrelated subsegments, *her candy* and *killed Mother See,* both of whose local grammatical relations are then directly contained in the grammatical relations of the entire sentence (10)b. Note that in the limiting case, all the words in a certain segment may be unrelated, as in (11)a, for example:

a . . . to by for . . .
b The one that he should be spoken to by for God's sake is his
    supervisor.   (11)

In this limiting case, the representation of grammatical relations of the segment can simply be identified with the lexical representations of the unrelated words. In sum, the order-free composition constraint asserts that sentential context may determine the *choice* of one of a set of locally computed grammatical relations for a segment, but the computation of grammatical relations for a segment may not involve the computation of the grammatical relations of the context. In other words, this postulate severely constrains the role of context sensitive operations in the syntactic mapping.

The *order-free composition* property is equivalent to requiring that the mapping m_G be *monotonic*.[8] In other words, the mapping m_G from segments to sets of grammatical relations must preserve inclusion: under any valid interpretation, if one string of words (s_1) is contained in another (s_2), then the (compatible) grammatical relations of the one ($m_G(s_1)$) must be included in those of the other ($m_G(s_2)$). Intuitively, monotonic functions are incremental. Thus, the monotonicity constraint requires the operations of m_G to add increments of local information from the string to the global representation of information and not subtract from the global representation.

---

[8]Recall that a function $f:X \to Y$ is called 'monotonic' if X and Y are ordered sets and $X_1 \leq X_2$ implies that $f(X_1) \leq f(X_2)$.

Universality

The fifth assumption that we shall adopt to constrain the syntactic mapping problem is the *universality* of the mapping procedure. It is plausible to assume that the procedure for grammatical interpretation, m_G, is the same for all natural language grammars G: that is, there is a universal m_U such that for any G, $m_G = m_U$. This constraint is motivated by the universality of the system for mentally representing natural language. We assume that the grammar G representing mature knowledge of a particular language is induced by a universal learning function. As in Pinker (1982), we assume that data for the learning function are sets of pairs (s,r), where s is the perceived surface string of words and r is the mental representation of meaningful grammatical relations of s; the learning function maps these data onto hypothesized grammars. To test a hypothesized grammar G^*, there must be some universal effective procedure for constructing the mental representations r of lexical strings s given G^*; call it m_U. Although it is conceivable that this universal procedure is different from the one that the language learner normally uses in comprehending language (m_G), the simplest, strongest, and most plausible assumption is that the procedures are the same. If so, the acquisition process must depend just as much upon the ability to comprehend language as the growing comprehension of language depends upon the acquisition process. The interaction of these two information-processing systems, linguistic acquisition and linguistic comprehension, motivates the universality constraint that $m_G = m_U$ for all G.

IMPLICATIONS FOR SYNTACTIC KNOWLEDGE REPRESENTATION

These processing constraints on the syntactic mapping problem—*creativity, finite capacity, reliability, order-free composition,* and *universality*—impose important limitations on the possible forms of syntactic knowledge representation, ruling out many possible systems of grammar—even apparently descriptively adequate ones—as systems of the mental representation of language. The creativity and finite capacity constraints require knowledge representations consisting of finite systems of explicit rules that are capable of characterizing infinite sets of natural language sentences. The reliability constraint further requires that these rule systems specify languages within the class of recursive sets, thereby ruling out unrestricted rewriting systems (Chomsky, 1963), arbitrary augmented transition networks (Woods, 1970), and arbitrary standard transformational grammars (Peters & Ritchie, 1973) as representations of linguistic competence. The remaining two contraints impose still stronger conditions on possible grammatical theories.

Because of these constraints, the syntactic mapping decomposes into a finite grammar G that generates the language and a finite set of operations m_G that recursively apply the rules of G to extract the meaningful grammatical relations of the language. In general, it is the syntactic (as opposed to the lexical) rules of generative grammars that specify infinite sets of structures; hence it is these rules that m_G must recursively apply in extracting the grammatical relations of a string. Recall that the order-free composition constraint states that under any valid interpretation of a string *xsy* consisting of adjacent segments *x,s,y,* the grammatical relations in $m_G(s)$ must be included in $m_G(xsy)$. Clearly, grammars whose syntactic rule components consist only of context-free phrase structure rules satisfy this requirement: in such grammars, a noun phrase is a noun phrase regardless of its external context. For these grammars, the order in which the segments are analyzed in the mapping from the data to the representation will therefore be irrelevant. However, non-context-free syntactic rules do not in general have this property: such rules may alter the analysis of the grammatical relations of a phrase according to the analysis of its context, and the resulting representation may therefore depend upon the order of application of these "context-sensitive" rules to different segments. Consequently, if m_G has the order-free composition property, the grammar G must be such that information about the surface-to-predicate argument structure associations given as the ouptut of m_G is independent of the order in which any "context-sensitive" nonlexical grammatical rules are applied in the derivation of the representation from the data.[9] For reasons we have already seen, transformational grammars do not have this property. Transformational grammars explicitly operate by changing the grammatical relations for each region of the string. It is of course possible to modify transformational grammars so that local grammatical relations in the transformational derivation are always preserved in surface forms, and this formal effect is partially achieved by the projection principle (Note 2). But for reasons we have already discussed in Note 2, transformational grammars with this provision still do not possess the order-free composition property: the order dependencies merely migrate to other components of the mapping onto surface forms. Augmented transition networks also lack the order-free composition property, for they enable register re-assignments to modify previously determined grammatical relations.

The universality constraint implies that grammatical relations must be encoded in a form that allows a universal decoding process for all natural languages. One might think that one of the various well known algorithms for context-free phrase structure parsing might provide just such a universal decoding process for natural languages, but recent work in theoretical linguistics forces

---

[9]This 'order-free' restriction does not refer to the absence of an 'extrinsic' ordering on these rules (as in Chomsky & Lasnik, 1977), but rather to the stronger requirement that the order of application of these rules in the derivation of the representation of grammatical relations from the data cannot alter that representation.

us to reject this possibility. Bresnan et al. (1982) have shown that there is no context-free phrase structure grammar that can correctly characterize the parse trees of Dutch. The problem lies in Dutch cross-serial constructions, in which the verbs are discontinuous from the verb phrases that contain their arguments. The phenomenon of 'discontinuous constituents'—that is, noncontiguous constituents defining single functional units—is pervasive in natural language, occurring in its most extreme forms in Australian aboriginal languages (Hale, 1981; Pullum, 1982). The results of Bresnan et al. (1982) show that context-free grammars cannot provide a universal means of representing these phenomena.

The processing constraints on the representation of syntactic knowledge may now seem too strong: on the one hand, G has to have the order-free composition property of context-free phrase structure grammars, but on the other hand, G cannot be a context-free phrase structure grammar because context free grammars cannot universally characterize the correct surface constituent structures of natural languages. But a solution to the syntactic mapping problem does exist. For any language L, where S is the set of strings over the lexical vocabulary of L, let us take G to be a *lexical functional grammar* for L, and R (the representations of grammatical relations) to be the set of *functional structures* of the language as defined by G. Then a map m: (S,G)↦R exists which satisfies all of the given constraints. This result is based on the mathematical characterization of lexical functional grammars in Kaplan and Bresnan (1982). It provides the foundations of a computational theory for investigating the mental processes that construct representations of grammatical relations.

The work on Lexical Functional Grammar represents a new approach to the study of the mental representation of grammatical relations, based on the cognitive theory outlined here. The universality of the LFG theory of grammatical representation is supported by detailed research (Bresnan, 1982b) on grammatical relations in such languages as English, French, Russian, Icelandic, and Malayalam (a Dravidian language). The order-free composition property is illustrated by the f-structure solution algorithm given in Kaplan and Bresnan (1982). Finally, the suitability of the LFG theory for modeling syntactic processes is shown in studies by Pinker (1982), Ford (1982), and Ford, Bresnan, and Kaplan (1982). These are psychological studies of competence-based models of language acquisition, comprehension, and production, incorporating LFGs as representations of our internal knowledge of language.

ACKNOWLEDGMENTS

This material is based on work supported in part by the National Science Foundation under Grant No. BNS 80-14730 and in part by the Cognitive and Instructional Sciences Group, Xerox Corporation Palo Alto Research Center.

REFERENCES

Baker, C. L. Syntactic theory and the projection problem. *Linguistic Inquiry,* 1979, *10,* 533–582.

Baker, M. Objects, themes, and lexical rules in Italian. In L. Levin, M. Rappaport, & A. Zaenen (Eds.), *Papers in LFG.* Bloomington, In.: Indiana University Linguistics Club, 1983.

Berwick, R., & Weinberg, A. The role of grammars in models of language use. *Cognition,* 1983, *13,* 1–61.

Bresnan, J. *Towards a realistic model of transformational grammar.* Paper presented at the MIT-AT&T Convocation on Communications, MIT, 1976.

Bresnan, J. A realistic transformational grammar. In M. Halle, J. Bresnan, & G. A. Miller (Eds.), *Linguistic theory and psychological reality.* Cambridge, Mass.: MIT Press, 1978.

Bresnan, J. *Bounded context parsability and learnability.* Paper presented at the Workshop on Mathematics and Linguistics, Hampshire College, December 1979.

Bresnan, J. Control and complementation. *Linguistic Inquiry,* 1982, *13,* 343–434. (a)

Bresnan, J. (Ed.) *The mental representation of grammatical relations.* Cambridge, Mass.: MIT Press, 1982. (b)

Bresnan, J. *On the nature of grammatical representation.* Short-term Humanities Council Visiting Fellow Lecture, Princeton University, May 1982. (c)

Bresnan, J. The passive in lexical theory. In J. Bresnan (Ed.), *The mental representation of grammatical relations.* Cambridge, Mass.: MIT Press, 1982. (d)

Bresnan, J. Polyadicity. In J. Bresnan (Ed.), *The mental representation of grammatical relations.* The MIT Press, Cambridge, Massachusetts, 1982. (e)

Bresnan, J., Kaplan, R., Peters, S., & Zaenen, A. Cross-serial dependencies in Dutch. *Linguistic Inquiry,* 1982, *13,* 613–635.

Chomsky, N. *Syntactic structures.* The Hague: Mouton, 1957.

Chomsky, N. Formal properties of grammar. In R. D. Luce, R. Bush, & E. Galanter (Eds.), *Handbook of mathematical psychology.* New York: Wiley, 1963.

Chomsky, N. *Current issues in linguistic theory.* The Hague: Mouton, 1964.

Chomsky, N. *Aspects of the theory of syntax.* Cambridge, Mass.: MIT Press, 1965.

Chomsky, N. Remarks on nominalization. In R. Jacobs & P. Rosenbaum (Eds.), *Readings in transformational grammar.* Boston, Mass.: Ginn and Co., 1970.

Chomsky, N. On wh-movement. In A. Akmajian, P. Culicover, & T. Wasow (Eds.), *Formal syntax.* New York: Academic Press, 1977.

Chomsky, N. *Rules and representations.* New York: Columbia University Press, 1980.

Chomsky, N. *Lectures on government and binding.* Dordrecht: Foris Publications, 1981.

Chomsky, N., & Lasnik, H. Filters and control. *Linguistic Inquiry,* 1977, *8,* 425–504.

Church, K. *On memory limitations in natural language processing.* MIT/LCS/TR–245; also Bloomington, In.: Indiana University Linguistics Club, 1980.

Fassi Fehri, A. *Complementation et anaphore en arabe moderne: Une approche lexicale fonctionelle.* These de Doctorat d'Etat, Universite de Paris III, 1981.

Fodor, J., Bever, T., & Garrett, M. *The psychology of language.* New York: McGraw-Hill, 1974.

Ford, M. Sentence planning units: Implications for the speaker's representation of meaningful relations underlying sentences. In J. Bresnan (Ed.), *The mental representation of grammatical relations.* Cambridge, Mass.: MIT Press, 1982.

Ford, M., Bresnan, J., & Kaplan, R. A competence-based theory of syntactic closure. In J. Bresnan (Ed.), *The mental representation of grammatical relations.* Cambridge, Mass.: MIT Press, 1982.

Greeno, J., Riley, M., & Gelman, R. Young children's counting and understanding of principles. *Cognitive Psychology,* in press.

Grimshaw, J. Subcategorization and grammatical relations. In A. Zaenen (Ed.), *Subjects and other subjects: Proceedings of the Harvard Conference on the Representation of Grammatical Relations.* Bloomington, In.: Indiana University Linguistics Club, December 1981.

Grimshaw, J. On the lexical representation of romance reflexive clitics. In J. Bresnan (Ed.), *The mental representation of grammatical relations.* Cambridge, Mass.: MIT Press, 1982.

Hale, K. On the position of Walbiri in a typology of the base. Bloomington, In.: Indiana University Linguistics Club, 1981.

Halvorsen, P. K. Semantics for lexical-functional grammar. *Linguistic Inquiry,* in press.

Hintikka, J. Quantifiers vs. quantification theory. *Linguistic Inquiry,* 1974, *5,* 153–177.

Hintikka, J. Quantifiers in natural languages. In E. Saarinen (Ed.), *Game theoretical semantics.* Dordrecht: D. Reidel, 1979.

Hintikka, J. On the any-thesis and the methodology of linguistics. *Linguistics and Philosophy,* 1980, *4,* 101–122.

Jackendoff, R. Morphological and semantic regularities in the lexicon. *Language,* 1975, *51,* 639–671.

Jackendoff, R. *X̄ syntax: A study of phrase structure,* Linguistic Inquiry Monograph Two. Cambridge, Mass.: MIT Press, 1977.

Kaplan, R. Augmented transition networks as psychological models of sentence comprehension. *Artificial Intelligence,* 1972, *3,* 77–100.

Kaplan, R., & Bresnan, J. Lexical-functional grammar: A formal system for grammatical representation. In J. Bresnan (Ed.), *The mental representation of grammatical relations.* Cambridge, Mass.: MIT Press, 1982.

Klavans, J. *Configuration in nonconfigurational languages.* Proceedings of the First Annual West Coast Conference on Formal Linguistics, Stanford University, January 1982.

Levelt, W. J. M. *Formal grammars in linguistics and psycholinguistics,* (3 vols.) The Hague: Mouton, 1974.

Levin, L., Rappaport, M., & Zaenen, A. (Eds.) *Papers in LFG.* Bloomington, In.: Indiana University Linguistics Club, 1983.

Matthews, R. Are the grammatical sentences of a language a recursive set? *Synthese,* 1979, *40,* 209–224.

Matthews, R. *Knowledge of language in a theory of language processing.* Paper presented at the conference on Constraints on Modelling Real-Time Processes, Marseille, France, June 21–26, 1982.

Miller, G. A. Some psychological studies of grammar. *American Psychologist,* 1962, *17,* 748–762.

Miller, G. A. Semantic relations among words. In M. Halle, J. Bresnan, & G. A. Miller (Eds.), *Linguistic theory and psychological reality.* Cambridge, Mass.: MIT Press, 1978.

Miller, G. A., & McKean, K. A chronometric study of some relations between sentences. *Quarterly Journal of Experimental Psychology,* 1964, *16,* 297–308.

Mohanan, K. P. Grammatical relations and clause structure in Malayalam. In J. Bresnan (Ed.), *The mental representation of grammatical relations.* Cambridge, Mass.: MIT Press, 1982.

Neidel, C. Case agreement in Russian. In J. Bresnan (Ed.), *The mental representation of grammatical relations.* Cambridge, Mass.: MIT Press, 1982 (a)

Neidel, C. *The role of case in Russian syntax.* Doctoral dissertation, MIT, 1982. (b)

Peters, S. On restricting deletion transformations. In M. Gross, M. Halle, & M. P. Schutzenberger (Eds.), *The formal analysis of natural language.* The Hague: Mouton, 1973.

Peters, S., & Ritchie, R. On the generative power of transformational grammars. *Information Sciences,* 1973, *6,* 49–83.

Pinker, S. A theory of the acquisition of lexical-interpretive grammars. In J. Bresnan (Ed.), *The mental representation of grammatical relations.* Cambridge, Mass.: MIT Press, 1982.

Pinker, S., & Lebeaux, D. *Language learnability and language development.* Cambridge, Mass.: Harvard University Press, in press.

Pullum, G. Free word order and phrase structure rules. In J. Pustejovsky & P. Sells (Ed.), *Proceedings of NELS 12,* Graduate Linguistics Students Association. Amherst, Mass.: The University of Massachusetts Press, 1982.

Putnam, H. Some issues in the theory of grammar. *Proceedings of Symposia in Applied Mathematics* (Vol. 12). American Mathematical Society, 1961.

Rappaport, M. The derivation of derived nominals. In L. Levin, M. Rappaport, & A. Zaenen (Eds.), *Papers in LFG*. Bloomington, In.: Indiana University Linguistics Club, 1983.

Rothstein, S. *On preposition stranding*. Department of Linguistics, MIT, 1982.

Simpson, J. *Control and predication in Warlpiri*, Doctoral dissertation, MIT, 1983.

Simpson, J., & Bresnan, J. *Control and obviation in Warlpiri*. Presented at the First Annual West Coast Conference on Formal Linguistics, Stanford University, January 1982.

Slobin, D. Grammatical transformations and sentence comprehension in children and adulthood. *Journal of Verbal Learning and Verbal Behavior*, 1966, *5*, 219–227.

Wanner, E., & Maratsos, M. An ATN approach to comprehension. In M. Halle, J. Bresnan, & G. A. Miller (Eds.), *Linguistic theory and psychological reality*. Cambridge, Mass.: MIT Press, 1978.

Wasow, T. On constraining the class of transformational languages. *Synthese*, 1978, *39*, 81–104.

Williams, E. Semantic vs. syntactic categories. Cambridge, Mass.: MIT Press, 1982.

Woods, W. Transition network grammars for natural language analysis. *Communications of the ACM*, 1970, *13*, 591–606.

6 Deductive Vs. Pragmatic Processing in Natural Language

T. Givón
University of Oregon
and
Ute Language Program
Southern Ute Tribe

METHODOLOGICAL PREFACE

This paper is primarily concerned with substantive issues arising out of the study of natural language. Whatever methodological import it may carry is suggested only obliquely. For the purpose of this volume, however, it may be both helpful and necessary to outline the methodological context within which it may be viewed. There are three elements that bind a scientific discipline together:

1. A common body of data;
2. A common body of methodology for sifting the data;
3. A common body of theoretical goals via which the sifted data find its explanation or meaning.

My favorite metaphors for doing science comes from deep-rock mining, where the mountain is your data, your digging tools and techniques are your methodology, and the ore or its derivative precious metal is your goal. Most of us tend to agree that a unification of the cognitive sciences into one discipline is at the moment both premature and unwise. They can support each other best by using their separate methodologies to sift through distinct data bases, although perhaps it may already be useful to develop a tentative body of common theoretical goals and explanatory parameters. What cognitive scientists outside linguistics tend to underestimate, I believe, is the degree to which linguistics as a scientific discipline is split right through its middle, to the point where it is not very clear to me that there is any common grounds remaining for the discipline as a whole.

Taking the two extreme poles of this schism, one may characterize them some-what schematically as follows:

	Formal Linguistics	Functional-Typological Linguistics
(a) *Data base:*	—Isolated utterances	—utterances in context
	—derived from typologically limited "competence"	—derived from natural-speech
		—cross-linguistic
	—non-variational	—variational
	—non-developmental	—developmental (diachronic, on-togenetic, phylogenetic)
(b) *Methodology:*	—deductive	—inductive/abductive/eclectic
	—grammaticality	—interpretability/functionality
	—formal/closed system	—open-ended system
(c) *Theory:*	—Characterizing language an a closed, formal, consistent system of rules of "competence" respon-sible for "generating" all and only the "grammatical" utterances of a language;	—Characterizing language as an open-ended functional system of communication and information processing;
	—Defining the formal universals of structure;	—explaining structural/formal/cod-ing properties of language in terms of functional requirements;

Although this characterization is admittedly abbreviated and may not always do justice to claimed ultimate goals, it does characterize actual practices (for further detail see Givón, 1979b).

Given such a fundamental split within linguistics, perhaps the most useful thing one could do in making a contribution—as a linguist—to the study of cognition, is to exemplify as clearly as possible one's own brand of linguistics *in practice*. With this in mind, I now turn to the substantive bulk of the chapter.

INTRODUCTION[1]

In the study of language and meaning, a persistently anti-empirical logic-bound tradition has plagued Western epistemology for over 2 millenia. The origins of this tradition may be traced back to one post-Socratic giant, Plato. Its real foundations, however, are buried in a facile and superficial analysis of an ar-

---

[1]This introduction is a personal overview, as is the entire chapter. I am quite aware that what I have to say may have been said, in parts, by others before me. While not wishing to short-change anyone, I prefer to follow Wittgenstein's (1918) precedent in this regard: ". . . I do not wish to judge how far my efforts coincide with those of other philosophers. Indeed, what I have written here makes no claim to novelty in detail, and the reason why I give no sources is that it is a matter of indifference to me whether the thoughts that I have had have been anticipated by someone else . . ." [p. 3].

tificially narrow range of language facts, culled from a narrower yet range of natural languages, and forced into an analytic mold that has little regard for the burden of empirical validation. This tradition has persisted in one guise or another via the two extremist schools of Western epistemology—Rationalism and Empiricism—through the medieval Modistae and their Tomist and Anselmic descendents, on through the *Port Royal* school and the *Age de Raison,* then onward through the first formal logicians of the 19th century, eventually coming into full bloom in the Logical Positivists of the early 20th century and their anti-linguistic bias. Modern American linguistics, although joining the frey relatively late in the early 1960s, fell squarely into an established formal logico-deductive tradition in its approach to the analysis of meaning, on both sides of the so-called Great Debate.

Challenges to the logico-deductive analysis of meaning begin with C. S. Peirce in the second half of the 19th century, via a tradition that traces itself back to Kant. A similar challenge was mounted later on by Wittgenstein, roughly contemporaneous with the ascent of Logical Positivism. These early challenges were quickly neutralized by the growing power and prestige of the "more rigorous" formal logicians. Within Positivism itself, the early broader-scoped Russell and Whitehead slowly gave way to narrower formalists such as Carnap, Tarski, and Montague. By then Peirce had been effectively co-opted by social philosophers and the impact of his pragmatism on epistemology thus largely obscured. The early Wittgenstein was admired as a brilliant gadfly, and the late Wittgenstein was dismissed by the Positivists as impressionistic mush. Formal logicians had of course, by then, been engaged in a rear-guard battle against the encroaching shadow of context-dependence in human language, reducing the open-ended complexity of reference and definite description to neatly-packaged deductive-logic formulae, and ruminating recursively upon old-time favorites such as quantification and predication. The vast field of language meaning and communicative use was left largely undisturbed. Thus, when American linguistics was at last ready to outgrow the Bloomfieldian strictures that relegated semantics to either the natural sciences or to mathematics,[2] the seductive rigor of formal logic was the only game in town. And so, elaborate deductive-looking monsters such as the Katz and Fodor "model" soon sprang all over the countryside, purporting to represent meaning as a closed system of atomic "features" and formal "rules."

When pragmatics finally reared its ugly head in American linguistics, it did so with a gentle whimper. The Sapir-Whorf hypothesis had been around for decades by then, a cogent observation[3] concerning the cultural relativity and thus context dependency of language and meaning. Both early and late structuralists were not

---

[2]Bloomfield (1933, Ch. 9, pp. 154, 157).

[3]The original hypothesis was two-edged, allowing either that language was culture-dependent or that culture was language-constrained (Whorf, 1956).

impressed, and soon managed to restrict the impact of the observations to "cultural vocabulary" and then relegate the whole phenomenon to the realm of Anthro-Linguistics at the very margins of the discipline. When Wittgenstein raised the very same issues in his Investigations, citing "games," "family resemblance," and "meaning through use," American linguistics was too busy comtemplating its post-Bloomfieldian navel. Of the post-Wittgensteinian Philosophers of Ordinary Language, Austin's forray into the realm of Speech Acts and Illocutionary Force[4] finally registered upon a new and semantically-oriented generation. But once again the impact was restricted to this "specialized" area of the grammar, and even there it was soon absorbed into the pseudo-formalism of "higer verbs."[5] In the same vein, Grice's observations were automatically couched in terms of Logic of Conversation,[6] and soon pressed into Gordon and Lakoff's forbidding Conversational Postulates,[7] where pragmatically open-ended entities such as "belief," "intent," "judge," "sincere," "reasonable," or "relevant" were construed as atomic primes. It would be of interest some day to speculate why the sweet siren song of formal, deductive logic recurrently proves so irresistible to linguists investigating pragmatics, who then proceed to force the subject matter into a mold that robs it of its very meaning.[8] The very soul of pragmatics may be given initially as the four properties below, contrasted with deductive logic:

(1) | *pragmatic systems* | *logico-deductive systems* |
|---|---|
| open-ended | closed |
| context-dependent | context free |
| continuous/non-discrete | discrete |
| inductive/abductive | deductive |

The implications of Pragmatics to the study of human language and human cognition are immense, but the Positivist tradition in logic and its bobsie-twin, the Generative tradition in linguistics, seem to force all pragmatic incursion into one or two blind alleys: Toward strangulation in irrelevant formalism, or toward marginal existence at the neglected edges of the linguistic system. And it seems that no amount of empirical evidence to the contrary, culled from psycho-linguistics, discourse studies, developmental psychology or perception, has so far succeeded in denting this doctrinaire delusion of modern linguistics.

In this chapter I propose to perform an epistemological *coup de grâce* upon this stale and misguided tradition. I first tackle some of the more habitual pre-

---

[4]Austin (1962).
[5]Ross (1970).
[6]Grice (1968).
[7]Gordon and Lakoff (1971).
[8]See for example Gazdar (1979) or Karttunen and Peters (1979), *inter alia*.

serves of the deductive logician, such as reference, definite description, and presupposition, showing how a deductive-logical analysis, whereas attractive and on occasion self-consistent, has very little to do with the actual meaning system used in natural language. I then proceed to discuss more complex cases that formal logicians have—for presumably good reasons—consistently avoided. The main argument I present, that the meaning system in natural language is inherently of a pragmatic nature, is made cumulatively and with an eye to one fundamental truth of science—and a fundamental feature of pragmatics: That *complete* proof is a deceptive mirage, and that science is bound to accept with grace the pragmatically-tainted *preponderance of evidence*.[9]

EXISTENCE AND REFERENTIALITY

On the face of it, this old bastion of logic-bound analysis seems altogether context-free, depending in no way on entities outside the bounds of the atomic proposition—or even outside the bounds of the quantified argument (NP) itself. Logicians could thus with impunity posit an "existential quantifier" that would instantiate an individual argument into some "real world," or so it seems. One may of course raise superficial arguments concerning pronouns whose co-referents are non-referential, as in:

(2)   I am looking for *a horse,* and *it* better be white

But this can be "handled" via the modal logic of "possible worlds." And, whereas such a treatment involves the tacit assumption that existence and reference do not involve mapping into *the* real world, but rather into a *universe of discourse,* the logician could still consider the "bulk" of reference to involve *this* real world, and relegate modal areas to the margins of the system.

There are other cases, however, that make the predicament involved in defining existence in logical terms more acute. They involve the reference-coding properties of many languages, perhaps most, where existence is not taken to be a mapping into *the* world, nor into a hypothesized/imagined universe of discourse. Rather, existence depends upon *communicative intent* of the speaker uttering the discourse, specifically on whether a particular individual argument (NP) is going to be *important* enough in the *subsequent discourse,* that is, whether its *specific identity* is important, or only its generic *type membership.* This behavior is illustrated, first with data from Israeli Hebrew, but identical examples may be cited from Turkish, Mandarin, Hungarian, Sherpa, Persian, old-Spanish, and all

---

[9]While this work deals primarily with epistemology, it is only to be expected that whatever valid conclusions emerge at the end will be equally applicable to the philosophy of science. Thus, to the extent that a scientific method aims at obtaining new knowledge, it must abide by the constraints suggested by Peirce (1955) and Wittgenstein (1918).

Creoles.[10] In such languages, the numeral 'one' has just become the marker of *referential* ('existing') *indefinite* NP's, so that in grammatical/semantic environments that allow non-referential interpretations, one observes the following contrast:

(3)   a. REF: ani mexapes sefer-*exad* she-neevad li
               I looking-for book-*one* that-lost to-me
               'I am looking for a book that I lost'
       b. NON-REF: ani mexapes sefer bishvil ha-yeled shel-i
               I looking-for book for the-boy of-me
               'I am looking for a book for my boy'

The numeral 'one', further, cannot mark the object of a negative sentence, where the object may either be *non*-referential or *referencial-definite*.[11] Thus consider:

(4)   a. REF-DEF: lo karati et ha-sefer ha-ze
               NEG read-I ACC the-book the-this
               'I didn't read this book'
       b. NON-REF: lo karati (af) sefer ha-shavua
               NEG read-I (any) book the-week
               'I didn't read (any/a) book this week'
       c. REF-INDEF: *lo karati sefer-*exad* (ha-shavua)
               NEG read-I book-*one* (the-week)

Unlike the scope of negation or the modal *look for,* most verbs in any language *imply the existence* of their object. Thus, one would expect in affirmative sentences using such verbs that the only possible contrast would be between referential *definite* and *indefinite,* and that further, in Hebrew the numeral 'one' would mark the referential-indefinite object in such a context, as is indeed the case in:

(5)   a. REF-INDEF: karati sefer-*exad* etmol, ve- . . .
               read-I book-*one* yesterday and-
               'I read a (certain) book yesterday, and . . .'
       b. REF-DEF: karati et ha-sefer ha-ze etmol
               read-I ACC the-book the-this yesterday
               'I read this book yesterday'

So far, then, the marking-system in Hebrew seems rather straight-forward: If the speaker refers to an indefinite NP that he/she actually *assumes to exist,* that NP is marked with 'one.' However, if the speaker does not *have in mind* any particular individual, but rather is talking about a *member of the type* regardless of specific

---

[10]For background and many details, see Givón (1973a, 1978, 1981a).
[11]For discussion of the *pragmatic* motivation for this restriction, see Givón (1979b, Ch. 3).

identity, then the NP is not marked with 'one.'[12] But this seemingly clear and logically-coherent picture now gets clouded, because one could also find a *logically-referential* object in this grammatical environment marked as a non-referential noun, that is, without the numeral 'one.' Thus consider:

(6)   karati sefer etmolt, ve- . . .
       read-I book yesterday and-
       'I read a book yesterday, and . . .'

In logical terms, both (5a) and (6) must refer to an individual, specific book, because if one has read it, that individual book must have perforce existed and been identified. But in terms of discourse-pragmatics, the two usages are radically different. In order to make this difference explicit, let us introduce a little more context into the narrative, and consider the two story-fragments below:

(7)   *LOG-REF, PRAG-REF:*.
       . . az axarey ha-avoda halaxti la-sifriya ve-yashavti sham
          so after the-work went-I to-the-library and-sat-I there
       '. . . So after work I went to the library and I sat there
       ve-karati sefer-*exad,*
       and-read-I book-*one*
       and I read a book,
       ve-ze haya sefer metsuyan . . .
       and-it was book excellent
       and it was an excellent book . . .'

(8)   *LOG-REF, PRAG-NON-REF:*
       . . . az axarey ha-avoda halaxti la-sifriya ve-lo haya li ma la-asot,
          so after the-work went-I to-the-library and-NEG-was to-me what to-do
       '. . . So after work I went to the library and I had nothing to do,
       az karati sefer, ve-karati shney itonim, ve-axar-kax halaxti ha-bayta . . .
       so read-I book and-read-I two papers and-after-that went-I to-home
       so I read a book, and a couple of newspapers, and then went
       home . . .'

In (7) above the logically-referential "book" becomes *an important topic,* where its specific (token) identity *matters.* It is thus marked with the referential-indefinite marker 'one.' In (8), on the other hand, an equally logically-referential "book" does *not* become an important topic, and its specific identity *did not*

---

[12]I am deliberately expressing these conditions here in terms of the speaker's intent or belief. Logicians could of course easily convert them into truth-relations between propositions, I suppose via modal logic and possible-worlds.

matter. Rather, what mattered was that it was a *member of the type "book."* The speaker was doing some book-reading, not really caring what specifically he/she read. And the marking system that seemed in (3) and (4) to code the *logical* property of referentiality, now turns out in (7) and (8) to code a *pragmatic* contrast, that is, whether referential vs. type identity is what *mattered,* from the *speaker's point of view,* in the given discourse context, most specifically in the *following* expansion of the narrative.[13]

If it is the case that one objects that this phenomenon is restricted to the languages using the numeral 'one' to mark referential-indefinites, let me illustrate briefly how languages using other devices to code referentiality/genericity exhibit exactly the same behavior. Consider first Bemba, a Bantu language, where referentiality contrasts of exactly the same type are coded by the VCV/CV shape of the noun prefix.[14] In environments allowing logically-non-referential interpretations of the noun, the VCV/CV contrast marks this logical opposition, as in:

(9)   a. REF: n-dee-fwaaya *ici*-tabo
                I-PROG-want REF-book
                'I want a *specific* book'
       b. NON-REF: n-dee-fwaaya *ci*-tabo
                I-PROG-want NREF-book
                'I want a book , be it any'

(10)   a. REF: n-shi-a-soma *ici*-tabo
                I-NEG-PAST-read REF-book
                'I didn't read *the* book'
       b. NON-REF: n-shi-a-soma *ci*-tabo
                I-NEG-PAST-read NREF-book
                'I didn't read *a/any* book'

But where the logical contrast does not exist, the *pragmatic* one manifests itself just like in Hebrew. Thus consider:

(11)   a. *LOG-REF, PRAG-REF:* n-a-soma ici-tabo
                I-PAST-read REF-book
                'I read *a/the* book (and its identity
                mattered)'

---

[13]This point was made, with respect to existential-presentative constructions that introduce referential-indefinite arguments into discourse for the first time, by Hetzron (1971). As Hetzron points out, it is the relative importance in subsequent narrative that determines whether a presentative construction is to be used.

[14]See details in Givón (1973a), and further discussion in Givón (1978). In the earlier paper, I noticed only the logical contrast, but not the pragmatic one. We live and learn.

b. *LOG-REF, PRAG-NON-REF:* n-a-soma *ci*-tabo
I-PAST-read NREF-book
'I did some book-reading (any old
book would have done)'

A similar situation may be seen in Ute, a Uto-Aztecan language that marks the
contrast of referentiality, in environments where a logically non-referential reading is allowed, by contrasting *incorporated* object nouns (NON-REF) vs. *inde-*
pendent ones (REF). Thus consider:[15]

(12) a. REF: pọ'ọ́qwa-tụ 'ásti'i
book-OBJ want-PROG
'I want a (*specific*) book'
b. NON-REF: pọ'ọ́qwa-' ásti'i
book-want-PROG
'I want a book (be it any)'

(13) a. REF: kacụ́-n pọ'ọ́qwa-tụ ('urú) pụníkya-na
NEG-I book-OBJ (that) see-NEG-PAST
'I didn't see *the* book'
b. NON-REF: kacụ́-n pọ'ọ́qwa-pụníkya-na
NEG-I book-see-NEG-PAST
'I didn't see *a/any* book'

But again where the logical contrast is absent, the *pragmatic* one manifests itself
and is coded by exactly the same device:

(14) a. *LOG-REF, PRAG-REF:* pọ'ọ́qwa-tụ pụníkya-pụgá
book-OBJ see-PAST
'I saw a book (and its identity mattered)'
b. *LOG-REF, PRAG-NON-REF:* *hl*pọ'ọ́qwa-pụníkya-na-pụgá
book-see-HAB-PAST
'I did some book-seeing (and the
specific identity of the book
didn't matter)'

The facts, thus, are not limited to a particular morphological paradigm, but are
rather widespread. In fact, the same *semantic* facts exist in English, except that
English is *under-coded* in the area of marking referential vs. non-referential
indefinites, often using the indefinite article for both. Thus, consider:

---

[15]For details see Givón (1981, Ch. 17).

(15) *Opacity-allowing context:* I am looking for *a* book
          (a) A *particular* book
          (b) Any member of the *type* "book"

(16) *Logically transparent context:*
       . . . *later on I read a* book . . .
          (a) . . . and it was terrific (i.e. its identity mattered)
          (b) . . . and then I went home (i.e. the book's identity didn't
              matter)

Now, the logic-bound semanticist may wish to argue that what we have here are two *separate* distinctions, a logical one that appears in enviornments allowing referential opacity, and a pragmatic one that appears in enviornments were nouns are obligatorily *logically*-referential. There are two reasons why this argument carries little weight. First, if it were accepted, then the fact that so many languages use exactly the same device to mark the two distinctions is an inexplicable accident. Second, it is really not true that the two distinctions are totally separate, but rather, they *do* have an inclusion relation with each other. That is, the *pragmatic* contrast of referentiality may appear by itself, in logically-transparent environments. But whenever a noun is logically non-referential, it is also *pragmatically* non-referential. Thus, one looks for "a book be it any" only when the specific identity of the book does not matter and has no importance for the speaker in this particular discourse context. Similarly, one "doesn't read any book" in contexts where the issue of whether one has read *any* book (rather than if one has read a *particular* book). What we have here, is a typical language situation whereby the *general* semantic feature of referentiality is the *pragmatic* one, distributing in all environments, playing the *unmarked* case. Whereas in some environments the contrast is *augmented* by a *logical* contrast, that has a more limited distribution, is not independent of the wider contrast, and thus plays the *marked* case.

The pragmatics of referentiality/existence clearly involve the components of "being important," "speaker's judgment," "relevance," and "following discourse context." None of these are closed, deductive, objective factors. Rather, they are open-ended, non-discrete, graded, and speaker-dependent. The judgment involved is inductive/probabilistic, taking into account a great number of factors that bear upon "thematic importance/relevance within the discourse." The logical-seeming distinction, of the speaker either having in mind or not having in mind a specific individual object, seems to be merely *the extreme case* within an inherently pragmatic space. The fact that logicians—and logic-bound linguists—have traditionally recognized only the extreme case, while neglecting the general phonemenon, is a reflection on the methodological bias they have been harboring in their approach to human language.

CO-REFERENCE AND DEFINITE DESCRIPTION

Logicians have traditionally tried to interpret the use of the definite article 'the' (and with it anaphoric pronouns such as 'he'/'she' and proper names) as a problem of establishing *unique referential identity*.[16] When that didn't quite match the facts of language, they proceeded to introduce the concept of *presupposition*, construed as a logical property pertaining to either propositions or parts of them, and again expressible in terms of *truth*.[17] And various later refinements and elaborations still retained the notion that definiteness is a matter of logic, propositions, and truth, all the while skirting the intrusion of the *speaker*, the *hearer*, and their respective *beliefs*.[18] Logicians, and logic-bound linguists, are thus fond of classical examples such as:

(17)   The king of France is bald

where the truth-value of the sentence may be discussed from two different perspectives. First, going along with the presupposition that "there is a king in France," (17) may be judged either true or false in case the king is either bald or not bald, respectively. Alternatively, if one does not subscribe to that presupposition, then (17) presumably "has no truth value." In this way, logicians reduced definiteness to the realm of presuppositionality.[19] It would take a relatively small expansion of the data-base, however, to illustrate how definiteness in human language is inherently a *pragmatic* phenomenon, involving the speaker's belief about what the hearer is *likely* to know or believe in, and about *how easy* it is going to be for the hearer to uniquely identify the referent under consideration.

Logicians have never been forced to wrestle with the great diversity of language devices all employed in making unique reference to definite NP's. These devices, or at least a commonly recognized sub-group of them,[20] may be hierarchized according to a continuous, non-discrete property that may be termed as either:

---

[16]Russell (1919, Ch. 16).

[17]Strawson (1950).

[18]Donellan (1966).

[19]Logicians proceed to identify two spearate interpretations of the negation of (17), i.e., 'The king of France is not bald,' an *internal* one that accepts the presupposition but denies the assertion, and an *external* one that denies the assertion *because of* failure of the presupposition (Keenan, 1969). The fact that language users tend to admit only the internal interpretation of such a negative sentence (Givón, 1979b, Ch. 3) seldom bothers logicians.

[20]It may be shown (Givón, 1979b, Ch. 2, and 1980, Ch. 17) that other constructions, such as passivization, Y-movement, indefinites, and focus-constructions, as well as the VS/SV word-order variation observed in Spanish (Silva-Corvalán, 1977) and Hebrew (Givón, 1976a, 1977) belong on the same continuum.

1. "The degree of difficulty that the speaker assumes the hearer will experience in identifying the referent uniquely"; or
2. "The degree of continuity/discontinuity in the referent-tracking sub-system in discourse"; or
3. "The degree of surprise/predictability in discourse."

Massive cross-language evidence supports the following hierarchy:[21]

(18)   *Easiest identification/most continuity/least surprise value*
Zero anaphora
Unstressed pronouns and verb-agreement
Stressed/independent pronouns
DEF-NP with right-dislocation
DEF-NP (with no dislocation)
DEF-NP with left-dislocation
Hardest identification/most discontinuous/most surprise value:

Because the evidence supporting this hierarchy is massive and complex, I will illustrate it with some examples from English. Consider first the relative ranking of zero anaphora and unstressed pronouns (or verb agreement) by comparing the following examples:

(19) a. He came into the room, saw Mary, pulled a chair and sat down.
      b. He came into the room, saw Mary, pulled a chair and sat down. *She* seemed tired, *he* thought. *He* relaxed.
      c. He came into the room, turned and saw Mary. *He* pulled a chair and sat down.

In (19a) above we find the highest degree of continuity/predictability:

1. *Topic continuity/identifiability:* The referent of 'he' retains control, and even though 'Mary' is introduced, she remains a mere *object* and does not break the continuity chain;
2. *Action continuity/sequentiality:* The actions are given in their natural sequence, without a break;

---

[21]For the relative position of zero anaphora vs. pronouns/agreement, see Hinds (1978), Li and Thompson (1979), and Givón (1980, Ch. 17), among others. For the relative position of stressed vs. unstressed pronouns, see Givón (1979b, Ch. 2), Linde (1979), Kunene (1975), Hinds (1978), or Givón (1980, Ch. 17), among others. For the relative position of pronouns vs. DEF-NP's, see Bolinger (1979), Hinds (1978), Duranti and Ochs (1979), or Givón (1979b, Ch. 2), among others. For the relative position of right-dislocated DEF-NP's, DEF-NP's and left-dislocated DEF-NP's, see Duranti and Ochs (1979), Keenan and Schieffelin (1977) and Givón (1979b, Ch. 2, 1980, Ch. 17), among others.

3. *Thematic continuity/tightness:* Somehow the whole passage is given as a single event, a thematic 'breath unit.'

In (19a), the topic is continuous and no interference from other topics exists. Therefore, except for the first introduction of the referent 'he,' all other references are given as *zero anaphora.*[22]
In (19b), we start with the same type of maximal continuity. But then a break occurs where 'Mary' takes over as an important topic/subject. And again when the topic/subject switches back to 'he.' All three continuities are broken in these cases, and zero-anaphora cannot be used any more. Rather, the *unstressed pronoun* must be used.[23] Finally, the sequence given in (19c) illustrates the fact that the discontinuity does not *have* to be due to topic switch, as in (19b), nor to an overtly-manifest discontinuity in the action sequence. It can be purely thematic, simply a way of organizing the 'same' sequence into different, more discontinuous *event groups.*
Let us next illustrate the relative ranking of unstressed vs. stressed pronouns:

(20) a. John told Bill that *he* was sick,
     b.   and that *he* couldn't come
     c. John told Bill that *he* was sick,
     d.   and that *hé* couldn't come
     e. John told Bill that *he* was an idiot,
     f.   and that *he* couldn't come
     g. John told Bill that he was an idiot,
     h.   and that *hé* couldn't come
     i. John hates Bill, and *he* hates Mary
     j. John hates Bill, and *hé* hates Máry

In (20a) the first referent of 'he' could be either John or Bill, but the second unstressed 'he,' in (20b), must be coreferent to the 'he' in (20a). On the other hand, the *stressed* 'he' in (20d) cannot be coreferent to the unstressed 'he' in (20c). Thus, unstressed pronouns are used when chains of topic–identity are broken-disrupted. In (20e), the mere semantics of 'telling somebody that you/he is an idiot' militates strongly toward assuming coreference of 'he' with the object

---

[22]In English, zero anaphora covers a much narrower range of the continuity scale, and is customarily referred to as *phrasal conjunction,* either of NP's or of VP's. In Japanese and Chinese (see Hinds, 1978), on the other hand, zero anaphora covers a wider range, perhaps most of the functional range covered by unstressed pronouns in English or subject agreement in Spanish. In the latter, further, subject agreement covers the functional range of zero anaphora as well, because it is obligatory.

[23]The functional range of unstressed pronouns in English is covered in Spanish by obligatory subject agreement.

rather than with the subject. And again, the unstressed 'he' in (20f) must be coreferent with 'he' in (20e). But again, the *stressed* 'he' in (20h) could *not* be coreferent with 'he' in (20g). Finally, 'he' in (20i) must be coreferent with the subject 'John.'[24] But the *stressed* 'he' in (20j) must be coreferent with the object 'Bill,' and thus codes *switch reference*.

The use of independent/stressed pronouns to mark *switch*-reference—vs. unstressed pronouns/subject agreement to mark *same* reference—is attested in other languages as well.[25] But the discontinuity associated with stress need not arise only from switch reference/topic, but from other *counter expectations*. Thus consider:

(21)   Everybody hated the draft, but John told Bill *hé* wasn't so sure

Thus, as in the contrast between zero-anaphora and unstressed pronouns, here too discontinuity/surprise may arise either from expectations about topics/participants *referential* continuity, or from other—*thematic*—sources.

The relative ranking of pronouns vs. DEF-NP's and right-dislocated DEF-NP's. In general, DEF-NP's are used when not enough topic continuity exists in the discourse to enable the hearer to identify the referent uniquely with a pronoun. This may involve either the presence of other NP's within the same pronominal gender, or also a major *thematic* break.

(22)   a. He gave presents, to the King, the Prince and the General. *The king* thanked him profusely . . .
        b. He gave presents to the King. *Hé* thanked him profusely . . .
        c. He came into the room, looked around and sat down. *He* was tired and confused; *he* slumped in his chair and waited.
        d. He came into the room, looked around and sat down. *The man in the blue suit* was tall and sprightly, close to middle age but still young.

In (22a) the presence of 'prince' and 'general' preclude referring to 'king' as 'he,' so 'the king' is used. In (22b) only 'king' is involved, so a pronoun may be used, although switch-reference requires it to be a stressed pronoun. In both (22c,d) no referential discontinuity is involved. Nevertheless, in (22c) two minor

---

[24]The verbs 'tell(to)' and 'hate' have different topic-orientation characteristics, because the subject of 'tell(to)' is *dative-recipient,* but that of 'hate' may be *inert* and *uninvolved* (i.e., 'patient'). This is the reason why it is easier to have an unstressed pronoun referring to the *object* of 'tell(to).' Thus, it is not merely the property of *subjecthood* that controls continuity of topic, but rather the property of *topciality,* where datives outrank accusatives (Givón, 1976b).

[25]Spanish (Silva-Corvalán, 1977), Lango (Noonan, unpublished), Ute (Givón, 1980).

thematic breaks occur, with pronouns being used to mark them. But the thematic break in (22d) is a more major, more disruptive one, removing the narrative line from the immediate action sequence (as in being tired or slumping, in (22c)), and introducing *general background* instead.[26] Thus, it is more appropriate to refer to the uniquely-identifiable referent with a DEF-NP expression.

The use of right-dislocated DEF-NP's may be likened to an "afterthought" device,[27] whereby the speaker first assumes that there is enough continuity in the discourse to warrant the use of a pronoun, then changes his/her mind and adds the full DEF-NP as "insurance." To illustrate this consider:

(23) a. The king danced and sang. Later on *he* retired . . .
      b. The king danced, then other people performed. Then *he* retired, the king did . . .
      c. The king danced, and the prince and the general sang. Then *the king* retired . . .

In (23a), with full referential continuity, a pronoun is used. In (23b) technically there is no referential confusion, but there is already a one-clause gap and the intrusion of other participants in the subject position. There is nothing "ungrammatical" with using only a pronoung here, but still it is a typical context for using right-dislocation in conversation.[28] Finally, in (23c) there is no reason to even hedge and use a pronoun first, because referential continuity/predictability is broken. So a DEF-NP is used.

Finally, left-dislocated DEF-NP's are used primarily over long gaps of absence, where a referent/topic is re-introduced into the discourse. To illustrate this in contrast with normal DEF-NP's, consider:

(24) a. There once lived a king and a queen in an enchanted forest. *The king* was fat and ugly . . .
      b. There once lived a king in an enchanted forest. He was married to a beautiful queen, and she was the real power in the realm. Near the forest lived a poor prince, and the queen used to visit him and have lunch. Now *the king, he* didn't like the guy . . .

One must bear in mind, however, that in tightly-planned, written text the use of left-dislocation is not common, though it is very common in informal speech and conversation.[29]

---

[26]Background descriptions obviously disrupt discourse continuity (see Givón, 1977; Hopper, 1979; Hopper & Thompson, 1980).

[27]The explanation couched in the term "afterthought" is due to Hyman (1975).

[28]Both right and left-dislocation are used primarily in conversation and informal speech (Duranti & Ochs, 1979; Givón, 1979; Ochs, 1979).

[29]See Ochs (1979), Duranti and Ochs (1979), Keenan and Schieffelin (1977).

The preceding—albeit encapsulated—survey of the major devices used in definite expressions in human language firmly establish that we are indeed dealing with *a scale*, and that the scale is sensitive to either of the three major factors of continuity:

1. Topic/referent continuity and identifiability,
2. Action continuity in a narrower, sequential sense,
3. Thematic continuity in a larger sense.

The logician could of course claim that all three are *objective properties* of *propositions*, or even grant us a point and make them objective properties of *discourse*. He could even go a step further and agree that these are properties of the *speaker's universe of belief*, and thus must be handled via a modal logic of "possible worlds." But even that concession won't do, because the data suggests that in all areas what is involved is the *speaker's judgment/belief* about *probabilities* concerning the *hearer's beliefs* and *capabilities*, involving *inductive reasoning* of extreme complexity and subtlety, where many different contributory factors constitute the input to the speaker's judgment, and where the use of the same syntactic device may be motivated by degree of uncertainty/ indeterminancy in *different* sources of the input. The major areas contributing to the speaker's decision are:

1. Assessment of the hearer's ability to identify referents unambiguously;
2. Degree of interference from other referents near and less-near to the actual locus where a definite-reference device is used;
3. Degree of action-continuity and sequentiality in the discourse, and the speaker's ability to follow that continuity;
4. Degree of higher thematic continuity/tightness in the discourse, and assessment of the hearer's ability to follow it;
5. The speaker's knowledge of the hearer's expectation about themes and topics/referents appearing in the discourse, grounded in either the speaker's assessment of the ongoing discourse, his knowledge of previous encounters with the hearer, his knowledge of the hearer's personality and computational abilities, or his knowledge of specific facts concerning the hearer's mind and its contents;
6. The speaker's telepathic ability and thus "direct access" to the thought processes of the hearer.[30]

---

[30]There is nothing "improper" for two people who have access to each other's thoughts either via intimate, extended day-to-day contact or telepathy, to meet each other after a year of absence, whereby one says: "*He*'s finally dead, right?" and the other replying: "*He* is. Wonder who's next . . ." .

One could of course go on and on, but it is not really necessary. It seems quite clear that definite description is a *pragmatic* matter, involving *gradations, open-endedness,* and *inductive/probabilistic judgment* of the speaker about the *hearer* and about the *discourse.* One could of course grant that when the data-base is sufficiently reduced and sanitized, some *corners* of the system may appear to the logician as logical-deductive in nature, involving truth relations between atomic propositions. But such reduction, whatever its motivation may be, has relatively little to do with the facts of natural language.

NON-DISCRETENESS OF REFERENCE AND DEFINITENESS

Most linguists follow logicians in assuming that an argument/NP is either refer-ential or generic/non-referential. But there is enough evidence from natural languages to suggest that while those extreme poles are indeed common, one may also find gradations and intermediates, something like "semi-referen-tiality" that is bound to make logicians shudder. For example, consider the following contrast:

(25) a. Did you see *anything* there?
      b. Did you see *something* there?

Officially, both expressions in (25) are non-referential, but (25b) is somehow "a little bit *more* referential" than (25a). The distinctions between the two involves, roughly, the *degree* to which the speaker is willing to commit himself/herself to a *specific* individual they have in mind that they *suspect*—but are not sure—may have in fact been involved. When a speaker uses (25a), he/she is *less* committed to having an individual in mind, whereas the use of (25b) suggests *stronger* commitment. But the gradation is even more extensive. Thus consider:

(26) a. Did you see *any* man there?
      b. Did you see *some* man there?
      c. Did you see *a* man there?
      d. Did you see *a* man there wearing a blue tie with green polka-dots and twirling a silver baton on his right-foot toe?

In (26), (26d) is practically a unique referential description, at least as it is likely to be used in natural language.

One may also show the phenomenon of "degree of genericity," as in the following example from Spanish:[31]

---

[31]A similar distinction is observed in Mandarin (Sandra Thompson, in personal communication), where the contrast is between "be" and "exist."

(27) a. María siempre habla con *brujos*
'Mary always talks to sorcerers'
b. María siempre habla con *los brujos*
1. *'Mary always talks to (the) sorcerers' (not fully identified)*
2. *'Mary always talks to the sorcerers (that I have identified)'*

Sentence (27b) has two readings, one (27b(2)) fully definite. But the reading (277(1)) is "semi-generic," roughly involving a *smaller generic pool* from which any smaller-yet group of "brujos" *may* be pulled out to fit the description. Thus, whereas the characterization of the generic expression in (27a) is that "any member of the type 'sorcerer' may fit the description," (26b(1)) narrows the generic pool—but not completely, not up to unique definite description.

A somewhat similar gradation may be shown for English, as in the following example:

(28) The man who killed Smith was insane
a. 'I know who exactly killed Smith, and he was insane'
b. 'Somebody killed Smith, I don't know his full identity, but one thing I know about him—he was insane'

It was Donellan's original observations[32] that the sense (28(a)) involves a *referential* use of the definite NP, whereas the sense (28(b)) involves an *attributive* use. This observation is correct, but it reduces the facts to a discrete, categorial interpretation. But one could easily show that the "attributive" use can be graded toward a less and less attributive, more and more referential use. Thus consider:

(29) a. The man who killed Smith was insane, 25-years-old, blue eyes, brown hair, married to a Russian aristocrat and lives in a suburb of Sheffield . . .

b. The man who killed Smith was insane, etc . . . , and he lives next door to me and his wife baby-sits for us . . .

Unique definite description can thus be a matter of *fine gradation,* approaching full referential identification in the smallest of increments, each one of which slightly increases the *probability* that the speaker is capable of unique identification.

Consider, finally, the following example from Ute, where in WH-questions one overtly codes the difference between referential (token) and non-referential (type-membership) questions, as in:[33]

---

[32]See Donellan (1966). This distinction is actually overtly coded in some languages, see Keenan and Ebert (1973).
[33]See Givón (1980, Ch. 12).

(30) a. *REF-identity:* 'áa-'ará 'ín ta'wá-ci?
         WH-be this-ANIM man-SUBJ
         'Who is this man?'
   b. *TYPE-identity:* 'íni-'ará 'ín?
         WH-be   this-ANIM-SUBM
         'What kind (of an animate) is this one?'

The answer to (30a) may be properly 'John,' 'my teacher,' 'the doctor,' and so on, that is, by some uniquely-refering DEF-NP. While the proper answer to (30b) may be 'a doctor,' 'a teacher,' 'a horse,' and so on, that is, by an attributive/non-referential indefinite expression. But the non-referential question—as in English—may show further gradation, involving roughly the *degree of certainty* of the speaker about the exact type-membership of the referent. Thus consider:

(31) a. *Normal TYPE-identity:* 'íni 'ará-'ay 'ín?
         WH be-PROG this-ANIM-SUBJ
         'What kind (of animate) is this one?'
   b. *Uncertain TYPE-identity:* 'íni-kwa 'ará-'ay 'ín?
         WH-DOUBT be-PROG this-ANIM-SUBJ
         'What kind (of an animate) could this one possibly be?'

Thus, not only can human languages treat "definite identification" and "having unique reference" as scalar properties, they can also treat "having unique type-membership" as a scalar property. Thus, whereas it is true that natural languages tend to *code* major *extreme* ends of this system in a seemingly discrete fashion, there is inherently nothing particularly discrete about the cognitive space of definiteness, referentiality, and genericity. Rather, it seems to be a continuous, non-discrete space involving certainties and probabilities.

TRUTH, FACT, AND PRESUPPOSITION

The early history of the notion of "presupposition" in linguistics was a direct and slavish outcrop of an earlier deductive tradition.[34] Several supposed "pragmatic" formulations still strive to represent the system as closed, tight, and deductive-looking.[35] I will now show, by citing language-data from various

---

[34]See eg. Keenan (1969, 1971), Horn (1972), *inter alia.* Karttunen (1974) has suggested that one may formulate presupposition for human language in *pragmatic* terms, that is, with reference to the *speaker's belief* rather than to atomic propositions. But he still formulates presupposition in terms of "truth relations" between various beliefs/propositions held by the speaker. This is a rear-guard attempt to salvage some deductive properties for the system.

[35]See Karttunen (1974), Grice (1968), Gordon and Lakoff (1971), or Gazdar (1979), *inter alia.*

sources, that the logicians once again have misrepresented the *general* thrust of a system that is in essence *pragmatic,* by rigorously eliminating from their database vast quantities of relevant evidence. I attempt to show, then, that presupposition in language does not involve truth relations between propositions, but rather probabilistic assumptions that the speaker makes about what the *hearer* is likely to be *familiar with* or *likely to accept without challenge*—on whatever grounds. Only in small sub-domains of the entire field does presupposition appear to be a logical phenomenon.

Negation

As has been shown elsewhere,[36] negative sentences in human language are "presuppositional" in terms of discourse-pragmatics, so that in order for the negative to "make sense" or "be used appropriately," either a previous *overt mention* of the corresponding affirmative must have taken place in the previous discourse, or otherwise the speaker must *assume* that the hearer is either *familiar* with the affirmative, *inclined to believe* in it, has some *probability* of tipping the *scale of belief* even slightly toward it, and so on. As a simple illustration, consider the following example:

(32) a. *Context:* "Hi, how are you, what's new?"
      b. *Reply-AFFIRM:* "Well, my wife is pregnant."
      c. *Reply-NEG:* "Well, my wife is not pregnant."

Reply (32b) is used appropriately on the background of: (1) total ignorance of specifics; and in addition (2) the culture-based assumption that women (at least nowadays) are more likely to be un-pregnant than pregnant. On the other hand, reply (32c) is appropriate only if the corresponding affirmative is "a pragmatic/discourse background" for the negative. But being "pragmatic discourse background" could not possibly mean "being logically presupposed," because that would lead us to the absurd conclusion that "NEG-p presupposes p."

The logician may now presumably claim that we are dealing here with two different phenomena, logical presupposition on one hand, and discourse-pragmatic background-ness on the other. But if that is so, then several facts of human language become accidental and hard to explain. For example, many languages group the major clause-types in syntax into two groups:[37]

(33) *Asserted:* main, declarative, affirmative, active clauses
      Background/presupposed: negatives, REL-clauses, ADV-clauses, questions

---

[36]See Givón (1979b, Ch. 3).
[37]See discussion in Givón (1979b, Ch. 2).

Thus, for example, in Bemba[38] every assertion is morphologically marked according to whether the verb itself is included or excluded from the *scope of new information* being asserted. But the following constructions are automatically marked for *excluding* the verb from the scope of new information:

(34) REL-clauses, WH-question, CLEFT-sentences, Negatives, ADV-clauses
      (including IF-clauses)

Most of these constructions are "logically presuppositional." IF-clauses presumably "have no truth value", whereas negation is "discourse/pragmatic presuppositional or 'backgrounded'." On what basis then are they grouped together? Again, I suggest, what is involved here is that there is a *general* property that is inherently pragmatic, something like:

(35) "The speaker assumes that a proposition p is *familiar* to the hearer,
      likely to be *believed* by the hearer, *accessible* to the hearer, within the
      reach of the hearer, etc. *on whatever grounds*"

And only within a limited sub-set of contexts does the grounds for accessibility to or belief by the hearer involve something resembling logical presupposition, obviously still couched—in natural language—in terms of the speaker's belief about the hearer's belief.

Restrictive relative clauses

Restrictive relative clauses are one of the strongest bastions of logical-presupposition in language. On the surface they seem quite solid, so that given the embedded sentence in (36a) below, the full sentence in (36b) must be "presupposed":

(36) a. The man *I saw yesterday* left
      b. I saw a/the man yesterday

Things begin to blur when one realizes that restrictive relative clauses may modify non-referential head nouns. Thus, consider:

(37) a. I didn't see anybody *who wore a blue shirt*
      b. ?Somebody wore a blue shirt

Logically, there is no way in which (37a) could presuppose (37b) in the *same* sense as (36a) presupposed (36b). Nonetheless, pragmatically something like

---

[38]See details in Givón (1975).

(37b) must have been the *discourse-background* for a proper use of (37a), perhaps a question such as:

(38) Did you see anybody *wearing a blue shirt?*

This property of backgrounded-ness is thus *shared* between (36a) and (37a), whereas the seemingly "logical" presupposition applies only to (36a). Consider next:

(39) a. A man *I had never seen before* came into my office and . . .
      b. ?I had never seen a/the man before

Again, there is no way (39b) could be presupposed by (39a) logically. But is it then *asserted* by (39a)? Seemingly, but not really. It is, obviously, a piece of *totally-new information* to the hearer. Nevertheless, its status is not the same as the predicate in (39a) (". . . came into my office . . ."). That predicate is *open to challenge* from the hearer. But the relative clause in (39a) is deliberately put in a *background position*—thus in a sense it is *shielded from challenge*. The speaker uttering (39a) is much more open to the following challenge directed at the "exposed" predication:

(40) Now hold it, I was with you all day and nobody came in at all!

He/she is much less open to the challenge:

(41) Now hold it, you *must* have seen him before!

Thus, it seems that the general contrast underlying relative clauses does not involve "presupposed" vs. "asserted," nor "old" vs. "new" information, nor even "accessible" vs. "inaccessible" to the hearer. What seems to unify *all* cases of restrictive relative clauses is rather the contrast between "information that is shielded from challenge by the hearer" vs. "information that is open to challenge." This is of course the discourse-pragmatic distinction between *background* and *foreground*, respectively. Now, the hearer may of course choose to challenge *anything* just for the heck of it. So the speaker simply assigns probabilities about what is less-likely to be challenged. Obviously, something the speaker knows that the hearer believes in (thus, "logically presupposed") is least likely to be challenged. So logical presupposition is an extreme case within the inherently pragmatic scale of probabilities. But beyond that the speaker exercises his/her judgment about the likelihood of challenge. In (39a) he presents a piece of totally new information in a "shielded," backgrounded position. There is no logical reason why the hearer couldn't challenge it. The mere putting of the information in that background position simply makes the challenge a bit less likely.

IF-clauses and BECAUSE-clauses

As mentioned earlier, IF-clauses and all other subordinate-adverbial clauses often tend to be grouped in language with the ''logically'' presuppositional clauses. But certainly an IF-clause cannot be presuppositional, logically it even doesn't have truth value. If some ''presupposition'' is involved here, it could not be logical, but must be discourse-pragmatic. To illustrate this, consider first:

(42)   a. *Context:* What will you do if *I give you the money?*
     b. *Reply:*   If you give me the money, I'll buy this house.
     c. *Context:* Under what conditions *will you buy this house?*
     d. *Reply:*   I'll buy this house if you give me the money.

The replies in (42b,d) are appropriate in their context, but they are *not* interchangeable. Logically, the IF-clause is presupposed in neither. But pragmatically it is established as *background* in (42a), whereas the *main* clause is established as background in (42c). Such variation is possible in English, where adverbial clauses may either precede or follow their main clauses. But it is impossible in some languages, where ADV-clauses—as well as all pragmatically-background clauses—can only *precede* asserted/main clauses.[39] Such languages pre-pose all ''topic/background clauses,'' including all ADV-clauses, REL-clauses, V-complements (of 'know' or 'think') and other ''topic'' clauses. In other languages, ADV-clauses in text are overwhelmingly preposed.[40]

Unlike IF-clauses, BECAUSE-clauses are ''logically presuppositional.'' But they too exhibit the same discourse-pragmatic variation as IF-clauses. Thus, consider:

(43)   a. *Context:* What did she do because *he insulted her?*
     b. *Reply:*   Because he insulted her, she slapped him.
     c. *Context:* Why *did she slap him?*
     d. *Reply:*   She slapped him because he insulted her.

Here logically the BECAUSE-clause is presuppositional in *both* (43b) and (43d). Pragmatically, however, it is backgrounded in (43b) but foregrounded in (43d). Logical presupposition, involving ''truth values,'' is thus a more limited phe-

---

[39]For an example from Cuave, a clause-chaining Papua-New Guinea language, see Givón (1982).

[40]Greenberg (1966) has observed that in the language in his survey, regardless of word-order type, the dominant tendency was roughly 80% of ADV-clauses in text pre-posed, and only 20% post-posed.

nomenon, often corresponding to—but never identical with—extreme cases of pragmatic backgrounded-ness.

Verb Complements, Fact, and Evidentiality

The semantics of cognition verbs such as 'know' or 'think' was an early preserve of logic-based presuppositional analysis in linguistics, where one could proudly point out to a neat dichotomy between "factive" (presuppositional) verbs such as 'know,' 'forget,' 'find out,' and so on, vs. "non-factive" (non-presuppositional) ones such as 'think,' 'believe,' 'guess,' and so on.[41] But this comfortable discreteness/categoriality begins to break down when one turns to consider the data of *evidentiality*. Rather than being an "exotic" phenomenon, evidentiality is in fact all-pervasive, although its surface coding differs from one language to the next.[42]

I have discussed elsewhere the facts of KinyaRwanda,[43] where the verb 'know' and 'think' are the same lexical verb, but only a complementizing particle tells a presupposed from non-presupposed complement clause. In addition, however, Rwanda has an "intermediate" subordinating particle, creating a contrast of *evidentiality* in the complements interpretable as the English 'know': If you know something from *direct experience*, you use one particle. Whereas if you know it via *hearsay* or *inference*—you use another. This is, inherently, a *scalar*, non-discrete space once again, grading the degree of *subjective certainty of the speaker* concerning proposition, according to the *source of evidence*. Again, logical presupposition—or "extreme certainty"—seems to be merely the *extreme edge* of a much wider, non-discrete pragmatic scale.

The situation is even more complicated. Superficially, one would expect a correlation between "direct experience," "higher certainty," and "fact/truth." But such a correlation does not always obtain. For example, in Sherpa (Sino-Tibetan), the truest and most certain of stories, the Life of the Buddha, is told in the *hearsay* mode. There is no doubt whatever that the story-teller considers the story to be true in the most absolute sense of the word.[44] Nevertheless, he himself didn't witness it, he has it from other story-tellers. Hence, the use of the hearsay mode.

SUBJECTS AND PREDICATES

Logicians have always taken for granted the traditional notion of "subject." Having contended themselves mostly with analyzing one-place predications,

---

[41]See eg., Kiparsky and Kiparsky (1968), Karttunen (1971a,b) or Givón (1973b), *inter alia.*
[42]See Chafe and Nichols (eds, in preparation).
[43]See Givón and Kimenyi (1974) for details.
[44]I have this story from a religious *lama*, himself a minor *tulku* (reincarnated being) of a minor lamaic chain of transmission.

they could usually get away with this informality, because the predicate ('F') is normally well defined, so that the rest of the proposition, bound by either an existential or universal quantifier, could be referred to as "subject." Things become more complicated, however, when the predicate itself is nominal. If it is generic, a logician may yet resort to some type-theory argument purporting to explain why one of the two nominals in the sentence is the subject and the other the predicate, and thus explain the contrast in:[45]

(44) a. John is a farmer
      b. *A farmer is John

But it is doubtful if logical criteria alone could decide which nominal is the subject and which is the predicate in cases where both are equally referential and definite, as in:

(45) a. The teacher is the cook
      b. The cook is the teacher

To linguists and speakers, however, (54a,b) are not interchangeable, but they depend on the rather nebulous (from the logician's point of view) criterion of "what one is talking about," a notion that cannot be defined within the bounds of isolated propositions, but is rather *discourse-context sensitive*.

In handling two-place predicates, the logician could easily develop case-role notions such as AGENT and PATIENT, then translate "John killed Bill" into "KILL (John)$_A$ (Bill)$_p$. But this notation will not differentiate between the active and its corresponding passive. The logician could of course add the "atomic" notion SUBJECT, then characterize the active as KILL (John)$_{AS}$ (Bill)$_p$ and the passive as KILL (John)$_A$ (Bill)$_{PS}$. But as linguists and speakers would know, "subject" is not an atomic notion, but rather is a discourse-pragmatic entity closely related to the notion 'topic.'

I have already discussed the use of both subject-agreement and left-dislocation in discourse, and suggested that subject-agreement is used to refer to the *continuing* clausal topic, whereas left-dislocation is used to mark the switching/disruptive clausal topic. Consider now:

(46) . . . now *John, I* saw him leaving a while back . . .
           TOP SUBJ

In normal conversational context, a construction such as (46) will be used when 'John' is being re-introduced into the discussion after a considerable gap, while 'I' represents the continued topic. Clearly, then, the logician could not posit an

---

[45]All other things being equal, the referential scope of the subject must be narrower than that of the predicate. See discussion in Givón (1973a) and Keenan (1976).

atomic, context-free notion "subject" to deal with language data, because that notion is tied together with many other devices along a *scale* that we have already shown to be discourse pragmatic in nature. The fact that some *corners* of an inherently pragmatic sub-system can be described in seemingly-logical terms is again interesting, but it certainly does not justify viewing the entire sub-system as deductive-logic based.

SPEECH ACTS

One would expect that the impact of post-Wittgensteinian ordinary language philosophers such as Austin would give impetus to a truly pragmatic linguistic analysis of speech acts, one of the most transparently pragmatic areas in human language. In fact, nothing of the kind happened. Rather, the material has been repeatedly forced into various straight-jackets of seemingly formal "logic of conversation," "conversational postulates," "maxims," "conventional implications," and their ilk, all trying to represent the speech-act foundations of language and communication as an axiomatic system.[46] Linguists seem to forever withdraw from the brink of the non-discrete, open-ended pragmatic precipice to which honestly-gathered data invariably lead them. In this section I suggest that there are some fundamental beliefs concerning the discreteness of speech-acts that require serious re-evaluation in light of language facts. The discussion is prompted by the scalar, non-discrete nature of Japanese speech acts.[47]

In Japanese, the speech-act valuation of a clause is marked by verb suffixes, and one can easily observe an eight-point continuum between the two extreme points of declarative and interrogative, as given in (47) below. The difference between the points on this continuum may be expressed in terms of the *speaker's certainty* as well as the *degree of response* sought from the hearer.

Particle	*traditional label*	*degree/type of sought response*
declarative		
1. Ø	neutral	high certainty; not expecting a challenge
2. -yo	assertive	high certainty; expecting challenge and being emphatic in order to counter it

---

[46]See Grice (1968), Gordon and Lakoff (1971), *inter alia*.

[47]The data is taken from Nakau (1973, p. 30), and was brought to my attention by Charlie Sato (in personal communication).

3. -na -kanaaa	exclamatory	emotional load; perhaps ex- pressing surprise (and thus less certainty)
4. -ne	confirmative	low certainty; inviting chal- lenge or comment
5. -desyoo -daroo	presumptive (polite) presumptive (plain)	more presupposed question; an- swer is expected as either yes or no
6. -kadooka	rhetorical	rhetorical question when one is fairly sure of an answer
7. Ø	informal	lower certainty in the answer, more reliance on hearer's response
8. -ka	formal	lowest certainty; formal and deferent to the hearer, and thus by implication more de- pendent on response

interrogative

Traditionally, points 1.–4. above were considered "declarative," whereas points 5.–8. were considered interrogative. But it is easy to see that we are dealing with a continuum that is quite complex, involving at least the following three mutually-dependent factors: (1) the speaker's confidence/certainty in his/her knowledge/information; (2) the speaker's willingness to admit challenge to his/her knowledge; and (3) the speaker's solicitation of confirmatory or corrective response. In addition to these three factors—which by themselves make the traditional dichotomy between declarative and interrogative a hopeless mirage—one must note that the speaker, in deciding which point on the continuum to use, exercises delicate *pragmatic judgment* concerning various *probabilities*, such as his/her own certainty and validity of source of information, the hearer's knowledge, the hearer's willingness to respond benevolently, and the hearer's disposition to attack or challenge. And in Japanese at least, subtle computations of the *social gradient* between the speaker and hearer must also be taken into account, and those are not fully independent of the more epistemic considerations.

Of course, one does not have to stray far to find similar gradations. Thus, *tag-questions* in English are clearly an intermediate grade between declaratives and interrogatives. Further, the normal yes-no question pattern in English is not neutral, but rather is systematically *biased* toward either an affirmative or negative response.[48] So that only the explicit construction below represents total 'neutrality':

(48) "Given p and non-p, please tell me which one is true"

---

[48]See Bolinger (1975) for details and argument.

One could also show that the data of *indirect speech acts* points toward another continuum, between declarative and imperative, where at least the following graded dimensions must be involved:

1. The degree of the speaker's attempt to elicit action from the hearer,
2. The degree to which the information volunteered by the speaker is relevant to the hearer's general and specific context (as assessed by the speaker).

Here again, the social-status gradient is tangentially relevant to the more epistemic factors.

Further, one could show a wealth of evidence supporting a continuum between interrogatives and imperatives, in the data area of indirect speech acts again. And there again a number of graded dimensions of the type mentioned earlier must be involved, with social-status factors again intruding. The entire semantic space of speech acts is thus a veritable mire of non-discreteness and pragmatic-probabilistic judgment. Only at the very extreme edges of the system do phenomena appear to be discrete-like and amenable to logical-deductive analysis.

TOPIC RELEVANCE

So far, we have dealt only with topics that, in one way or another, have been the province of philosophers and logicians for quite a while. But some of the more complex problems arising from the use of human language in communication have never even been touched by logicians. One of those is the propriety of using one expression—an NP or a phrase or a larger chunk—as a *topic raiser* to invoke a "comment" in a subsequent sentence. As a fairly rudimentary example, consider the following left-dislocated example in a descending order of "topic relevance":

(49) a. Speaking of politics, I like Reagan.
      b. Speaking of Reagan, I like politics.
      c. Speaking of Reagan, how do you like Haig?
      d. Speaking of Reagan, this country'll never be the same.
      e. Speaking of Reagan, the prime rate is going up again.
      f. Speaking of Reagan, are you sure Nixon liked cheese too?
      g. Speaking of Reagan, the sky is blue.

Without belaboring the obvious, one could say that *in principle* anything could be the invoking topic of anything, anything that is within the cognitive network is in principle "related to" anything else within the network, and the only empirical question to be determined is "by what degree?", "by how many computations?'', ''in what specific way?''[49] Now, if the cognitive network were a closed, deductive system, then in principle one could specify the degree of connectedness in an exact way, pointing out to a discrete number of specific computational steps. But it is clear that the kind of considerations underlying relevance judgment in the examples above are totally open-ended, they are sensitive to anything the speaker might suspect the hearer has access to, including trivial and accidental information that belies systematic formalization. In principle, then, the problem of topic relevance, that plays a crucial role in the use of language in communication—including the more ''syntactic looking'' processes disscussed earlier—cannot be dealt with within a logical-deductive system, without trivializing it or masking its explosive implications.[50]

THE CONTEXT-DEPENDENT, RELATIVISTIC NATURE OF LEXICAL MEANING

In the analysis of the ''purely semantic'' lexical meaning, the same Positivist tradition of *radical reductionism* observed by Quine[51] in philosophy has largely prevailed in linguistics. One is thus conditioned to talk about ''semantic features'' as if they are atomic, absolute, primal units of meaning in the grand tradition of Carnap[52] and Katz and Fodor.[53] Even people far removed from this tradition, such as the Diverians,[54] proceed upon the assumption that it is possible to segregate the ''core meaning'' of words or morphemes, which is invariant and context-free, from ''contextual inferences'' which are contingent and context-sensitive.

Wittgenstein challenged this tradition of logical atomism from two separate perspectives, and I deal with his second challenge first.[55] In his Investigations, Wittgenstein attacks logical atomism by citing *games, tools, family-resemblances,* and in general vocabulary items that are more obviously *culturally-dependent.* The Positivists and their conscious or unconscious followers in linguistics are forced to concede the force of Wittgenstein's argument, but confine

---

[49]AS we shall see later, the same argument holds for the problem of *semantic* relatedness in general, and for the description of semantic-cognitive networks and semantic change. Example (49g) merely stretches the system towards its less-likely, absurd end, and is reminiscent of the ''Russian-jokes'' I used to hear in my childhood, such as: ''There were once two brothers, the first one was tall and handsom, and the second one liked cheese, too.''

[50]Thus, Grice (1968), for example, includes ''relevance'' in one of his maxims without further amplification.

[51]See Quine (1953, pp. 20–46).

[52]See Carnap (1947).

[53]See Katz and Fodor (1963).

[54]See for example García (1975) or Kirsner (1979), *inter alia.*

[55]See Wittgenstein (1953).

its scope to what they consider to be "marginal" areas of the semantic system, while persisting to interpret the "bulk" of the system in terms of a closed deductive system of semantic prime, worrying about "projection," "analyticity," and "contradiction" within such a system. What I hope to do in this section is demonstrate, with a few rather simple-minded examples, how the shoe is on the other foot, and how what Wittgenstein demonstrated to be the "less objective" margins of the semantic system of natural language is in fact true *with a vengeance* for the entire system.

Suppose you and I were taking a walk in pre-Columbian Southwestern Colorado, and suppose we both saw the following event:

(50) "A man is slowly walking up a hill; he reaches the top, then he kneels down in front of a pile of stones and raises his arms to the sky."

And suppose that a friend asked us both: "What was that man doing?", and we had a conflict in reporting the event, with you claiming (51a) and I claiming (51b):

(51) a. "He was praying to the Great Spirit."
      b. "He was mourning his dead mother."

It would be a trivial endeavor to show how both "praying" and "mourning" are largely non-objective, culture-dependent meaning constructs. But how about the "more objective" parts of (50)? Consider first "slowly," and it is easy to see how its meaning is totally relativistic, *gestalt*-determined, founded upon the notion of *norm/average* which is, further, not fixed but rather depends upon the entity and the type of change/event/movement involved.[56] How about "walk" then? And how objective are the criteria differentiating "walking" from "loping" from "jogging" from "running?" Well, how about "up"? At what point of departure from the absolute "physical" horizontal plane does one begin to move "upward"? And is that absolute horizontal plane defined in terms of gravity? In terms of the earth's circumference? In terms of our visual perception of the horizon? Next, take "hill," and wonder how it is to be differentiated from "mound," "heap," "pile," "peak," or "mountain." Size has obviously something to do with it, but then there's nothing absolute about size in language, where a big mouse is much smaller than a small elephant. Take "reaching" next, and ponder about the point where one reaches the top of something, especially something such as a hill that does not have a discrete, right-angled apex but rather curves gradually. Now proceed to "kneel"—how bent does one's knee need to be to qualify for real kneeling? And how elevated does it

---

[56]Viz. the difference between "a fast horse" and "a fast jet-plane."

contact-point on the ground need to be? And is knee-contact necessary, and if not what proximity of the knee to the point of contact/approach is required to qualify for kneeling? And—horror of horrors—what exactly is the "knee" itself?

At this point the Positivist should be unable to contain him/herself and burst in a frustrated harangue about the objective, scientific nature of our anatomy. The knee is indeed that bend between the femur and the tibia/fibula, capped by the knee-cap. But from what perspective are the femur and tibia/fibula straight and the knee a bend/joint between them? Let us consider three perspectives:

(52) a. *"Normal"*:

b. *"Micro"*:

c. *"Macro"*:

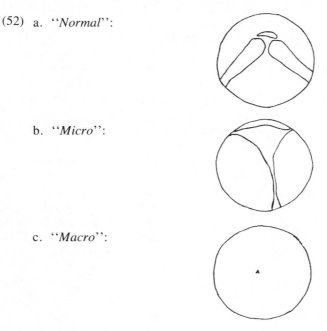

The Positivist may now howl about optical tricks and the *finess* of scientific measurement, and one may indeed wish to defer the discussion of whether scientific truth is in any sense subject to less-pragmatic and more-objective criteria of meaning.[57] But the cognitive map represented in language clearly judges whether objects are straight or bent in a *frame-dependent* way, given a particular perspective and a *utilitarian context*. Let us go on to investigate the notion of "in front," and we find that different cultures construe it differentially with respect to the position of the observer/reporter, the physical characteristics of the two objects involved, and their relative position vis-a-vis each other.[58]

---

[57]See discussion in Givón (1979b, Ch. 8).

[58]Hill (1974) illustrates such differences between English and Hausa in their construction of spatial relation terms.

And each one of these three dimensions is potentially a continuum. We may go on to "pile" now, and face the same relativistic problems as with "hill." And on to "stones," and worry how different they are from "rocks" or "pebbles." And on to "raising," where we must worry about what angle of rise is criterial for "raising," where we must worry about what angle of rise is criterial for "raising" to hold. And on to "arm" where similar problems as seen above for "knee" may be raised. And on to "sky" and criterial angles, visual field, and so on.

The last refuge of logical atomism and "objective meaning" is of course science, and it is an interesting—and for the moment, separate—issue if scientific "truth" is at all relevant to the cognitive map represented in language. Still, how objective is the Linnaean classification of flora and fauna? Of "horse," "zebra," and "donkey"? Of "goat," "sheep," and "bharal"?[59] Of "simian," "primate," and "human"? Honest biologists have long ago owned up to the arbitrariness—in principle—of taxonomic boundaries between species and genera. Those boundaries are *useful*—in very much the same way as the taxonomic map of language. But from a purely logical point of view, that of criteria, they are just as contingent and relativistic.

The most seemingly objective science is physics, whose wide perspective and relatively unclattered subject matter is the envy of both biologists and behavioral scientists. But even there it seems that the finer and more sophisticated our observation techniques—our *framing*—become, the more and finer sub-atomic particles we "discover." Science, an "objectivist" would retort then, is obviously dependent on calibrated instruments, but that should not impinge upon our faith that ultimately, out there, an "objective" reality exists independently of the instrumentation. That reality may indeed exist, but what is the source of our faith in its being in any sense "objective"? Obviously, only from The Ultimate Perspective—that is in principle denied to us—[60] could one make such a judgment, support such a faith. The *specifics* certainly have not been shown to be frame-independent.

Having once established that meaning is in principle a pragmatic matter, a frame-dependent entity, one nevertheless must concede the great areas in our cognitive map where *relatively stable* frames have been established by the organism, most obviously in the areas of our construing the physical universe. There is nothing wrong in owning up to those areas of illusory stability and seeming context-independence, so long as one remembers that they are forever perched over the pragmatic abyss, and that their stability is relative in two different senses: (1) relative to the more obviously relativistic areas in our cognitive map; and (2) relative in terms of the speed of change and context-sensitive interaction

---

[59]Viz. Gordon Schaller, a famous biologist, and his quest for the Bharal in Matthiessen (1978).
[60]I owe this discussion of The Ultimate Frame to stimulating exchanges with Tom Bikson and Martin Tweedale, both of whom probably disagree with my views.

they do indeed retain. One could, in addition, hazard a guess why the emergence of relative areas of stability and "objectivity" within our otherwise shifting pragmatic map of reality is a useful feature of *survival,* a functional *modus operandi* of bio-sentient systems: No rapid *decision making, action schemata* and *categorial yes/no choices* for survival are possible otherwise. The objectivization and categorialization of reality is a prime survival-linked feature of such systems, it keeps manifesting itself repeatedly at all levels of perception and cognition; it keeps re-creating itself—out of the inherently pragmatic, chaotic abyss of non-discreteness, by re-framing or re-adjusting existing frames over time. As we see directly below, such re-adjustment is a fundamental fact of language and language change.

PRAGMATICS AND LANGUAGE CHANGE

Ever since Saussure, linguists have tended to segregate diachronic from synchronic study, pretending that it was possible to appreciate language "as if at any given point it is a fixed system." Although it is true that such idealization is a necessary methodological-heuristic step,[61] it is still the fact that language—within the minds of speakers, rather than as some abstract system of *langue*—is *always* in the middle of change in lexicon/meaning, syntax, morphology, and phonology. Language as a cognitive map is thus not only a system of coding knowledge, but perhaps primarily a system of re-coding, modifying, and re-structuring existing knowledge and integrating into it newly-acquired knowledge. Within the generative linguistic tradition, such modification is often represented as a purely formal process of permuting and re-combining a fixed inventory of atomic primitives, or juggling the order of relatively-fixed formal rules, thus totally trivializing the process of language change,[62] and also making diachronic linguistics much less of a challenge to the logical-deductive dogma.

There are three aspects of diachronic change that would be incompatible with a formal/deductive approach to meaning:

1. *The system-parts dependency:* In a logico-deductive closed system, the primitives are fixed, and any "change" in the system is basically trivial, involving formal deductions from existing propositions.[63] Further, the primitives in such a system have an absolute, atomic value and are thus not system-dependent. Language change, however, shows rather clearly that in human language "primitives" may be added or subtracted, split or merged, and that further, a change in sub-parts of the system requires re-definition of the entire system.

---

[61]See discussion in Givón (1979b, Ch. 6).
[62]See for example Kiparsky (1971) or Lightfoot (1976a,b).
[63]Cf. Wittgenstein (1918).

2. *Open-endedness:* Aside from Goedel's observations *in principle,* deductive systems are closed and can absorb no new primitives, nor new propositions that are not deducible from existing ones. The human cognitive map as represented in language change, however, clearly attests to the open-endedness of the system and its ability to absorb new knowledge without short-circuiting the existing network.

3. *Analogy, similarity, and relevance:* The definition of "primitives" in any deductive system precludes their bearing any "similarity" to any other primitives. In natural language—and particularly in language change—one most commonly deals with *scales/clines* where A resembles B, B resembles C and so on up to Z, but where the "resemblance" relation does not have to be of the same kind across the entire scale, but may be *of any kind,* thus recalling Wittgenstein's *Family Resemblance.*[64] Most commonly, change occurs along such clines by *analogy* or *abduction,*[65] which may be based on semantic, structural or pragmatic similarity. Now, logic-bound linguists may strive to express similarity in discrete terms, that is, "number of discrete features being shared," but this simply defers the problem to another level, that is, "similarity between features."

In this section I cite a number of typical examples of diachronic change in meaning, morphology, word-order, and syntactic structure. In each case I point out how a logico-deductive analysis of what goes on is *in principle* bankrupt, and that only a pragmatic characterization is indeed compatible with the facts. The overall force of the argument will remain close to Peirce and Wittgenstein: So long as the system of knowledge representation is elastic, open-ended, and changeable, it could not possibly be a deductive-based system. The consideration of language learning and language evolution will be brought to bear later on as correlates to the central argument.

Lexical-semantic change

Consider the following semantic split, where both English 'know' and 'can' come from the same PIE and PG root, **gno* and **knāw-an,* respectively, with the meaning 'know' being older, and 'being able to' being later-derived.[66]

---

[64]Cf. Wittgenstein (1953).

[65]For the role of analogy and abduction in language change, see Anttila (1977) and Anderson (1973).

[66]Raimo Anttila (in personal communication) has sketched roughly the following history for the PG stem **knāw-an:* First, a variation between the aorist-perfect *knāw-an* which yielded eventually *know,* and the present form *gṇne.* From the latter form, via **kuna,* came the OHG for *kann.* The addition of the causative suffix *-ja* to *kann* produced the G. *kennen* and E. *cennan,* and from the latter *ken* 'know' is a direct development. OHG *kann* is the stem preceding E. *can.*

Initially, the sense ''being able to'' was merely a *contextual inference* from ''know,'' a *pragmatic*-based inference of roughly the following kind:

(53)   ''If one *knows how to do* something,
       then the probability is higher that one *can do* it''

That the inference in (53) is probabilistic/inductive rather than deductive is easy to see, because ''ability to do'' involves not only ''knowing how to do,'' but also possibly ''having the physical power to,'' ''having the will-power to,'' or ''being physically/mentally un-restrained.'' Now, if the sense ''be able to'' can be pragmatically/contextually inferred, then presumably ''know'' at that historical stage in English may be described as *polysemous,* roughly along the following lines:

(54)   KNOW - - → 'be able to' / 'know how to perform an act'
              - - → 'know' / elsewhere

In (54), then, 'know' is the ''core meaning'' and 'can' a special case in a narrower context, still pragmatically inferable as in (53). Next what must have occurred is a *generalizing inference,* again pragmatic/inductive in nature, roughly on the line:

(55)   ''If one *can* do something because one *knows* how to do it, perhaps one
       can do it for other reasons as well, such as (i) physical/mental power,
       or (ii) being unrestrained''

Deductively such an inference is absolutely unwarranted, but *inductively* it proceeds along a ''family resemblence'' cline, noting the ''similarity'' between the three senses of 'ability' (knowledge how, power, and lack of outside restraint). Once such an inference has occurred, however, the structure of the lexical item KNOW as in (54)—with 'ability' defined as a contextual variant of 'knowledge'—is now disrupted, because the other senses of ability have nothing to do with knowledge. At this stage speakers face a conflict, between the semantic domain of ABILITY and that of KNOWLEDGE, both of which overlap at one point, namely the sense ''knowing how to.'' The conflict, in this particular case, was resolved by taking advantage of the *phonological variant kann/can,* which (presumably by some inferential/inductive stages) then lost all its 'know' senses except 'know how to do'—and absorbed the other 'can' senses, of power and lack of restraint. This restructuring may be thus summarized as:

(56)   KNOW →   'know'
       CAN → (i) 'be able to because of knowledge'
            (ii) 'be able to because of power'
            (iii) 'be able to for lack of restraint'

But the fortuitous appearance of an *etymologically*-related stem is not a prerequisite for the restructuring. Rather, a *semantically*-related stem could do just as well. This was presumably the case in German, where the 'know that' and 'know NP' senses of *können* were transferred to *wissen*, with *können* retaining largely the domain of 'be able to.' But *wissen* comes from the IE *woida*, which is a perfect-resultative form, and is attested in the L. *vidē-re* 'see,' a stative form, with the most likely scenario being that 'see' was the original meaning, retained in Latin,[67] which the perfect-resultative 'having seen' became re-analyzed as 'having seen and thus know.' And this analogical extension from ''perception'' to ''cognition'' is again itself the product of pragmatic/inductive inference, whereby the original situation may be given as a contextually-predicted submeaning of 'see':

(57) *WID- - - → 'know'/PERF-'having seen'
        - - → 'see' / elsewhere

The following pragmatic/inductive inference must have then occurred:

(58) ''If one *has perceived* something, then one must be in possession of *knowledge/cognition* of/about it''

The inference from physical perception to mental cognition is of course not shocking, and must be a human universal.[68] Nevertheless, it is not a *deductive* inference, but rather inductive-probabilistic. Presumably, the next development from 'see' to 'know' in *WID- involved a *generalizing inference* again, something like:

(59) ''If one understands something because of prior perception/seeing, maybe then one understands it for other reasons as well, such as hearing, explanation, introspection, divine inspiration, etc.''

Once again we have an overlap between two semantic domains, of ''cognition'' and ''visual perception,'' with the point of overlap that cements the *family* resemblance between the two domains being ''understanding due to seeing.'' And once again a phonological/morphological variant form, the perfective stem *woida*, was exploited as the specialized form to carry the restructured meaning

---

[67]Raimo Anttila (in personal communication).

[68]Thus, for example, in most core-Bantu languages the verb 'hear' -*umfwa* is also used for 'understand.' Or note the English expression 'I see your point.' But even more concrete expressions of *possession* may be semantically bleached toward understanding/cognition. Thus, in Amharic ''it has entered me'' means ''I understand it,'' and in English ''I get/got it'' has a similar sense. In Hebrew ''I caught it'' may mean ''I've understood it,'' etc.

'know,' unloading (in those languages) the sense of 'see' and loading the non-perceptual senses of 'know.'[69]

Metaphoric Extension of Meaning

Consider the following gradual extension of the meaning-scope of "sweet," originally a physical-taste adjective:

(60) a. sweet apple (physical-taste, beneficial/rewarding)
      b. sweet music (audio/abstract, beneficial/rewarding)
      c. sweet victory (abstract, beneficial/rewarding)

At first glance, one may argue that we have here a straight case of *semantic* bleaching, whereby 'sweet' has the *core meaning* "beneficial/rewarding," and the extensions are *contextual inferences* from that core meaning, gradually bleaching the more physical/concrete aspects of meaning. Thus, a Diverian linguist would argue rather straneously that no semantic change has occurred at all between (60a), (60b), and (60c), but only a widening of the range of contextual inferences.[70] But the concept of "core meaning" becomes nebulous when one considers the fact that the very same lexical item may mutate in different directions, using different components of its total meaning cluster as the so-called "core meaning." In this connection, see the discussion of the mutations of the verb 'sit.' In other words, *anything* within the semantic field/domain of a lexical item can be considered either "core" or "contextually inferred" meaning, given particular *contexts*. But the role of context in this case is *to define what is relevant* within the domain, what will be held constant and thus considered "core," as against what will be mutated and thus considered "contextual inference." This determination is *in principle* non-deductive, but rather involves the pragmatic judgment of 'relevance' and 'similarity.' Further, much like the cases discussed earlier—which are after all metaphoric in very much the same sense—we deal here with *family resemblance* rather than with objective/core/fixed meaning features.

I would like to claim here that all semantic change is *in principle* of the same kind as metaphoric extension, and that the availability of "related stems" or "semantically related words" simply conspires to mask this generality in the eyes of some scholars. Thus, consider the extension of the meaning of the Hebrew root *šb* 'sit':

---

[69]In Germanic the sense 'see' was unloaded upon *seen,* itself etymologically coming from IE *sekw-* 'follow' via exactly the same type of pragmatic inference, i.e., "if one follows with the eyes (special context of 'follow') then one sees." Etc. Etc. Etc.

[70]See eg. arguments in García (1975) and Kirsner (1979), *inter alia.*

(61) a. *Early stage: šb* 'sit'
     b. *Later stage: y-šb* 'sit'
          šb-t 'rest' - - → 'rest from work'- - → 'strike'

The split between 'sit' and 'rest' is obviously a contextual inference from 'sit,' as are the later extension of 'rest' toward 'rest from work' and eventually 'strike.' The added phonological differentiation most likely arose from the imperfect third-person ms. sg. prefix -*y* for 'sit' and the nominalizing suffix -*t* for 'rest.' For this particular extension, the ''core'' is presumably ''not be at motion.'' But that is only one possible *inference* arising from the physical fact of sitting. But 'sit' may also develop into the the verb 'be' and eventually into the *progressive aspect,*[71] and those extensions are not based on the ''core'' 'not be at motion,' but rather on the ''core'' 'be at location,' later bleached into 'be/exist' or into 'be in time' in the case of the progressive aspect. The distinction between ''core meaning'' and ''contextual/pragmatic inference'' is thus fundamentally bankrupt. Any portion of a semantic domain can be either, pending the *context* where a particular semantic/metaphorical change takes place—a context given in terms of the other lexical items participating *by analogy.* The group of semantic/lexical items assembled *contingently* for a particular purpose is thus the *semantic/functional* domain for that particular contingency, and their relation to each other is in principle a Wittgensteinian *family resemblance.*

Functional Re-analysis in the Morphology

Consider next the following change, where an erstwhile locative-directional pre/post-positions such as 'to' becomes next the dative/benefactive marker 'to/for' and eventually the *human-accusative* marker, as in the case of the Spanish *a.* Such a change, initially, proceeds by bleaching along a well-established cline of verbs, and the move from locative to dative-benefactive may be summarized as:[72]

(62) a. ''*move* X-object to Y-location''   (locative)
     b. *give/send/bring* X-object to Y-person''   (human-locative)
     c. ''*give/tell/show* X-abstract to Y-person''   (Human-dative)
     d. ''*do* X-act to/for Y-person''   (human-benefactive)

---

[71]In Juba-Arabic, a Creole language from the So. Sudan, the Arabic *gaad* 'sit' has become the progressive marker *ga/ge/gi* (Mahmoud, 1979). The bleaching of locational to temporal expressions is universally observed, see Traugott (1974), Clark (1973), or Givón (1973b), *inter alia.* The bleaching of locational 'be' to existential 'be' is again widely documented (see summary in Givón, 1979b, Ch. 8).

[72]See discussion and many details in Givón (1976b). Similar changes, up to at least the dative/benefactive point, are documented for English *to,* Bantu *ku-,* Hebrew *le-,* Sherpa *-la,* among others.

The four inductive/pragmatic steps of inference involved in (62a–d) may be given roughly as:

(63)   (i)   "If a person is the *locative-goal* to which an object moves, that person must be a *conscious-recipient* of that object."

    (ii)   "If a person can be a conscious-recipient of *concrete* objects, that person could also be a conscious/recipient of *abstract* ideas."

    (iii)   "If a person is the *recipient* of anything, changes are he/she is the *possessor* of that thing."

    (iv)   "Whoever *possesses* an object most likely also *benefits* from that object."

The probabilistic nature of these inferences, and the family-resemblance character of the verb-continuum along which they progress, need no further elaboration. But how did the inference progress further from dative/benefactive to human-accusative? Three facts about dative/benefactives are involved here:

1. They are in text overwhelmingly *human*[73]
2. They are in text overwhelmingly *definite;*[73]
3. They have a high probability in discourse/text of undergoing a "dative-shift" and thus being promoted to *direct object/accusative.*[74]

Now, in English dative-shifting involves the loss of the preposition 'to/for' but in Spanish it does not. Thus, compare:

	Direct-Object variant	*Indirect-Object variant*
(64) a.	He gave me a book	He gave the book to me
b.	He played me a record	He played a record for me
c.	Dio el libro a María	Le dio a María un libro
	gave the book to Mary	her gave to Mary a book
	'He gave the book to Mary'	'He gave Mary a book'

The analogical, "family-resemblance" nature of this inferential space next gave rise in Spanish to the following inferences:

(65)   a.   "If an object is *dative,* it has a high probability of also being *human* and *definite*"

---

[73]See discussion in Givón (1976b, 1979b, Ch. 2).
[74]See discussion in Givón (1979b, Ch. 4).

b. "If an object is *dative-human-definite*, it has a high probability of also being *direct object*"

This purely probabilistic reasoning is responsible for re-analyzing *a* as, among others, a *human-accusative* marker. But the very same inferential space could lead to another resolution. In Swahili, for example,[75] the object agreement/pronoun first marked dative objects, then was extended to definite direct objects, and finally was further extended—within one gender only—to be the marker of *human direct objects*, definite, and indefinite alike. The more general extension in Swahili, then, went from dative to *definite*-direct object, rather than from dative to *human*-direct object.

Reanalysis of Syntactic Constructions

I have suggested elsewhere[76] that the functional domain of passivization involves three separate sub-domains that overlap in the passive construction:

(66) 1. *The topic-assignment domain:* Here a non-agent gets 'promoted' to the clausal subject/topic position. Other members of this domain are, for example, anaphora, pronouns, definitization, right and left dislocation, etc.:

2. *The impersonalization domain:* Here the identity of the agent is suppressed in various ways. Other members of this domain are various impersonal/neutral constructions;

3. *The stative/detransitive domain:* Here an event is construed as a resulting state, its "active" properties thus suppressed. Other members of this domain are stative-adjectivals, reflexives, reciprocals, perfective-resultatives, etc.

Now, in diachronic change giving rise to passive constructions of the "classical" English type, one which reflects properties of all three functional domains of passivization, one finds cases where the passive arises from constructions within each domain that are *not* per se passive constructions, but rather "belong to the passive family" via one of the three domains. Undergoing change, they then gradually acquire some of the properties characteristic of the *other* two domains. These changes involve the following inductive/probabilistic inferences:

(67) a. "If the *identity of the agent* is *suppressed,* the next most-likely participant in the clause will be the *topic* of the clause."

b. "If the *stative/resultative aspect* of the action is focused upon, then it is likely that the *status/identity of the agent* is less important."

---

[75]See discussion in Givón (1976b).
[76]See Givón (1981b).

c. "If the *clause-topic is a non-agent,* then it is most likely that the patient-related properties of the event, such as its being a *resulting state,* are focused upon."

Inference (67a) leads from the agent-identity-suppression domain to the topic-identification domain. Inference (67b) leads from the stative-intransitive domain to the agent-identity-suppression domain. And inference (67c) leads from the topic-identification domain to the stative-detransitive domain. Each is in principle pragmatic, and each is supported by a wealth of cross-linguistic data.

Word-order Change

In Early Biblical Hebrew[77] (EBH) the "unmarked" word order was VSO and a special marked order—involving either an "anterior" break in the action sequence or a topic-switch—was SVO. In Late Biblical Hebrew (LBH), the two orders changed their valuation, with SVO becoming the unmarked word order—in context of both action and topic continuity—and VSO assuming specialized, oft semi-frozen, values (though not the original perfect/anterior/topic-switch value of the EBH SVO). This re-analysis also involves a crucial change in the tense-aspect system, from a perfect-aspect system to a past-tense system, whereby the EBH "perfect" changed its valuation from "anterior" to "past," whereas the EBH "imperfect" changed its valuation from "preterit" to "future/nonpunctual." Both word-order and tense-aspect changes may be characterized as "over-use" processes of de-marking:

(68) a. Over-use of the more marked left-dislocated SVO word-order to identify subjects that are *easier* to identify and represent *higher predictability/continuity* than those normally marked by left-dislocation[78]. This de-marked the SVO word-order;

   b. Over-use of the anterior "prior to" value that first involved a highly-specific *look-back function* of "perfect" or "pluperfect," making it a more general "in the past" marker. This de-marked the perfect aspect and made it a past tense.

Both changes are well attested elsewhere, mostly independent of each other.[79] Both represent an "over-kill" communicative strategy, whereby the speaker decides that—just for safety's sake—he will use a more marked device to insure beyond a shred of doubt that the hearer got the message, thereby de-marking or

---

[77]See Givón (1977).

[78]See earlier discussion.

[79]See Givón (1980) for the perfect-to-past change, and Givón (1976b) for the de-marking of left-dislocated constructions.

devaluating that device. A similar decision is involved in de-marking the right-dislocated VS word-order and eventually changing a language from SV to VS "neutral" word-order.[80] Now, the input into such a decision by the speaker presumably involves monitoring a variety of probabilities, concerning syntactic, semantic, discourse-pragmatic, and social-interactive factors all pertaining to the hearer's ability to decode the message without inordinate difficulty. There is no way in the world that such a decision procedure could be characterized deductively, within a closed system. It is inherently a context-sensitive, pragmatic process.

Diachronic, Ontogenetic, and Phylogenetic Language Change

The kind of gradual, step-by-step changes occurring during the child's acquisition of his/her first language is fundamentally of exactly the same kind as the diachronic changes described earlier.[80] Thus, the type of modification and enlargement of both the knowledge-coding system of the growing child and of his/her knowledge-communicating system is pragmatically based, necessarily open-ended and in principle is incompatible with a logico-deductive description. Further, as I have suggested elsewhere,[81] there are strong grounds for believing that the evolution of human language proceeded roughly along the same lines seen in diachronic and ontogenetic change. Thus, not only is a deductive-based account incapable of characterizing the constant, ongoing modification language is undergoing at all times, but it is a most unlikely model to account for the way language has evolved. In this sense, thus, Chomsky's profoundly anti-evolutionary view of human language is certainly compatible with his view of grammar as a closed algorithmic system.[82]

INTERMEZZO: KNOWLEDGE SYSTEMS, WITTGENSTEIN AND PEIRCE

Bertrand Russell, in his foreword to Wittgenstein's *Tractatus,* acknowledges the force of Wittgenstein's argument concerning logico-deductive systems, yet ruefully dismissed the main point, noting that for someone who argued that there was nothing new to be said, brother Ludwig surely managed to say quite a book-full. Russell's comment was of course fundamentally misdirected, because W.

---

[80]See discussion in Slobin (1977) and Givón (1979b, Ch. 5), *inter alia.*

[81]See discussion in Givón (1979b, Ch. 7).

[82]To wit: ". . . It is quite senseless to raise the problem of explaining the evolution of human language from more primitive systems of communication that appeared at lower levels of intellectual capacity. . ." (Chomsky, 1968, p. 59).

was not talking about language but about logic (though Russell's positivism of course took it for granted that logic was the only worthy model of language). Wittgenstein's point is of course not new, having been made some decades earlier by Peirce.[83] Both saw clearly that a deductive, axiomatic system cannot express new information, but only *tautologies* (totally familiar knowledge) or *contradictions* (totally strange and un-integrable knowledge). Peirce went on to coin the term "abduction" to characterize the kind of non-deductive inference, induction or *intuitive leap* that must underly the acquisition of new information. Still, Wittgenstein, perhaps inadvertently, laid down the foundations for a general *theory of information* within any communicative system—certainly within natural language—where "new information" always operates somewhere between the two extremes of tautology and contradiction: It cannot be *totally* new and thus incompatible with all previous knowledge (i.e., "contradictory"). And it cannot be *totally* old and thus redundant and of no interest (i.e., "tautological"). Within such extreme bounds, one could conceive of the seemingly *logical* impropriety of tautologies and contradictions such as:

(69) a. Joe is a teacher; *he is a teacher* (tautology)
      b. Joe is a teacher; *he's not a teacher* (contradiction)

as being merely the extreme margins of the system. But the bulk of the system of linguistic communications operates somewhere *between* these two extremes, where "degree of redundancy" and "degree of newness/surprise" cannot be ruled upon by deductive means, but must be inferred *pragmatically*. There are three empirically-based arguemnts that militate for such conclusions:

1. The context-dependency argument.   The logician would consider (69a,b) above closed systems, and thus pretend that the contradictory or tautological second proposition in each is that way because of the first proposition. But the use of language in communicative context is *wide open* in two distinct senses: First, *generically,* any item of shared knowledge within the culture/lexicon can potentially by an *implicit* part of the context for *any* proposition. Second, any span of the "specific/immediate" context the speaker judges to be *within reach* of the hearer can be legitimate context for a proposition. Decisions on both grounds are *in principle* probabilistic and pragmatic, given the total open-endedness of "context."[84]

2. The speaker-hearer argument.   For logicians, context is "objective," overtly-listed premises. They are never comfortable with "I" and "you." But in

---

[83]See Peirce (1955), and discussion in Anttila (1977).

[84]Logicians, of course, create a sanitized notion of context by listing a finite number of propositions as "premises."

language and communication, it is the "I" that makes judgments about what—generically and specifically—can be taken as the *likely* context for "you," and this horror of subjective, inductive inference would bust any deductive system wide open. In addition, sooner or later the self-inclusion problems associated with the "I" create the classic infinite-meta-regress problem (cf. Russell's paradox), and while a logician may *legislate* against it, as Russell has done,[85] that does not make the problem go away.

3. *The topic-relevance argument.* As we have seen earlier, even agreeing to reduce the notion of context to immediate, visible, propositional context is not going to get the logician off the hook. This is so because *anything* within that visible, immediate context is judged by the speaker as to its potential *relevance* as context. And as we have seen earlier, such a judgment is in principle open-ended and pragmatic.

To sum up, then, whereas deductive logic may define the upper bounds of information transfer systems, the extreme cases, the bulk of the actual transactions are perforce pragmatic in nature, where "new" and "old" information are judged *relative* to open ended generic and specific contexts.

SYNONYMY, POLYSEMY, AND CONTEXT

Whereas lexicographers seem to take *synonymy* for granted as a matter of practice, a linguist would find it extremely hard to identify a bona fide case of synonymy in either lexicon or grammar. Marchand refers to the seeming absence of synonymy in the lexicon as "the economy principle,"[86] and that may indeed be a correct explanation, at least in part, because one would be hard put to explain why speakers should store two forms coding *exactly* the same meaning or function. Apparent cases of synonymy tend to dissolve, on closer examination, into more subtle semantic, pragmatic, social, stylistic, dialectal, or idiolectal variation. On the other hand, *polysemy*—and grammatical *ambiguity* in sentences out of context—are much easier to document, being indeed one of the most pervasive facts of lexicon and grammar. The types of diachronic changes discussed earlier clearly demonstrate this, so that in fact language change at all levels virtually depends on contextually-hinged proliferation of senses and functions.

Synonymy and polysemy are the two extreme poles of the system of *coding* in language. Synonymy is the extreme case of *over-coding*, where the very same

---

[85]See discussion of Russell's paradox and its implications for epistemology and pragmatics in section 13, below.
[86]See Marchand (1964).

message unit is coded by more than one code unit. Polysemy is the extreme case of *under-coding,* where several message units—presumably along a continuous portion of a semantic/functional domain—are coded by only one code unit. Why does language seem to avoid the one extreme so consistently and indulge in the other just as consistently? The answer, I believe, must lie—perhaps paradoxically—in the idealized semiotic imperative of language, the one demanding that "there ought to be a one-to-one correspondence between code and message units." On the synonymy side, language clearly abides by this dictum. How come then it violates it so consistently on the polysemy/ambiguity end?

If language were a closed, deductive system dealing largely in context-free meaning, this question would indeed be baffling. But the discussion here has, I believe, already established that lexical and grammatical meaning/function in language is *open-ended* and *context-sensitive.* Thus, the reason why speakers can tolerate the seemingly high incidence of polysemy/ambiguity in language must be that they constantly make use of the context to disambiguate. Thus, whereas a small core or "margin" of the semantic system indeed "behave as if they are absolute and context free," the bulk is context-dependent, at least when one considers the actual facts of language use. Given such a concept of language, it seems that it does indeed "strive to operate" by the principle of one-to-one correlation between code and message. Either total synonymy (i.e., an infinite number of code-units expressing the same message) or total ambiguity (i.e., an infinite number of messages coded by a single code-unit) would be a communicative nightmare. The first would impose an enormous and non-functional *memory burden,* whereas the second would impose *total dependency on context.* It seems to me that language is in fact a *mixed, compromise* system, relying to some extent on memory burden—where items and rules can be memorized in a relatively "atomic," context-free fashion, while to some extent relying on disambiguation via context, where items and rules shift their meaning/usage depending upon the pragmatics of context. That such a compromise should be rooted in the neurological capacity of the human organism seems too obvious to require further comment.

ONTOGENY, PHYLOGENY, AND THE RISE OF AUTOMATED PROCESSING ROUTINES

When one considers the ontogeny and phylogeny of human language, one is struck by the fact that earlier stages resemble more the extreme pole of total polysemy/ambiguity and thus, total context-dependence in communication; and that from those earlier stages onward the system grows gradually *less ambiguous* and *more richly coded.* It is certainly true of both early child communication and primates communication that they are context-dependent in the extreme,[87] revolving primarily around here-and-now, visible topics and I-and-you. Thus, one may view the evolution of language, both phylogenetically and ontogenetically, as the slow rise of syntax out of discourse, "core meaning" out of contextual meaning, semantics out of pragmatics, and thus—in very much the same sense—deductive logic out of inductive/probabilistic/context-dependent inference. One could, further, view such developments as the rise of *routinized, automated information coding and processing,* in the sense I have defined elsewhere for the rise of syntax out of discourse-pragmatics.[88] When the human organism gives rise to such growing "islands" of context-free, deductive-seeming, semi-automated processing routines, it must perforce increase its reliance on memory, which I believe is the case both ontogenetically and phylogenetically. The rise of context-free information coding and processing must thus be correlated with an increase in the *neurological complexity* of the organism's brain.

Biological sub-systems seldom evolve in a functional/teleological vacuum. The rise of automated information-processing routines in the human organism was to some extent a functional imperative, enabling the organism to carry on increasingly fast and complex decision procedures, action schemata, and complex planning, all based partially on context-free taxonomies of the experiential input. But one must remember that what makes these islands of relative cognitive firmament functional and successful in executing the tasks for which they are suitable, is their fundamental connection and constant interchange with the great submerged iceberg of context-sensitive and open-ended pragmatics.

One could, further, comment on the likely division of labor, within a cognitive system, between context-free automated routines and context-sensitive pragmatic inference. The first are used along established, routine pathways, moving between *major classificatory* nodes of the system, where predictability is high and indeterminacy is low. On the other hand, pragmatic, context-sensitive processing is used along less-established, trail-blazing areas of the system, where predictability is low and insecurity high. Context-free, deductive processing may thus be likened to the *automatic pilot* of an airplane. Thus, between major airports, when problems are routine and well understood, the automatic pilot is switched on. In areas of greater uncertainty, however, where automated decision-making is not possible, *analytic piloting* is a survival imperative.

One may also speculate about what the analytic scanning capacity is doing when the communicative device is on automatic pilot. One possible answer is that it is scanning and analyzing at a higer organizational level—at the *theme* and *story* level as well as at the interactive level with the *hearer*—in order to decide when the switch-back from automatic to analytic processing is required.

---

[87]See discussion in Givón (1979b, Ch. 7).
[88]See discussion in Givón (1979b, Ch. 5).

DISCUSSION

Language, Communication, and Pragmatics vs. Logic

I think an empirical study of language data, without idealizing, pre-sanitizing, or abstracting use, context, and communicative function beyond recognition, points clearly to the fact that 3000 years of deductive bias of whatever stripe can achieve little in either describing or illuminating language and cognition. This is not to say, however, that a careful look at deductive systems, contrasting them with pragmatic ones from the perspective of information/communication theory, cannot yield some interesting insight about the parameters that bound cognition and communication. The major differences between the two systems may be summarized as follows:

		pragmatic processing	*deductive processing*
(a)	*Space:*	continuum	categorial/discrete
(b)	*Context relation:*	context dependent	context free
(c)	*System bounds:*	open-ended, changeable	closed, fixed
(d)	*Mode of inference:*	inductive/probabilistic	deductive
(e)	*Mode of proof:*	preponderance of evidence, open	deductive proof, closed
(f)	*Mode of data-processing:*	analytic	automatic, algorithmic
(g)	*Functional distribution:*	processing at areas of low certainty and high complexity	processing at areas of high certainty and low complexity
(h)	*Speed of processing:*	low speed	high speed
(i)	*Memory/hardware dependence:*	low dependence	high dependence
(j)	*Program/software dependence:*	low dependence	high dependence

The fact that a fundamentally pragmatic-based cognitive system as that of the human organism nevertheless gives rise, repeatedly and in various sub-systems, to islands of relative firmament/deductability, is an important evolutionary and epistemological fact. The hybrid system that is thus created is capable of adapting to two major parameters of environment and survival:

1. *Flexibility, change, and interdeterminancy:* This is a fundamental fact of reality, and only an organism capable of dealing with it could survive in a real universe.

2. *Speed of decision making, planning, and action:* This is a fundamental requirement of competing and surviving in a potentially hostile universe, where time of reaction is of the essence.

There are few if any organisms that do not reveal a measure of routinized behavior. And clearly the human organism in its cognitive evolution has been developing larger areas of deductive, algorithmic processing. Nevertheless, one must never forget that those areas rest upon a vast foundation of pragmatic processing, a foundation that is *not* fossilized but remains *functional*, constantly interacting with the deductive outcrops to which it gives birth, contributing crucially to their modification and re-structuring, and ultimately remaining the silent, analytic, watchful, conscious pilot over the entire organism.

Routinized vs. Analytic Processing in Other Parts of the Organism

When I stumbled upon the distinction between the syntacticized vs. the pragmatic mode of communication (Givón, 1979b), it seemed to me then that a more general phenomenon must be involved, one reflective of general processing of perceptual and cognitive input in higher mammals and probably in live organisms in general. Over the last few years, it has been gratifying to find out how well that initial hunch indeed pans out. The psychological literature is replete with works on the automatization/routinization of perception, cognition, and memory. A vast literature exists concerning the routinization/schematization of motor skills. Similarly, in the neurological literature the discussion of routinized processing and the rise of feedback-free circuits is by now standard.[89] One may of course choose to regard all these phenomena as separate and distinct from what I have tried to suggest holds for the processing of language. But such compartmentalism is bound to be self-defeating for our eventual understanding of both what cognition is now, as well as—most intriguing of all—how it evolved out of so-called "lower" functions of the living organism. Presumably, the same adaptive advantages accrue to the organism from the routinization/automation of behavior at *any* level: increased processing speed, the ability to free the analytic aparatus for other—occasionally *higher*—functions. Presumably again, the same survival advantages would militate for the retention of analytic processing at some—at the very least the *highest*—levels of processing:

---

[89]For routinization and automatic vs. analytic processing in perception and cognition, see Posner and Keele (1970). Posner and Klein (1973), Atkinson and Shiffrin (1968) and Schneider and Shiffrin (1977), *inter alia*. For automatization and routinization of motor behavior, see Keele (1968), Schmidt (1975, 1980), Shapiro and Schmidt (in press), Shapiro, Zernicke, Gregor, and Diestel (1980), or Carson and Wiegand (in press), *inter alia*. For the neurological basis for routinization and automated schemata, see Galambos and Morgan (1960) or Paillard (1960), *inter alia*. For genetically-coded, evolutionarily older neurological routines/reflexes, see Smith (1980), *inter alia*.

the retention of flexibility, the ability to react to feedback and fast-changing feedback, the coping with indeterminacies and imponderables.

Ultimately, I think, we must elaborate not only a theory of human cognition as it is now, but also a theory explaining the rise of human cognition as an *evolutionary*, biological process. What I believe will emerge as part of such a theory is a *hierarchic* view of the evolution of organismic functions, whereby simple, "lower" functions get routinized via repetition and the eventual categorization of tokens of both stimulus and behavior as "belonging to a more general type X." This categorization makes routinized processing of the response/ behavior possible, thus freeing the analytic, feedback-dependent capacity to seek, create, and pursue higher levels of organization. But eventually those higher levels get similarly categorized and eventually routinized, and so on, perhaps potentially ad infinitum.

Categorization into hierarchic token-type systems may thus be viewed as the *prerequisite* for routinization, part and parcel of the same general process of *systems creation*. And routinization of processing is what makes it possible, in turn, to keep increasing the *depth* of the processing system, by freeing the analytic capacity to pursue the next level.

The Rise of Order Out of Chaos: Carnap and Wittgenstein Revisited

To some extent, the rise of deductive, closed, well-ordered, discrete, and algorithmic systems out of the non-discrete, chaotic mire of pragmatics must remain a fundamental mystery of the sentient organism. Philosophical extremists since time immemorial, be they Western epistemologists and latter-day logical atomists or mushy mystics and late-Wittgensteinian cop-outs, have all striven to represent the roots of cognition as one of the exteme poles. If epistemology is ever to become relevant to and compatible with the empirical study of cognition and behavior, it must reject both extremes with equal vigor, both Carnap's fear of coping with the pragmatics of reality in the study of mind as well as in 'hard' science, and Wittgenstein cathartic rejection of the deductive mode in his late years. The central fact that epistemology must eventually contend with, I believe, is the rise of *temporary, illusive* but nevertheless "real" islands of relative firmament and order out of the inherently chaotic universe of experience. Ultimately, I believe, figure-ground pragmatics—the idea that the picture is stable and "real" only as long as the *frame* remains fixed, must play a central role in such an enterprise. The rise of order out of chaos, and thus the rise of consciousness itself, must be ascribed to the evolution of *creative framing and re-framing*, through which portions of experience—those which are *repetitive, recurrent, predictable,* and *routine*—are made into background or "frame," and thus *held constant* just long enough to produce, within the frame, the illusion of a stable, coherent picture.

Knowledge, Self-Knowledge, and Russell's Paradox

There is enough linguistic evidence to support the view that the concept of "knowledge" in language involves *inclusion* of the known by the knower. Another cardinal fact of language and cognition is that the *self* can also be the object of knowledge. And further, that the fact and process of self-knowledge may in turn be the object of knowledge, and so on, *ad infinitum*. Russell attempted to contend with the infinite-meta-regress, which such self-inclusion and self-reference raise, by rigorous prohibitions in his *Theory of Types*,[90] banning the switch from one meta-level to another "within the same system." Such legislative *fiat* may indeed answer to the needs of the logician and his concern with closure and consistency, but it does not abolish the facts of language and cognition. A serious epistemology, it seems to me, must take account of and contend with the indispensibility of the "disallowed" self-inclusion. This is so because one of the most common modes of *re-framing* is indeed that of "self-perception," "self-knowledge," or "self-reference." The perennial problematic nature of the "*I*" to logicians notwithstanding, the process of *objectifying the "I"* has always been an essential component in the adaptive neuro-biological arsenal of consciousness, allowing individuation, planning, and the exertion of will in action and communication. In the struggle for survival, the "I" remains *outside* the frame, manipulating what is inside to its advantage. The very same "I" may, however, choose to include itself *within* the frame, allowing self-reflection, re-contextualizing itself increasingly as *part of The Whole*, subject to the same universals that may apply to other parts, perhaps ultimately subject to Grace itself. The possibility of consciousness exercising such double—or multiple—vision makes equally accessible to use the abyss of *schizophrenia* but also the bliss of *satori*, the ultimate meta-level of them all. The predicament of sentient beings and their ultimate reward are thus seen as emanating from the same fundamental pragmatic source.

Finally, one must note that Russell's injunction against crossing meta-levels has reasonable precedence in the behavior of the cognitive organism itself. Thus, if everything is held at flux at all times, *nothing* will ever be processed. In viewing cognition as a map of the experiential universe, one often observes that certain areas of that universe seem to be held *more or less* constant, relatively speaking. Chief of those are: (a) the self; and (b) the physical world as the largest context for changes. Linguistic/epistemological evidence seems to suggest that on the background of these two "constants," more shifty and less stable facets of experience are then construed as changes, events, actions, or movements. Lacking the *Ultimate* frame, the cognitive organism thus seems to construct a second best one.

---

[90]See Russell (1919, Ch. 13) or Whitehead and Russell (1913, Ch. 12).

ACKNOWLEDGMENT

I am indebted to T. K. Bikson, Tom Bikson, Henning Andersen, Raimo Anttila, Martin Tweedale, John Haiman, John Verhaar, and Erica García for critical comments and much encouragement. Needless to add, the various blunders and overblown claims expressed here remain my own.

REFERENCES

Anderson, H. Abductive and deductive change. *Language*. 1973, *49*.

Anttila, R. *Analogy*. The Hague: Mouton, 1977.

Atkinson, R. C., & Shiffrin, R. M. Human memory: A proposed system and its control processes. In K. W. Spence & J. T. Spence (Eds.), *The psychology of learning and motivation: Advances in research and theory*, New York: Academic Press, 1968.

Austin, J. L. *How to do things with words*. Cambridge, Mass.: Harvard University Press, 1962.

Bloomfield, L. *Language*. New York: Holt, Rinehart & Winston, 1933.

Bolinger, D. Yes-no questions are *not* alternative questions. In H. Hiz (Ed.), *Questions*. Dordrecht: Reidel, 1975.

Bolinger, D. Pronouns in discourse. In T. Givón (Ed.), *Discourse and syntax: Syntax and semantics*. New York: Academic Press, 1979.

Carnap, R. *Meaning and necessity*, Chicago: University of Chicago Press, 1947.

Carson, L., & Wiegand, R. L. Motor schemata for motion and retention in young children: A test of Schmidt's schema theory. *Journal of Motor Behavior*, in press.

Chafe, W., & Nichols, J. Eds.*Evidentiality in language*. Norwood, N.J.: Ablex, in preparation.

Chomsky, N. *Language and mind*. New York: Harcourt, Brace and World, 1968.

Clark, H. Space, time, semantics and the child. In T. E. Moore (Ed.), *Cognitive development and the acquisition of language*, New York: Academic Press, 1973.

Donellan, K. Reference and definite description. *The Philosophical Review*, 1966, *75*.

Duranti, A., & Ochs, E. Left dislocation in Italian conversation. In T. Givón (Ed.), *Discourse and syntax: Syntax and semantics, Vol. 12*. New York: Academic Press, 1979.

Galambos, R., & Morgan, C. T. The neural basis of learning. In J. Field, H. W. Magoun, & V. E. Hall (Eds.), *Handbook of physiology, neurophysiology*, Washington, D.C.: American Physiological Society, 1960.

Garcia, E. *On the role of theory in linguistic analysis: The Spanish pronoun system*. Amsterdam: North Holland, 1975.

Gazdar, G. *Pragmatics: Implicature, presupposition and logical form*. New York: Academic Press, 1979.

Givón, T. Opacity and reference in language: An inquiry into the role of modalities. In J. Kimball (Ed.), *Syntax and semantics*. New York: Academic Press, 1973. (a)

Givón, T. The time-axis phenomenon, *Language*, 1973, *49*. (b)

Givón, T. Focus and the scope of assertion: Some Bantu evidence. *Studies in African Linguistics*, 1975, *6*.

Givón, T. On the VS word-order in Israeli Hebrew: Pragmatics and typological change. In P. Cole (Ed.), *Studies in modern Hebrew syntax and semantics*. Amsterdam: North Holland, 1976. (a)

Givón, T. Topic, pronoun, and grammatical agreement. In C. Li (Ed.), *Subject and topic*. New York: Academic Press, 1976. (b)

Givón, T. The drift from VSO to SVO in Biblical Hebrew: The pragmatics of tense-aspect. In C. Li (Ed.), *Mechanisms For Syntactic Change*, Austin: University of Texas Press, 1977.

Givón, T. Definiteness and referentiality. In J. Greenberg (Ed.), *Universals of human language* (Vol. 4), Stanford: Stanford University Press, 1978.

Givón, T. *Discourse and syntax: Syntax and semantics, Vol. 12.* New York: Academic Press, 1979. (a)

Givón T. *On understanding grammar.* New York: Academic Press, 1979. (b)

Givón, T. *Ute reference grammar,* Ignacio, Colo.: Ute Press, 1980.

Givón, T. On the development of the numeral 'one' as an indefinite marker. *Folia Linguistica Historica,* 1981, *1.* (a)

Givón, T. Typology and functional domains. *Studies in Language,* 1981. (b)

Givón, T. Tense-aspect-modality: The Creole prototype and beyond. In P. Hopper (Ed.), *Tense-aspect: Between semantics and pragmatics,* TSL # 1. Amsterdam: J. Benjamins, 1982.

Givón, T., & Kimenyi, A. Truth, belief and doubt in Kinya-Rwanda. *Studies in African Linguistics.* Supplement, No. 5, 1974.

Gordon, D., & Lakoff, G. Conversational postulates. *Papers from the Seventh Regional Meeting.* Chicago Linguistics Society, University of Chicago, 1971.

Greenberg, J. Some universals of grammar with particular reference to the order of meaningful elements. In J. Greenberg (Ed.), *Universals of language,* Cambridge, Mass.: MIT Press, 1966.

Grice, H. P. Logic and conversation. In P. Cole & J. L. Morgan (Eds.), *Syntax and semantics, Vol. 3.* New York: Academic Press, 1968, 1975.

Hetzron, R. Presentative function and presentative movement. *Studies in African Linguistics,* Supplement No. 2, 1971.

Hill, C. Spatial perception and linguistic encoding A case study in Hausa and English. *Studies in African Linguistics,* Supplement No. 5, 1974.

Hinds, J. *Anaphora in Discourse.* Edmonton: Linguistic Research. 1978.

Hopper, P. Aspect and foregrounding in discourse. In T. Givon (Ed.), *Discourse and syntax: Syntax and semantics, Vol. 12.* New York: Academic Press, 1979.

Hopper, P., & Thompson, S. Transitivity in grammar and in discourse. *Language,* 1980, *56.*

Horn, L. *Presupposition in language.* Doctoral Dissertation, UCLA, 1972.

Hyman, L. The change from SOV to SVO: Evidence from Niger-Congo, In C. Li. (Ed.), *Word Order and Word Order Changes.* Austin, University of Texas Press, 1975.

Karttunen, L. The logic of English predicate complement constructions. *Language.* 1971, (a)

Karttunen, L. Some observations on factivity. *Papers in Linguistics.* 1971, *4.* (b)

Karttunen, L. Presupposition and linguistic context. *Theoretical Linguistics.* 1974, *1.*

Karttunen, L., & Peters, S. Conventional implicatures. In C.-K. Oh, & D. Dineen (Eds.), *Syntax and semantics,* vol. 11: *Presupposition.* New York: Academic Press, 1979.

Katz, J. J., & Fodor, J. A. The structure of a semantic theory. *Language,* 1963.

Keele, S. W. Movement control in skilled motor performance. *Psychological Bulletin,* 1968, *70.*

Keenan, E. L. *A logical base for a transformational grammar of english.* Doctoral Dissertation, University of Pennsylvania, 1969.

Keenan, E. L. Two kinds of presuppositions in natural language. In C. Fillmore & T. Langendoen (Eds.), *Studies in linguistic semantics.* New York: Holt, Rinehart & Winston, 1971.

Keenan, E. L. Toward a universal definition of subject. In C. Li (Ed.), *Subject and topic.* New York: Academic Press, 1976.

Keenan, E. L., & Ebert, K. A note on marking transparency and opacity. *Linguistic Inquiry,* 1973, *4.*

Keᵉnan, E. O., & Schieffelin, B. *Foregrounding referents: A consideration of left dislocation in discourse.* Los Angeles: University of Southern California, 1977 unpublished manuscript.

Kiparsky, P. Historical linguistics. In W. O. Dingwall, *A survey of linguistic science.* College Park: University of Maryland, 1971.

Kiparsky, P., & Kiparsky. C. Fact. In M. Bierwisch & K. E. Heidolph (Eds.), *Progress in linguistics.* The Hague: Mouton, 1968.

Kirsner, R. Deixis in discourse: An exploratory quantitative study of the modern Dutch demonstrative adjectives. In T. Givón (Ed.), *Discourse and syntax: Syntax and semantics, Vol. 12*. New York: Academic Press, 1979.

Kunene, E. C. Zulu pronouns and the structure of discourse. *Studies in African Linguistics*, 1975, 6.

Li, C., & Thompson, S. Third person pronouns and zero anaphora in Chinese discourse. In T. Givón (Ed.), *Discourse and syntax: Syntax and semantics, Vol. 12*. New York: Academic Press, 1979.

Lightfoot, D. Syntactic change and the autonomy thesis. Paper read at the *Conference on Syntactic Change*, University of California, Santa Barbara unpublished manuscript. May 1976.

Lightfoot, D. The base component as a locus of syntactic change. In W. Christie (Ed.), *Proceedings of the 2nd International Conference on Historical Linguistics*. Amsterdam: North Holland.

Linde, C. Focus of attention and the choice of pronouns in discourse. In T. Givón (Ed.), *Discourse and syntax: Syntax and semantics, Vol. 12*. New York: Academic Press, 1979.

Mahmoud, U. *Process of language change in the southern Sudan*. Doctoral dissertation, Georgetown University, 1979.

Marchand, H. *The categories and types of present-day English word formation*, University of Alabama Press, 1964.

Matthiessen, P. *The Snow leopard*. New York: Viking, 1978.

Nakau, M. *Sentential complements in Japanese*. Tokyo: Kaitakusha, 1973.

Noonan, M. *Switch-reference in Lango*. State University of New York, Buffalo, unpublished manuscript.

Ochs, E. Planned and unplanned discourse. In T. Givon (Ed.), *Discourse and syntax: Syntax and semantics, Vol. 12*. New York: Academic Press, 1979.

Paillard, J. The patterning of skilled movement. In J. Field, H. W. Magoun, & V. E. Hall (Eds.), *Handbook of physiology*, vol. 3, section 1, *Neurophysiology*, Washington, D.C.: American Physiological Society, 1960.

Peirce, C. S. *Philosophical writings* Edited by J. Buchler, New York: Dover, 1955.

Posner, M. I. Attention and cognitive control. In R. L. Solso (Ed.), *Information processing and cognition: The Loyola symposium*. Hillsdale, N.J.: Lawrence Erlbaum Associates, 1974.

Posner, M. I., & Keele, S. W. Retention of abstract ideas. *Journal of Experimental Psychology*, 1970, 83.

Posner, M. I., & Klein, R. M. On the function of consciousness. In S. Kornblum (Ed.), *Attention and Performance IV*. New York: Academic Press, 1973.

Quine, W. van O. *From a logical point of view*. Cambridge, Mass.: Harvard University Press, 1953.

Ross, J. R. On declarative sentences. In R. A. Jacobs & P. S. Rosenbaum (Eds.), *Readings in English transformational grammar*. Watham, Mass.: Ginn and Co., 1970.

Russell, B. *Introduction to mathematical philosophy*. London: Allen and Unwin, 1919.

Schmidt, R. A. A schema theory of discrete motor-skill learning. *Psychological Review*, 1975, 82.

Schmidt, R. A. Past and future issues in motor programming. *Research Quarterly*, 1980, 51.

Schneider, W., & Shiffrin, R. M. Controlled and automatic human information processing, I: Detection, search and attention. *Psychological Review*, 1977, 84.

Shapiro, D. C., & Schmidt, R. A. The schema theory: Recent evidence and developmental implications. In J. A. S. Kelso' J. E. Clark (Eds.), *The development of movement control and coordination*. New York: Wiley, in press.

Shapiro, D. C., Zernicke, R. F., Gregor, R. J., & Diestel, J. D. Evidence for generalized motor programs. *Journal of Motor Behavior*, 1980.

Silva-Corvalán, C. *A discourse study of some aspects of word-order in the Spanish spoken by Mexican-Americans in West Los Angeles*. Masters thesis, UCLA, 1977.

Slobin, D. Language change in childhood and history. In J. Macnamara (Ed.), *Language learning and thought*. New York: Academic Press, 1977.

Smith, J. L. Programming of stereotyped limb movement by spinal generators. In G. E. Stellmach & J. Requin (Eds.), *Tutorials in motor behavior*. Amsterdam: North Holland, 1980.

Strawson, P. F. On referring. *Mind*, 1950, *59*.

Traugott, E. C. *Spatial expressions of tense and temporal sequencing: A contribution to the study of semantic fields*. Unpublished manuscript, Stanford: Stanford University, 1974.

Whitehead, A. N., & Russell, B. *Principia Mathematica*. Cambridge, England: Cambridge University Press, 1913.

Whorf, B. L. *Language, thought, and reality*. J. B. Carroll (Ed.), Cambridge, Mass.: MIT Press, 1956.

Wittgenstein, L. *Tractatus logico philosophicus*. Trans. by D. F. Pears & B. F. McGuinness, New York: The Humanities Press, 1918.

Wittgenstein, L. *Philosophical investigations*. Trans. by G. E. M. Anscomb, New York: MacMillan Co., 1953.

7 Psychological Constraints on Language: A Commentary on Bresnan and Kaplan and on Givón

Herbert H. Clark
Barbara C. Malt
Stanford University

Ever since Wundt's debates with the neogrammarians late in the nineteenth century, psychologists and linguists each have been interested in what the other has had to say about language. In Wundt's day, the study of language was a unified field to which both psychologists and linguists contributed. But as the two fields became more specialized, the study of language structure got cut off from the study of language use. Language structure was considered the province mainly of linguists, and language use the province mainly of psychologists.

Despite the split, language structure was still assumed to bear some relation to language use. The best known claim on this issue was Chomsky's (1965): "A reasonable model of language use will incorporate, as a basic component, the generative grammar that expresses the speaker-hearer's knowledge of that language [p. 9]." But for psychologists, it was difficult to see how the grammar Chomsky favored at the time could be incorporated into a theory of language use. There had been several attempts to do this, and all were failures. This was probably why Chomsky cautioned in his very next sentence: "But this generative grammar does not, in itself, prescribe the character or functioning of a perceptual model or a model of speech-production." This was an odd caveat. If the grammar is something that a person "puts to use in producing and understanding speech (Chomsky, 1970)," then shouldn't it partly prescribe, in the sense of constrain, the character or function of speech production and understanding?

The problem is clear. Linguists had generally come to study language structure with little regard for language use. Generative grammars were being devised without considering the role they would play in the models of speech production and understanding. This had to be short-sighted. It was as if one division of

General Motors were designing automobile engines without consulting the division designing the chassis into which these engines would be installed.

With the chapters by Bresnan and Kaplan, and by Givón, we have excellent examples of linguists bucking this tradition. More than most linguists, they are attempting to study language structure in its relation to language use—although their approaches are very different. Bresnan and Kaplan, like many of their predecessors, want to devise grammars of languages such as English. They differ from most predecessors in wanting the grammars to fit directly into models of speaking and listening. Givon, in contrast, wants to show how the functions to which language is put explain the form that various languages have evolved into.

What is common to Bresnan and Kaplan, and Givón, is the idea that the form and function of language are subject to cognitive, or psychological constraints. One of their aims is to describe these constraints and their consequences for language structure. For many cognitive psychologists, the constraints they offer may not look very psychological. They are sometimes couched in abstract formalisms in which the psychological content is obscure; they are generally not based on studies of psychological processes; they are sometimes based on psychological notions that are more speculative than proven. For these and other reasons, many psychologists may be wary of parts of Bresnan and Kaplan's and Givón's enterprises. Should they be?

The basic issue, we believe, is what constitues a "psychological" constraint. Psychologists and linguists expect very different things of such constraints. To bring out the differences, we first present a strong psychological view of cognitive constraints on language: we describe several constraints most psychologists would accept as psychological, see what makes them psychological, and apply these criteria to constraints proposed by Bresnan and Kaplan, and by Givón. Later, we describe what linguists would be more likely to consider a psychological constraint and see how these two views might be reconciled.

FOUR POSSIBLE PSYCHOLOGICAL CONSTRAINTS

In the past few years, there have been many proposals by psychologists and other non-linguists for how language is constrained by psychological factors. The constraints entertained range from perceptual to social constraints, as in this classification offered by Clark and Clark (1977):

> A priori, every human language must be susceptible of: (1) being learned by children; (2) being spoken and understood by adults easily and efficiently; (3) embodying the ideas people normally want to convey; and (4) functioning as a communication system in a social and cultural setting [pp. 516–517].

The same four types turn up in different guises in Slobin's (1979) "ground rules to which a communicative system must adhere if it is to function as a full-fledged human language. [p. 188].'' Let us call the four categories *learnability, processibility, expressibility,* and *social utility.*

Most of the constraints in these categories would strike psychologists as clearly "cognitive" or "psychological" in character. To see why, we will take up four proposed constraints, one for each category. The four examples, admittedly, aren't equally convincing or well founded, but our purpose is to see what they have in common.

Learnability: Regular paradigms

According to Slobin (1973), children find it difficult to learn morphological paradigms that aren't regular—that don't adhere to regular rules. In English, for example, the paradigm for possessive constructions is highly regular: To form the possessive of a noun phrase, add *-s* (in its various phonological realizations) to the noun phrase. There are a few exceptions to this rule, such as *my, your, his,* and *their,* but they are few. The paradigm for the past tense of verbs, in contrast, is highly irregular. The general rule is this: To form the past tense of a verb, add *-ed* (in its various phonological realizations) to the present tense form. But for this rule, there are many exceptions—the so-called strong verbs such as *broke, ate, had,* and *rang.* As evidence for his generalization, Slobin noted that English children make systematic mistakes in acquiring past tense verb forms, saying *breaked* for *broken,* and *ringed* for *rang.* He also noted that Turkish inflectional morphology, which is highly regular, is acquired very early compared to English inflectional morphology, which is much less regular. There is other evidence as well.

If regular paradigms are easier to acquire than irregular paradigms, we should find them in abundance in languages, and we do. And all other things being equal, we should find systematic changes in languages toward regular paradigms, and there is considerable evidence on this score too. As reviewed by Bybee Hooper (1979), languages tend to extend regular endings (so-called "analogical extension") and to get rid of alternative endings (so-called "analogical leveling"). In English, for example, forms such as *wrought, dreamt, spelt,* and *shone* have been or are being replaced by their regular counterparts *worked, dreamed, spelled,* and *shined.* (As we note later, there are independent forces that lead to irregular paradigms.) Here, then, is a well documented feature of language learning that has been argued to constrain the form an evolved language can have by constraining the direction in which languages can evolve.

Processibility: Constituent structure

Words and their meanings live only briefly in listeners' short term memories. Not only do the phonetic shapes of words die out very quickly, but so do the several "automatically" available senses of ambiguous words, which disappear within tenths of seconds (see, e.g., Swinney, 1979). If two words need their meanings to be accessed at the same time to be interpreted correctly, they should be processed more efficiently the closer they are in the sentence, and most efficiently when they are adjacent members of the same constituent. The words that most depend on each other this way are those that have the same referents. In the noun phrase *the small elephant,* for example, the interpretation of *small*—a relative adjective—depends on the noun being modified, here *elephant,* and so does the speed of that interpretation (Rips & Turnbull, 1980). Then, all other things being equal, understanding should be optimal when words that are mutually dependent in this way belong to the same constituent.

Very generally in languages, words mutually dependent in this way *are* likely to belong to the same constituent, as in the English noun phrase *the small elephant.* This tendency has long been known in what Vennemann (1974) has called Behaghel's First Law: What belongs together mentally is placed close together syntactically (see also Moravcsik, 1971). As a specific example, take the relations between classifiers (C), quantifiers (Q), and nouns (N), as in the construction *two head of cattle,* which has the constituent structure ((Q + C) + N). As Greenberg (1972, 1975) has argued, the quantifier is semantically dependent on the classifier and not on the noun; hence the quantifier ought to form a constituent with the classifier and not the noun. In many languages, such classifiers are required in every noun phrase that contains a quantifier. In an extensive survey of these languages, Greenberg found no examples of C + N + Q, or Q + N + C, in which C + Q don't form a constituent. All he found were the other four possible orders of C, N, and Q, in which C + Q formed an independent constituent. So it can be argued that the processing needs of the language user constrain the way words get grouped into constituents in all or almost all languages.

Expressibility: Color terms

The components of the human eye are each more sensitive to some colors than others (see Kay & McDaniel, 1978; Miller & Johnson-Laird, 1976). Very roughly, one part of the visual system is geared to represent grays—from black to white—and another part to represent hues. The visual system is particularly sensitive to the four primary hues—red, green, blue, and yellow. If speakers of a language develop words for denoting brightness, the argument goes, they should first code black and white, the two ends of the grayness continuum. If they develop words for hues, they should find it easiest to add red, green, blue, and yellow. And if they develop still more words for hues, they should add names for certain in-between hues, such as brown, pink, grey, purple, and orange.

As Berlin and Kay (1969) demonstrated—with later emendations—languages of the world acquire their basic color vocabulary in a highly regular order. All languages have at least *black* and *white,* or *light* and *dark.* Languages with only three basic color terms also have *red,* and the next three terms to be added are *green, blue,* and *yellow.* The last terms to be acquired are *brown, pink, purple, orange,* and *grey.* The evolution of color terminology, then, follows the psychological constraints put on it by human visual perception.

Social utility: Politeness

In his study on face-to-face interaction in public, Goffman (1967) noted that two people in social contacts try to maintain face—Goffman's notion of the public self-image that people claim for themselves. People try to maintain their freedom of action, their freedom from imposition; they also try to maintain their positive self-image—their desirability to at least some other people. Brown and Levinson (1978) argued that if maintaining face is universal, then there is a set of strategies every language should exploit in polite expressions. For example, every language should have evolved indirect means for making requests. In requests, the speaker is, by definition, trying to impose on the addressee, and this can be threatening to the addressee's face, or self-image. To be polite, the speaker needs to reduce the threat and hedge his request—for example, by giving the addressee the option of not complying.

According to Brown and Levinson's survey of three unrelated languages—English, Tzeltal, and Tamil—this sort of device is universally used to enhance politeness in requests. In English, it is more polite to ask *Can you open the door?,* which gives or appears to give the addressee the option of not complying, than to say *Open the door,* which doesn't give him that option. These two examples translate almost directly into Tzeltal and Tamil, with the identical consequences. There are parallel realizations in the three languages of many other face-saving and face-preserving strategies too. Thus, a psychological process is argued to lead to a universal class of devices for requesting favors and for other face-threatening speech acts.

WHAT IS A PSYCHOLOGICAL CONSTRAINT?

These examples, whatever their status, illustrate the strong position psychologists could take toward "psychological" or "cognitive" constraints on language. They bring out features they might see as desirable, even necessary, for a constraint to be considered truly psychological. We will draw out four such features under the labels empirical grounding, structure independence, theoretical coherence, and linkage.

Empirical Grounding

A psychological constraint ought to be grounded in empirical findings. In support of the claim that regular paradigms are easier to learn than irregular paradigms, Slobin referred to a plethora of evidence. For the primacy of black, white, red, blue, green, and yellow in the visual system, Kay and McDaniel, and Miller and Johnson-Laird, pointed to an even broader and more substantial range of evidence. The evidence for the fast decay of phonetic shapes and meanings in short-term memory is quite firm, with new evidence appearing yearly, and the evidence for the psychological process of maintaining face, though of quite a different kind, is substantial too.

All this evidence has been verified for a variety of cultures and languages—or can be assumed to hold across language groups. Slobin's evidence is explicitly taken from many language groups. The psychophysical evidence on colors and the properties of short-term memory shouldn't depend on the language a person speaks. Maintaining face, like other anthropological features, may not be universal, but the claim is only that in those cultures in which it occurs, certain features should appear in the language. (Note that if there were a race of colorblind people, we wouldn't expect the same color names either.) The point is that if we wish to maintain that something is a universal constraint on language, we must show that it holds cross-linguistically and cross-culturally.

Structure Independence

All four constraints we have reviewed were grounded on evidence derived not from facts about the structure of language, but from facts about processes in language use. The evidence was, as we shall put it, *structure independent:* it was independent of language structure per se. The constaint on color terminology was derived from physiological and psychophysical evidence that has nothing to do with language at all, and the constraint on maintaining face came from observations of people in face-to-face situations, with and without the use of language. The constraints on regular paradigms and short-term memory take a little more explanation.

In the evidence on learning regular paradigms, Slobin didn't appeal to facts about language structure per se; for example, he didn't appeal to the presence or absence of regular paradigms in various languages of the world. Rather, he appealed to facts about how children learned paradigms with and without exceptions—how often they made errors in learning various paradigms. He could even have appealed to experimental analogues of learning regular and irregular paradigms of nonsense words, which show results parallel to the findings on children (Palermo & Eberhart, 1968; Palermo & Howe, 1970; Palermo & Parrish, 1971).

The constraints from short-term memory too are based on structure independent sources. The evidence that the phonological shapes of words in short-term memory are transient comes from studies on genuine utterances, on isolated words, on nonsense phonological strings, and on digits. It doesn't come from evidence about what constitutes a sentence or other language structures. The availability of multiple word senses also comes from studies on utterances, on isolated words, independent of language structure. Even though this evidence pertains to language, it is structure independent if it is evidence of psychological processes behind speaking, listening, and acquisition.

The point of this criterion is to keep psychological constraints from becoming vacuous. Imagine a linguist—let us call her the autonomous linguist—who is not in the least interested in explaining language structure by way of psychological constraints. She is interested only in getting as general a description of language structure as possible by referring to other facts about language structure. In deciding how to write a rule in English, she might note that Finnish, Japanese, and Dyirbal all require a certain form of rule, hence the English rule would be most general if it were in the same form. She would be appealing here to facts about language structure as an "explanation" for her particular formulations of the English rule.

But as evidence for a *strong* psychological constraint (in the sense we have characterized it), this reasoning is circular. A feature found in all languages is prima facie evidence that there *may* be a psychological constraint leading to that feature, but the feature itself doesn't constitute the constraint. To claim it does would be to fall prey to the fallacy *post hoc, ergo propter hoc*. Suppose the autonomous linguist discovered Berlin and Kay's facts about the universality of *black, white,* and the other color terms in languages of the world. It would be circular for her to posit a psychological constraint expressed this way: "All languages must have color terms for black and white, and the next four terms acquired must be ones for red, green, blue, and yellow." This is surely just a redescription of the data. As Givón noted, the autonomous linguist's constraints are constraints on the *description* of language structure, not on its explanation. By this criterion, a good number of what passes for psychological constraints in the linguistic literature go by the board.

Theoretical Coherence

Most psychological constraints don't lead to single isolated features of language—like piano keys that each produce just one note. Constraints such as short-term memory, and color vision, generally have a host of consequences—like organ keys that trigger rank upon rank of pipes. One must show how each constraint fits within a coherent theory of psychological constraints.

Consider Brown and Levinson's theory for the expression of politeness in language. The constraint they posited wasn't intended to account just for indirect requests. It had many different consequences—in the form of different politeness "strategies": presuppose common ground, joke, use in-group identity markers, exaggerate interest in the addressee, give reasons, seek agreement, hedge, be pessimistic, show deference, apologize, and so on. Or consider color terminology. In their arguments, Kay and McDaniel (1978) could have restricted themselves to the primacy of black, white, red, blue, green, and yellow in the visual system, but they recognized that their account ought also to fit into a broader account of vocabulary acquisition. Reasoning this way, Brown (1977, 1979) has appealed to related principles in accounting for the order in which languages acquire the botanical terms for tree, shrub, grass, herb, and vine and the zoological terms for bird, snake, fish, bug, worm, and mammal. Rosch, Mervis, Gray, Johnson, and Boyes-Braem (1976) have expanded on the criteria that appear to apply in determining which categories of objects get "basic" names and which do not. So the psychological constraints on politeness and on color terminology aren't isolated constraints. They each belong to systems intended to explain a range of linguistic features.

Linkage

For a psychological process to explain a feature of language, it must be accompanied by a theory about how it is *linked* to the language feature. It isn't enough to say that languages have regular paradigms because children learn them more easily than irregular ones. One must specify how children's learning brings this about. Providing such a link isn't easy, and it is the part of the argument most often left unspecified. The linkage between constraint and language feature almost always requires a complex model about why languages have the features they do.

What, for example, *is* the linkage between children's learning and regular paradigms in language? Bybee Hooper (1979), Bybee and Slobin (1982), and Slobin (1977) have proposed a theory in which children's (and adults') over-regularizations get incorporated into languages through a process of language change. In support of their argument, they have appealed to evidence of how languages change in the directions predicted by their theory. Slobin has also appealed to evidence on the creation of pidgins and creoles. Most linkage theories will have this character. They must explain how a psychological constraint molds the creation of languages, as in the invention of pidgins, and how it shapes the evolution of languages both in ordinary language change and in the creolization of pidgins.

The linkage problem is complicated by the fact that language is influenced by many diverse, often conflicting, psychological constraints operating at once. If the push toward regular paradigms from children were the only influence on paradigms in languages, after so many years of evolution there should be only highly regular paradigms. But as Vennemann (e.g., 1973) has argued, another force on paradigms is phonological reduction. Over long periods of time, suffixes tend to get reduced phonologically and even disappear. This is the way English has lost its case and gender systems, and Dutch is in the process of losing them now. When this happens, languages also tend to find other ways of expressing what was lost. Suppose English eventually loses its plural suffix -s, as is already happening in certain non-standard dialects. Then English will have no way of marking singular versus plural, and if it is like other languages, it will evolve a motley collection of new devices to do the work of the old. In this way, English will have evolved a highly *irregular* paradigm for the plural, but because of a *different* set of psychological constraints—those that lead to phonological reduction and to semantic contrast. Vennemann has argued that these forces of language change are cyclic. So for the linkage theory to be complete, it must specify how the relevant constraints interact.

To summarize briefly, we suggest that a strong psychological constraint on language has four properties. It is grounded on firm psychological evidence. The evidence it is grounded in is independent of language structure. It belongs to a coherent theory that leads to a body of related predictions. And its linkage to language structure is explicit and theoretically sound. These four criteria, as ideals, will rarely be met in practice. Yet they provide a standard for examining any psychological constraint that is proposed.

BRESNAN AND KAPLAN ON PSYCHOLOGICAL CONSTRAINTS

In their chapter, Bresnan and Kaplan take up the issue of psychological reality in grammars: What is the relation of grammars to theories of speaking, understanding, and acquisition? Briefly, they argue that the grammar of a language should represent the linguistic knowledge people use in speaking and understanding it. Certain forms of grammar, such as transformational grammar, cannot in principle represent that knowledge. Indeed, if one looks at language processes, one can posit certain "theoretical" constraints that grammars must adhere to. So the relation between linguistic knowledge, as represented in grammars, and language processes, as represented in theories of speaking and understanding, is reciprocal: each constrains the form the other can take.

Bresnan and Kaplan's arguments are inventive and clear, and there is much in them we can only agree with. Yet as commentators, we will take a harder look at some of the assumptions behind the arguments, especially behind what they call "theoretical" constraints. In the original version of the chapter—the version for which we first prepared comments—these were called "cognitive" constraints, which is what piqued our interest. Despite the change in name, the issue remains the same: What is the grounds for positing these constraints?

The Competence Hypothesis

The superordinate constraint behind Bresnan and Kaplan's enterprise is, as they say, Chomsky's *competence hypothesis:* "A reasonable model of language use will incorporate, as a basic component, the generative grammar that expresses the speaker-hearer's knowledge of the language (Chomsky, 1965, p. 9)." This hypothesis can be taken in many ways, and they rightly argue against Chomsky's position that, as they say, "psychological reality [of grammars] is whatever linguistic theory is about." They go on to argue, "[The competence hypothesis] requires that we take responsibility not only for characterizing the abstract structure of the linguistic knowledge domain, but also for explaining how the formal properties of our proposed linguistic representations are related to the nature of the cognitive processes that derive and interpret them in actual language use and acquisition."

The competence hypothesis is clearly a psychological constraint. If grammars must be incorporatable into models of language use, then only certain forms of grammars, and languages, are possible, because people can only process certain types of utterances. Part of Bresnan and Kaplan's arguments against transformational grammars is based on this logic. But if the hypothesis expresses a psychological constraint, how does it fare against the four standards we listed? The problem is that it has rarely been held up to standards like these, because most work on grammars is done by linguists whose interests lie elsewhere. Still, Bresnan and Kaplan are explicitly concerned with it as a constraint on language, so it is appropriate to ask the question anyway.

The competence hypothesis is susceptible to test both as a whole and in part. We will take up just one part—central to the hypothesis—as expressed in the *single representation* assumption: the representation of grammatical knowledge is the same, or isomorphic, in speaking and understanding. Of course, "it is uncontroversial that stored knowledge structures underlie all forms of verbal behavior," as Bresnan and Kaplan say. But it isn't so uncontroversial that the *same* stored knowledge structure underlies all these forms.

There are a number of facts people take for granted that suggest quite the reverse. A native Californian can understand a range of accents of spoken English—Australian, Indian, Scottish, the American south—yet not have the slightest competence for producing them. He can understand syntactic forms in these dialects, as well as in Shakespeare, Joyce, and even Bellow, over which he has no productive control. He can understand a large number of words—much of his recognition vocabulary—that he couldn't use himself. His deficiencies in production lie in syntax, vocabulary, morphology, phonology, and semantics, suggesting that at all levels of language structure, the process of listening has access to more "knowledge" than does the process of speaking. There are also systematic differences between what children can understand and what they can produce (E. Clark & Hecht, 1983). As with adults, the ability to comprehend particular constructions precedes the productive control over the same constructions.

These observations can be viewed several ways. One is to suppose there is a body of linguistic knowledge people can access in understanding but not in speaking. This premise allows one to preserve the single-representation assumption. A more radical view is to suppose that comprehension and production access distinct representations of linguistic knowledge, even though, in normal people, the two representations code much the same information and are closely coordinated: people use their comprehension system to monitor and adjust what they produce, bringing production into line with comprehension. Under this view, the single-representation assumption is incorrect.

As evidence for the more radical view, consider phonology. The processes of hearing speech sounds—all the acoustic, phonetic, and phonological processes that investigators of speech perception have learned so much about—bear little resemblance at any level of abstraction to the processes of sound production—planning phonetic sequences, creating articulatory programs, and executing these programs. The first involves the ear and theories of auditory perception, and the second, the mouth and tongue and theories of motor movements. The two processes appear to involve distinct parts of the cortex as well. All that theories of phonetic perception and phonetic production need have in common is that the phonemes identified in perception, when veridical, are the same phonemes the speaker intended to articulate. Even the intention to produce a phoneme, and the recognition of that intention, need not make reference to the same representation, as long as they are coordinated in some way. In any case, the language representations that the two processes make reference to in realizing and recognizing these intentions don't need to look alike.

As an analogy, consider a television set that produces images and an eye that interprets them. The representation to which the television "refers" in its complicated chain of electrical activity may not look at all like the representation that people make reference to in recognizing the objects being imaged. Yet if we erected a window between the television set and the viewer, we could describe at many levels of abstraction the patterns of light passing through the window—according to the intuitions of the viewer. In a way, this is what grammar writers do. They describe, in a relatively neutral representation, the sound patterns that emerge from the mouth and strike the ear and how these are correlated with the speaker's meanings. That representation may not describe either what is in the producer of those sounds or in the perceiver. Indeed, they cannot describe both if the "knowledge" exploited in speaking is not the same in all respects as that exploited in listening. We can't assume that just because the production and comprehension systems coexist in the same brain, or are under the control of the same mind, they share a single representation system. The motor and perceptual systems in the brain are distinct, so why shouldn't their linguistic components—the production and comprehension systems—be distinct too.

So much for the empirical grounding of the competence hypothesis. How does it fare against the second, third, and fourth of our criteria? Any empirical grounding we could think of would be structure independent. Until there is an explicit model of how linguistic knowledge is put to use in models of speaking and understanding, it is impossible to ask about theoretical coherence. Probably the most challenging criterion is linkage, and on this, there has been some work. Several investigators have studied whether transformational grammar (Wexler & Cullicover, 1980) and Bresnan and Kaplan's lexical functional grammar (Pinker, 1980) are in principle learnable by children, and this is a first step in linking the competence hypothesis to the possible forms a language can take. Yet these investigations haven't been concerned with specific models of speaking and understanding, nor with theories of language change or language creation (in pidgins and creoles).

The fate of the competence hypothesis lies ultimately in the models of speaking and understanding that achieve practical success. The way most models are being developed today, they will probably *not* make reference to identical or isomorphic representations of linguistic knowledge, even though the form they take will be constrained by the grammar. If this happens, the competence hypothesis will be robbed of its central assumption.

Creativity and Finite Capacity

Creativity and finite capacity are two constraints that lead to important decisions about how to write grammars. Creativity specifies that the grammar must be capable, in principle, of producing an infinite number of grammatical strings. Finite capacity has two parts. Part 1 specifies that words and syntactic relations must be finite, and Part 2 specifies that people's *mental* capacity for storing knowledge must be finite. As a result of creativity and finite capacity, Bresnan and Kaplan argue, "the mapping [from representations to sentences] must consist of the recursive composition of finitely many operations that can project a finite store of knowledge of a particular language onto infinite sets of data [Bresnan & Kaplan, p. 123]."

As strong psychological constraints, these two constraints are circular. Creativity is really an observation about language structure—that there seems to be no principled limit on the amount of recursion possible in languages of the world. So is Part 1 of finite capacity, which expresses the observation that all languages consist of a finite vocabulary put together by a finite set of syntactic operations. A grammar would need to reflect both creativity and finite capacity Part 1 just to be an adequate description of a language. So creativity and finite capacity Part 1 fail on the criterion of structure independence.

Part 2 of finite capacity, that people's mental capacity for storing knowledge is finite, is clearly a psychological constraint with much empirical support. But its linkage with linguistic phenomena is problematic. Bresnan and Kaplan assume that if a language is creative and uses finite means, it must be recursive. This clearly needn't be so. In American sign language, signers could in principle exploit the ability to mimic, through infinitely gradable gestures, any movement they wished to denote. In English, too, we could use vowel and fricative length in an infinitely graded way to represent, say, the physical extent of some object. With such analogue devices, signers and speakers could get infinite expressibility from finite means without the use of recursion. The point is that, in principle, a language could fit the creativity and finite capacity constraints and yet not have recursion.

The way Bresnan and Kaplan use finite capacity is to constrain the sets of elementary words and relations to be finite, not the possible length of sentences or depth of recursion. Bresnan and Kaplan could have claimed instead that the set of elementary words was infinitely expandable and used finite mental capacity to constrain sentence length and depth of recursion. Applying finiteness of mental capacity to elementary words and grammatical relations and not to sentence length or depth of recursion was a strategic decision made to fit the observed recursion in language. The decision is the right one, but it is based on adequacy of description—a structure dependent criterion—not on the constraints per se.

Reliability

According to this constraint, "the syntactic mapping can thus be thought of as reliably computing whether or not any string is a well-formed sentence of a natural language." As evidence, Bresnan and Kaplan suggest that "independently of knowledge of specific context, even independently of meaningfulness, speakers can reliably classify sentences as grammatical or ungrammatical," suggesting that "classification of strings as grammatical or ungrammatical is based on an automatic procedure."

The alternative, as Bresnan and Kaplan point out, is that language users use so-called heuristic strategies—strategies that do not constitute an "effectively computable characteristic function." It has long been argued, of course, that listeners use such strategies (see Clark & Clark, 1977; Fodor, Bever, & Garrett, 1974, for reviews), and computational systems have been implemented that understand almost entirely on the basis of non-syntactic information (see Birnbaum & Selfridge, 1981; Reisbeck & Schank, 1978). Even if listeners use non-heuristic procedures, they also use heuristic procedures. In some cases, contrary to Bresnan and Kaplan's claim, they even seem incapable of accessing the appropriate recursive procedure. How else are we to explain why many of Wason and Reich's (1979) informants could never see what was wrong with *No head injury is too trivial to ignore?* Or why many of Gleitman and Gleitman's (1970) informants consistently misinterpreted noun compounds like *bird-house boot?*

Bresnan and Kaplan, of course, don't deny people use heuristic procedures. But to maintain the reliability constraint, they must demonstrate that there exists *no* syntactic construction that is consistently interpreted by means of such procedures. Given how widespread, useful, and powerful such procedures are, that proposition seems implausible.

Order-free Composition

With order-free composition, the grammatical relations derivable from an arbitrary fragment of a sentence, like *not told that,* must be included in the grammatical relations derivable from the entire string, like *I was not told that she was coming.* When people are given *not told that,* they can compute all grammatical relations the fragment could ever have as part of a complete sentence. To figure out the possible relations, they don't need the complete sentence at once.

As empirical grounding for the constraint, Bresnan and Kaplan argue that "complete representations of local grammatical relations are effortlessly, fluently, and reliably constructed for arbitrary segments of sentences." Although this holds for many segments, Bresnan and Kaplan provide their own counterexample with *to by for,* which, as they note, could be a fragment of *The one that he should be spoken to by for God's sake is his supervisor.* It seems unlikely that people could compute, or conceive of, all the possible relations derivable from this segment. If we allow for unlimited embeddings, the possibilities are indeed infinite. It might be countered that people can't compute them because of "performance limitations"—that is, because they are limited by *other* psychological constraints. If so, the empirical grounding for order-free composition is incomplete: Some local grammatical relations are computed "effortlessly, fluently, and reliably," and others are not. Without a model of how the process is otherwise constrained, we can't tell whether the data support, or disconfirm, order-free composition.

Bresnan and Kaplan's real motivation in proposing the constraint appears to be computational and, therefore, not structure independent. It would be convenient for writing grammars if, as Bresnan and Kaplan argue, "sentential context may determine the *choice* of one of a set of locally computed grammatical relations for a segment, but the computation of grammatical relations for a segment may not involve the computation of the grammatical relations of the context." Like recursion, languages appear to allow this possibility in principle, so it is reasonable to require order-free composition of a grammar. That makes order-free composition a constraint on possible *descriptions* of languages, not a psychological constraint on the form languages can take.

So as desirable as order-free composition might be for writing grammars, it doesn't seem to be a *psychological* constraint by the strong standards set up earlier. It doesn't seem to express a mental capacity, or ability, or bit of competence so much as it characterizes a property of languages.

Universality

In the universality constraint, the procedure for grammatical interpretation is assumed to be the same for all natural language grammars. There is assumed to be a "universal effective procedure" for constructing mental representations for sentences. The idea is that there is a universal mental representation for natural languages that is induced by a universal learning function. It is also plausible, then, that the mapping induced by that learning function is also universal.

This constraint is motivated by psychological concerns. The autonomous linguist could, if she wished, design very different grammars for each language, proposing, for example, one kind of grammar for mainly free-word-order languages and another for fixed-word-order languages. But if grammatical representations play a role in speaking and understanding, Bresnan and Kaplan suppose, it is implausible that people's mental representations should take radically different *forms* depending on the language they speak.

The grounds for this constraint, then, is plausibility. But is the alternative so implausible? Even within English, different people could well process utterances differently. Consider the evidence, reported by Peters (1977), that different children learn their first language according to different styles. Some are analytic about word structure, whereas others treat words as Gestalts. If, as Bresnan and Kaplan suppose, the procedure for constructing mental representations is induced by the learning function, then these two groups might develop mutually exclusive strategies for handling certain structures, ending up with two distinct representations of language, both of which, however, fit English in use. Indeed, adults with varying amounts of spatial ability have been shown to use fundamentally different strategies in certain areas of comprehension, one type relying heavily on spatial abilities and the other type not (MacLeod, Hunt, & Mathews, 1978; Mathews, Hunt, & MacLeod, 1980). In speaking spontaneously, adults also appear to hesitate, monitor, and correct themselves according to consistently different styles (Maclay & Osgood, 1959). Both adults and children are known to vary tremendously in size of vocabulary, fluency, size of short-term memory, and spatial abilities, and languages have surely evolved to accommodate this variation. They could also have evolved to accommodate distinct types of mental representations, as may be needed to account for the differences among children, and among adults, in both comprehension and production. If it is reasonable to assume heterogeneity in language processing, it is also reasonable to assume heterogeneity in procedures for grammatical interpretation from one person to the next.

By the same argument, the principal procedures used for grammatical interpretation could vary from language to language, too. Speakers of a mainly free-word-order language might exploit a class of procedures that speakers of mainly fixed-word-order languages never use. If this turned out to be true, there would be little reason for constraining the procedures to be the same in the two languages.

So Bresnan and Kaplan's five theoretical constraints on grammars don't fare too well against the four standards set out earlier, and the explanation is obvious. They weren't designed to. They were motivated not so much by psychological concerns—by examining psychological theories to see how they might constrain grammar—as by linguistic concerns—by trying to rationalize the constraints that languages seemed obviously subject to. We return to this alternative approach to psychological constraints later.

GIVÓN ON PSYCHOLOGICAL CONTRAINTS

Givón, in his chapter, takes a very different tack to the study of language. His main goal is to contrast "formal-logical systems" of representation with pragmatic systems. He argues that, because formal-logical systems are fundamentally closed, context-free, discrete, and based on deductive inference, they are inadequate as representations of pragmatic systems of language, which are fundamentally open-ended, context-dependent, continuous, and based on inductive inference. To make this argument, he appeals to several case studies in "the meaning system in natural language." Unlike Bresnan and Kaplan, Givón doesn't explicitly state psychological constraints for his proposals. Yet the arguments he advances presuppose such constraints, which he exploits in many explanations.

What aspects of language is Givón talking about? One of the most basic distinctions in linguistics is between form and content, or between structure and function—that is, between the structure of what is produced (phonetic segments, words, constituents, sentences, and so on) and the function or use to which that structure is put (to refer, ask questions, denote, and so on). Although language function may be open-ended, context-dependent, continuous, and inductive, language structure certainly isn't. English, like apparently all languages, has a finite vocabulary of elementary words and a small number of syntactic devices for combining words. When Givón argues that language ought to be treated as open-ended, he can only be talking about *function*. Givón's complaint, then, comes down to this. The finiteness, context-independence, discreteness, and deductive nature of linguistic form have often been assumed to hold for linguistic function as well, and any model built on this assumption is necessarily incorrect.

The bulk of Givón's chapter is devoted to showing that language (read "language function") is at every point open-ended, context-dependent, continuous, and inductive. For each set of functions he considers, he argues: (1) here is a well-known property of language structure; (2) it reflects a continous, open-ended, context-dependent, or inductive set of functions; and (3) here is a plausible psychological process that would lead to 2. We consider this argument, because it is used repeatedly, against our four standards—empirical grounding, structure independence, theoretical coherence, and linkage. We select only a few of Givón's many examples to illustrate his appeal to psychological constraints.

Co-reference and Definite Description

The empirical grounding Givón appeals to are almost always facts about language use. In his discussion of co-reference and definite description, he uses a variety of informal examples to argue that reference devices vary along a dimension of "identifiability of referent" or "surprise value of referent." In English, the most identifiable and least surprising referents are introduced with null anaphora—ordinary ellipsis—and the least identifiable and most surprising with definite NP's in left-dislocation, as in *My sister, she just left*. Givón puts four other intermediate constructions between the two extremes. He uses the examples to show that the speaker's choice along this continuum depends on his belief about the hearer's ability to identify referents unambiguously, the hearer's ability to follow the thematic content of the discourse, the speaker's knowledge of the hearer's expectation about themes and topics in the discourse, and other such things. All these factors, he suggests, are characterized by the four properties of pragmatic systems he has given, which amounts to the claim that the referential devices lie on the continuum they do because speakers need to be able to make graded distinctions in ''identifiability'' or ''surprise value.''

Givón's evidence here is structure independent, because it was derived from observations about language use and not merely language structure. Further, Givón has offered what is really psychological evidence for each of the factors suggested. He provides fragments of discourse in which, say, stressed and unstressed pronouns are used, and he asks us, his readers, to go along with his intuitions about how these pronouns would be interpreted. Through a network of examples, he argues for the factors that go into his ''identifiability'' continuum.

Although identifiability and surprise value have some empirical grounding, they don't come from a clear psychological theory about how they combine to determine the form of reference. What Givón's examples show is that people make certain distinctions among the six syntactic forms he discusses, and that very roughly, they fall along a continuum of ''identifiability'' or ''surprise.'' But this continuum probably consists of a number of overlapping scales collapsed onto a single scale. For example, a speaker could use *the woman* instead of *she* for many reasons. He may believe that the woman being referred to isn't readily available in the listener's memory—isn't ''on stage in consciousness'' as Chafe (1974) would say. Or he may believe that, even though this woman is readily available in memory, the listener might confuse her with another woman also readily available in memory. Or, even though he believes the woman is readily available, he wants to indicate a change in topic. Each of these reasons is different, and they cannot all be called ''identifiability'' or ''surprise.'' The same goes for many other contrasts that are collapsed in this continuum.

The continuum-like appearance of these six forms may arise from their hierarchical nature. The main contrast on the scale is between pronouns (forms 2 and 3 on Givón's continuum) and definite descriptions (forms 4, 5, and 6). If the use of any pronoun presupposes ''ready availability in consciousness,'' then that single contrast accounts for the main break in the scale (between pronouns and definite descriptions). All other contrasts among the six forms must then presuppose at least this contrast. The scale only *looks* continuous, then, because the individual contrasts are hierarchically nested, with some presupposing others.

As for linkage, Givón says little, here at least, about how the speaker's desire to distinguish along the continuum of ''identifiability'' or ''surprise'' happens to map onto these six forms. Intuitively, the idea is probably that the more uncertain the speaker is that his addressee will be able to identify his reference, the more information he will include in his reference phrase. Indeed, the six forms lie on a continuum of how much information they express and at what point in the reference process. But how did English, for example, happen to evolve these six forms instead of two, or thirty, other forms, which might also vary this way in informativeness? For linkage to be complete, this question needs answers for languages in general.

Lexical Meaning

For quite a different attack on "formal-logical systems," Givón argues that word meanings are inherently context-dependent and open-ended. *Slowly* means different things depending on the type of change or event or movement involved, the norm or average speed for that type of movement, and so on. Words like *hill, mound, heap, pile, peak,* and *mountain* are applied according to inherently fuzzy criteria for dividing up the corresponding conceptual domain. From these and many other examples, Givón concludes, "Meaning is in principle a pragmatic matter, a frame-dependent entity." Yet there are "great areas in our cognitive map where *relatively stable* frames have been established by the organism, most obviously in the areas of our construing the physical universe [Givón's emphasis]."

Givón appears to suggest that these features of language—context dependence in the lexicon along with areas of stability—are a consequence of the way people are, of a set of cognitive constraints on language. We might state the main two constraints implicit in Givón this way: People don't think in discrete categories, yet they establish relatively stable categories in certain areas of thought. Givón doesn't discuss evidence for these notions except to take up the informal language examples and show how they suggest lack of discrete categories. As for any empirical grounding for the psychological claims here, then, Givón provides almost none.

What is striking about the lexicon of a language like English, given Givón's views, is how regular and stable it is. What *slowly* is used to mean on an occasion does depend on the context, but in a highly predictable way: if something is slow, it is below the average or normal speed expected of that type of object in this context. The same remarks apply to other words and their meanings. That is, most words have conventional meanings, presumably listed in people's mental lexicons, that are *not* open-ended and context-dependent; nevertheless, they can be used on particular occasions to denote meanings that are highly open-ended and context-dependent (Clark & Clark, 1979; Clark, 1978, 1983; Nunberg, 1979). There is something right about the long-held intuition that words like *dog* and *slowly* have sets of stable meanings. What needs to be explained is how they can have stable meanings, yet be used in such highly context-dependent ways.

With this added complication, it is hard to see how Givón will link his apparent assumption that people don't think in discrete categories with the way language is—both stable yet context-dependent. How should languages change because of this? So, however convincing Givón is on the context dependence of utterance meaning, he doesn't let us see beyond to a genuine theory of word meaning and word use, and how this theory may be a consequence of his assumptions.

Language Change

When Givón speculates on the history of certain changes in meaning, morphology, syntax, and word order, he appeals to a variety of psychological constraints. He suggests, for example, that *can* and *know* derive from the same Indo-European root, having got split from one another through reanalyses forced by such pragmatic inferences as "If one *can* do something because one *knows* how to do it, perhaps one can do it for other reasons as well, such as: (1) physical/mental power; or (2) being unrestrained." In metaphoric extension, he appeals to "contextual inferences" that involve "the pragmatic judgment of 'relevance' and 'similarity.'" In morphological and syntactic changes, he appeals to such inferences as "If an object is *dative,* it has a high probability of also being *human* and *definite."* and "If the *identity of the agent* is to be *suppressed,* the next most likely participant in the clause will be *likely* to become the *topic* of the clause [all emphases are Givón's]." In explaining a certain word order change, Givón appeals to an " 'over-kill' communicative strategy, whereby the speaker decides that—just for safety's sake—he will use a more marked device to insure beyond a shred of doubt that the hearer got the message."

The empirical grounding for most of these constraints doesn't exist. Many seem to have been assumed just to make Givón's analysis work; in that way, they are structure dependent. Others, such as the idea that people can and do judge "relevance" and "similarity," are surely correct, but in the form given don't constrain much. Still others, like the "over-kill communicative strategy," are at best doubtful. Contrary to the over-kill strategy, for example, speakers normally give no more information than is needed—it's impolite to provide too much information, which may implicate that the listener is incompetent—and, instead, let listeners *ask* for more information as they need it (see, for example, Sacks & Schegloff, 1979).

Givón would probably be the first to say that the theoretical coherence and linkage of these constraints with language universals has yet to be worked out. What he has given us is not so much a model of how pragmatics constrains language, but an illustration of how pragmatics might conceivably constrain language. Still, arguments of this kind would fare better if they stuck closer to highly plausible constraints related to well established psychological processes.

Givón, to summarize, appeals to psychological constraints of many types, but the constraints rarely live up the four standards of strong psychological constraints. Most are not well grounded in independent empirical evidence. Most do not belong to explicit theories about how psychological constraints shape language change, or mold creoles that emerge from pidgins. In Givón's approach, there remains a wide gulf between the "psychological constraints" appealed to and the "psychological constraints" most psychologists would want to call their own.

STRONG AND WEAK CONSTRAINTS

Although Bresnan and Kaplan's and Givón's ultimate aims are very different, they approach psychological constraints in much the same way. They examine language closely for features they can quite safely claim to be universal and then posit psychological or cognitive constraints that might plausibly explain them. Bresnan and Kaplan focus mainly on syntactic features, whereas Givón casts his eye over a range of features of both form and function. We have tried to point out how far these constraints are from those we have called strong psychological constraints. To keep them distinct, we might call the constraints proposed by Bresnan and Kaplan, and by Givón, *weak* psychological constraints.

Although we have been championing strong psychological constraints, there is a clear place in the study of language for weak psychological constraints too— so long as they are seen for what they are. What *are* they? If Bresnan and Kaplan's and Givón's constraints are any example, they are conjectures, proposed mostly on the basis of observations about language and its use, about how the mind must be constituted for language to be the way it is.

Reasoning from language universals to potential psychological constraints can lead to powerful conjectures. When Berlin and Kay (1969) discovered there was but a small set of color terms used by all languages, it was easy to conjecture that the color vocabulary was constrained by the nature of the visual system, and it remained for Kay and McDaniel (1978), Miller and Johnson-Laird (1976) and others to provide the psychophysical rationale. In this way, a weak psychological constraint was promoted to a strong one. Not all conjectures have been followed up so directly. Greenberg (1963), in his classic article on universals of word order, wrote informally of harmonic and disharmonic relations—for example, how an adjective-noun order in noun phrases was "harmonic" with a possessive-noun order in noun phrases. Greenberg clearly had in mind a type of psychological constraint: harmonic relations are more easily processed than disharmonic relations. Bartsch and Vennemann (1972) later raised Greenberg's harmony to the status of a "principle of natural serialization," which they clearly intended to be a psychological constraint—a weak one in our sense. Unfortunately, there have been no attempts to find structure-independent evidence for this constraint, though such evidence would take us a long way toward explaining both sides of the problem at once—the psychological processes that lead to harmonic relations, and the explanation of the harmonic relations. So weak psychological constraints may have their most important value as provocateurs: they goad us to look for strong psychological rationales for universal features of language.

Not everyone sees language universals this way. There have evolved two very different working assumptions about the origins of language universals, and these have led to a good deal of misunderstanding. Most linguists and some psychologists work from what might be called *Chomsky's wager.*

> It is highly likely that most aspects of language that are universal are a result not of general cognitive constraints, but of constraints specific to language functions—specific to an autonomous language faculty. It is therefore appropriate a priori to assume autonomous psychological constraints and to leave it to others to prove otherwise.

Many psychologists and some linguists, on the other hand, make the opposite bet, which might be called *Wundt's wager.*

> It is highly likely that most language universals are a result not of linguistically autonomous constraints, but of constraints general to other cognitive functions. It is therefore appropriate a priori to assume that language universals derive from general cognitive constraints and to leave it to others to prove otherwise.

Bresnan and Kaplan seem to accept Wundt's wager, although when they retreated from calling their constraints "cognitive" to calling them "theoretical," they may have been trying to hedge their bet. Givón seems to hold unwaveringly to Wundt's wager.

Chomsky's wager—which reflects Chomsky's own beliefs closely though probably not exactly—seems ultimately unsound. Its chief problem is that it encourages investigators not to look for structure-independent explanations of language universals, but to be satisfied with a linguistic *description* of a universal, assuming it is also a description of a feature of the human language faculty. If Kay and McDaniel had accepted Chomsky's wager, they would never have sought an explanation for color terminology in the workings of the human visual system. Many investigators appear to accept Chomsky's wager in syntax but Wundt's wager elsewhere in language. This too seems premature. It seems impossible a priori to distinguish those universals whose explanations probably lie within an autonomous language faculty, if there is one, from those whose explanations lie without. It is difficult even to see how one would draw a conceptual line between those processes that are strictly language autonomous and those that aren't. Our own wager is that as work continues, there will be fewer and fewer language universals that cannot be explained from outside such a faculty, and that the autonomous language faculty will go the way of the medieval humors—it will cease its scientific existence.

So we commend Bresnan and Kaplan, and Givón, for raising a number of weak psychological constraints that with time may be transformed into strong psychological constraints. The constraints they have offered are significant not because they represent psychological reality today but because they hold promise for psychological reality tomorrow.

ACKNOWLEDGMENTS

The preparation of this commentary was supported in part by grant MH-20021 from the National Institute of Mental Health. We thank many colleagues for their advice on this chapter, especially Eve V. Clark and Thomas A. Wasow.

REFERENCES

Bartsch, R., & Vennemann, T. *Semantic structures: A study in the relation between semantics and syntax.* Frankfurt, W. Germany: Athenäum Verlag, 1972.

Berlin, B., & Kay, P. *Basic color terms: Their universality and evolution.* Berkeley: University of California Press, 1969.

Birnbaum, L., & Selfridge, M. Conceptual analysis of natural language. In R. C. Schank & C. K. Reisbeck (Eds.), *Inside computer understanding.* Hillsdale, N.J.: Lawrence Erlbaum Associates, 1981.

Brown, C. H. Folk botanical life-forms: Their universality and growth. *American Anthropologist,* 1977, *79,* 317–342.

Brown, C. H. Folk zoological life-forms: Their universality and growth. *American Anthropologist,* 1979, *81,* 791–817.

Brown, P., & Levinson, S. Universals in language usage: Politeness phenomena. In E. Goody (Ed.), *Questions and politeness.* Cambridge, Eng.: Cambridge University Press, 1978.

Bybee Hooper, J. L. Child morphology and morphophonemic change. *Linguistics,* 1979, *17,* 21–50.

Bybee, J. L., & Slobin, D. I. Rules and schema in the development and use of the English past. *Language,* 1982, *58,* 265–289.

Chafe, W. L. Language and consciousness. *Language,* 1974, *50,* 111–133.

Chomsky, N. *Aspects of the theory of syntax.* Cambridge, Mass.: MIT Press, 1965.

Chomsky, N. Remarks on nominalization. In R. A. Jacobs & P. S. Rosenbaum (Eds.), *Readings in English transformational grammar.* Boston: Ginn, 1970.

Clark, E. V., & Clark, H. H. When nouns surface as verbs. *Language,* 1979, *11,* 430–477.

Clark, E. V., & Hecht, B. F. Comprehension, production, and language acquisition. *Annual Review of Psychology,* 1983, *34,* 325–349.

Clark, H. H. Inferring what is meant. In W. J. M. Levelt & G. B. Flores d'Arcais (Eds.), *Studies in the perception of language.* London: Wiley, 1978.

Clark, H. H. Making sense of nonce sense. In G. B. Flores d'Arcais & R. Jarvella (Eds.), *The process of language understanding.* New York: Wiley, 1983.

Clark, H. H., & Clark, E. V. *Psychology and language: An introduction to psycholinguistics.* New York: Harcourt Brace Jovanovich, 1977.

Fodor, J. A., Bever, T. G., & Garrett, M. F. *The psychology of language: An introduction to psycholinguistics and generative grammar.* New York: McGraw-Hill, 1974.

Gleitman, L. R., & Gleitman, H. *Phrase and paraphrase: Some innovative uses of language.* New York: W. W. Norton, 1970.

Goffman, E. *Interaction ritual: Essays on face-to-face behavior.* Garden City, N.Y.: Anchor Books, 1967.

Greenberg, J. H. Some universals of grammar with particular reference to the order of meaningful elements. In J. H. Greenberg (Ed.), *Universals of language.* Cambridge, Mass.: MIT Press, 1963.

Greenberg, J. H. Numeral classifiers and substantival number: Problems in the genesis of a linguistic type. *Working Papers in Language Universals* (Stanford University), 1972, *9*, 1–39.

Greenberg, J. H. Dynamic aspects of word order in the numeral classifier. In. C. N. Li (Ed.), *Word order and word order change.* Austin: University of Texas Press, 1975.

Kay, P., & McDaniel, C. K. The linguistic significance of the meanings of basic color terms. *Language*, 1978, *54*, 610–646.

Maclay, H., & Osgood, C. E. Hesitation phenomena in spontaneous English speech. *Word*, 1959, *15*, 19–44.

MacLeod, E. M., Hunt, E. B., & Mathews, N. N. Individual differences in the verification of sentence-picture relationships. *Journal of Verbal Learning and Verbal Behavior*, 1978, *17*, 493–507.

Mathews, N. N., Hunt, E. B., MacLeod, E. M. Strategy choice and strategy training in sentence-picture verification. *Journal of Verbal Learning and Verbal Behavior*, 1980, *19*, 531–548.

Miller, G. A., & Johnson-Laird, P. N. *Language and perception.* Cambridge, Mass.: Harvard University Press, 1976.

Moravcsik, E. A. Agreement. *Working Papers in Language Universals* (Stanford University), 1971, *5*, A1–A69.

Nunberg, G. The non-uniqueness of semantic solutions: Polysemy. *Linguistics and Philosophy*, 1979, *3*, 143–184.

Palermo, D. S., & Eberhart, V. L. On the learning of morphological rules: An experimental study. *Journal of Verbal Learning and Verbal Behavior*, 1969, *7*, 337–344.

Palermo, D. S., & Howe, H. E., Jr. An experimental analogy to the learning of past tense inflection rules. *Journal of Verbal Learning and Verbal Behavior*, 1970, *9*, 410–416.

Palermo, D. S., & Parrish, M. Rule acquisition as a function of number and frequency of exemplar presentation. *Journal of Verbal Learning and Verbal Behavior*, 1971, *10*, 44–51.

Peters, A. M. Language learning strategies. *Language*, 1977, *53*, 560–673.

Pinker, S. A theory of the acquisition of lexical-interpretive grammars. *Occasional Papers No. 6.* Center for Cognitive Science, MIT, 1980.

Reisbeck, C., & Schank, R. C. Comprehension by computer: Expectation-based analysis of sentences in context. In W. J. M. Levelt & G. B. Flores d'Arcais (Eds.), *Studies in the perception of language.* Chichester, England: Wiley, 1978.

Rips, L. J., & Turnbull, W. How big is big? Relative and absolute properties in memory. *Cognition*, 1980, *8*, 145–174.

Rosch, E., Mervis, C. B., Gray, W., Johnson, D., & Boyes-Braem, P. Basic objects in natural categories. *Cognitive Psychology*, 1976, *8*, 382–439.

Sacks, H., & Schegloff, E. A. Two preferences in the organization of reference to person in conversation and their interaction. In G. Psathas (Ed.), *Everyday language: Studies in ethnomethodology.* New York: Irvington, 1979.

Slobin, D. I. Cognitive pre-requisites for the acquisition of grammar. In C. A. Ferguson & D. I. Slobin (Eds.), *Studies of child language development.* New York: Holt, Rinehart & Winston, 1973.

Slobin, D. I. Language change in childhood and in history. In J. Macnamara (Ed.), *Language learning and thought.* New York: Academic Press, 1977.

Slobin, D. I. *Psycholinguistics* (2nd Edition). Glenview, Ill.: Scott Foresman, 1979.

Swinney, D. A. Lexical access during sentence comprehension: (Re)consideration of context effects. *Journal of Verbal Learning and Verbal Behavior,* 1979, *18,* 645–660.

Vennemann, T. Explanation in syntax. In J. Kimball (Ed.), *Syntax and semantics* (Vol. 2). New York: Academic Press, 1973.

Vennemann, T. Topics, subjects, and word order: From SXV to SVX via TVS. In J. M. Anderson & C. Jones (Eds.), *Historical linguistics I: Syntax, morphology, internal and comparative reconstruction.* Amsterdam: North Holland Publishing, 1974.

Wason, P. C., & Reich, C. C. A verbal illusion. *Quarterly Journal of Experimental Psychology,* 1979, *31,* 591–598.

Wexler, K., & Cullicover, P. *Formal principles of language acquisition.* Cambridge, Mass.: MIT Press, 1980.

IMPORTANT

Library Notice

ELMHURST COLLEGE
ELMHURST , ILLINOIS   60126
OVERDUE NOTICE 88/06/29
ITEM   DUE
001.535A366D
       C001   STX   880615
DESIGNING INTELLIGENT S
YSTEMS$ NEW YORK

PLEASE RETURN OR RENEW THE
ITEM(S) ABOVE. TO RENEW,
CONTACT LIBRARY AT
ELMHURST COLLEGE

DELIVER TO:

S. HARRISON WHITE
GRADUATED-1974
406 S. 6TH AVE.PO BOX 0072
MAYWOOD, IL.60153-0072

PATRON ID: . . .   22

PSYCHOLOGY

The chapters in this section deal with a number of important empirical and theoretical methodological issues. The first two chapters by Swinney and by Van Lehn, Brown, and Greeno have a common structure. The authors discuss several methodological issues and then illustrate their arguments with applications from their own work. Charniak, using historical examples, justifies the diverse empirical and theoretical methodologies from the various disciplines that make up cognitive science.

Swinney's chapter is a fine demonstration of the effective use of empirical methods to explore fundamental questions in language comprehension. Swinney provides a qualitative analysis of the processes of comprehension into a collection of subprocesses. The question he focuses on is whether or not the various subprocesses are weakly or strongly coupled. That is, are there significant semantic influences on the processes involved in lexical access. Whether or not a system is nearly decomposable, is a fundamental question in the analysis of any complex information-processing task. Swinney shows that empirical techniques can be developed to provide a very clean answer to this question. We view Swinney's chapter as supporting a contention made in Chapter 1 concerning the role of psychology's empirical methodology in cognitive science. We claimed that empirical methods have a central place in theory building and theory testing. Swinney's chapter is an elegant example of the successful application of such methodologies.

Van Lehn, Brown, and Greeno attempt to characterize the limitations of computational methodologies, primarily derived from artificial intelligence, as tools for constructing theories in cognitive science. They point out that it is often difficult to disentangle the contributions made by various mechanisms incorporated in a complex program in accounting for intelligent action in an environment. In addition, computational techniques are so flexible that they permit unprincipled fixes to a theory, for example, the kludges discussed by Lehnert in her chapter. Van Lehn et al. propose solutions to the difficulties inherent in computational theories and illustrate these solutions with their own work.

Charniak's discussion attempts to rationalize the methodological diversity so apparent in the chapters in this section and well illustrated by practically every contribution to this volume. Charniak argues that the heterogeneous methodologies that characterize cognitive science today are in fact necessary. He asserts that all mainstream methodologies have the potential to make important contributions to our further understanding of cognitive processes. Twenty-five or fifty years from now we may be able to look back and say that some subset of these methodologies were more profitable than others but at this point in time we cannot see far enough into the future to successfully pick the correct ones. Charniak states a theme that is developed further by the two general discussants, Suppes and Mandler, that the current methodological diversity is both necessary and healthy.

8 Theoretical and Methodological Issues in Cognitive Science: A Psycholinguistic Perspective

David Swinney
Tufts University

Introduction

The fundamental concern of cognitive science is to understand the nature of both the structural and functional properties of intellectual capacity and intellectual operation. And in so far as such a goal is maintained as a central focus in each of the various fields comprising cognitive science it will follow that roughly the same set of general theoretical issues and methodological approaches will be at stake in each. However, it is equally clear that any *real* catholicity of theory and methodology across disciplines can only exist when these fields (including, at least, aspects of psychology, linguistics, philosophy, neuroscience, and artificial intelligence) share *in detail* the same underlying assumptions about the nature of these issues and their possible solution sets. Thus, it is impossible to overstate the importance that such (typically unstated) assumptions play in constraining theoretical and practical approaches to any issue and in making approaches incorporating one constellation of assumptions *un*intelligible to those with a slightly different constellation.

It is obvious that some fields within cognitive science share more basic goals and approaches than do others, and it is certainly true that the interaction of some of these fields has produced a growing tendency for the development of mutually held theoretical assumptions. (Psychology and linguistics are prime examples: since its inception, psycholinguistics has borrowed extensively from linguistic theoretical foundations and, particularly in the past few years, linguistics has tended to expand its explanatory domain to include data on the *use* of linguistic knowledge.) However, there are great differences in the underlying approaches and basic theoretical goals in even these mutually assimilative areas, differences that have inevitably led to misunderstandings and a subsequent failure of those in one discipline to successfully use and appropriately apply knowledge gained by those in the other.

In what follows I examine a number of basic issues and assumptions underlying work in one domain of cognitive science, that of cognitive psychology. The goal of this presentation is to make explicit some of the considerations involved in a cognitive psychological approach to issues so that a reasonable basis for valid incorporation of data from this domain into other fields of cognitive science may be established. Further, as cognitive psychology is itself a diversified field, I focus discussion on a number of theoretical considerations (and their related methodological concerns) in one area of cognitive psychology—psycholinguistics. This focus on language processing seems appropriate in as much as a great many of the fundamental concerns in cognitive psychology have received their most extensive examination in the past decade from work in psycholinguistics.[1] However, it is my contention that the issues of perception and production discussed here are not different in kind from issues of perception and production in any other aspect of cognitive psychology, and thus the material presented here is taken to be representative of a fairly general set of concerns in and approaches to cognitive science.[2]

The discussion of theoretical and methodological issues in psycholinguistics is followed by an illustration of the issues at stake through an extended example of work in a specific area of language comprehension—lexical processing. Following this, a short discussion of the validity of utilizing such data to support claims in fields not sharing assumptions under which the data were gathered, is made.

BASIC THEORETICAL AND METHODOLOGICAL CONSIDERATIONS IN PSYCHOLINGUISTICS

The central concern of any theory of language perception is to establish the manner in which we assign interpretations to acoustic or visual language stimuli (i.e., understand words and sentences). Such a theory must, at the very mini-

---

[1]The reasons for this are fairly obvious: language has proven to be a somewhat more accessible domain for study than some others in cognitive psychology, and the systematic classification and observation brought by linguistics and philosophy to language have provided a convenient initial stepping stone for the investigation of the psychological properties underlying human language perception and production.

[2]It is worth noting that the issues that will be discussed all have *human* intelligence as their target of inquiry. This goal is shared by most, but certainly not all, approaches in cognitive science and provides a limiting criterion on all such work. The distinction between those fields that focus on human capacity and operation (to whatever degree) and those that entertain data from 'other' types of intelligent behavior appears to be that with perhaps the most profound consequences for mutual interdisciplinary understanding.

mum, detail the distinct sources of information that are involved in comprehension (including both the representational structure and the content of such information) and the processes by which this information is organized and integrated. As such, what distinguishes such psycholinguistic examination from many other approaches to language is its focus on the fact that language understanding (like many other cognitive functions) is a dynamic *process*. That is, theoretical descriptions of language from a *psychological* point of view must incorporate the fact that language understanding involves operations (on various sources of knowledge) that are distributed over time. A number of important issues, most of which are highly interrelated, follow from this point:

1. *Real-time vs. static analysis of language.*   The first of these is that *static* models of putative language structure are of questionable (and, at best, indeterminent) importance in the enterprise of understanding language performance. Models can be classified as static on two grounds: the first of these can be seen in the distinction drawn between those mental processes involved in real-time language perception (the *process* of understanding a sentence) and those involved in various post-perceptual language functions (recall, recognition, paraphrase, or some other such manipulation of already-perceived sentential/propositional material). Not only do quite different mechanisms likely underly perceptual and post-perceptual processes, but data taken from one of these two sources may not legitimately reflect processes in the other. Thus, it is important to distinguish between theory relevant to (and data derived from) on-line processes as compared to post-perceptual processes. The latter can, at best, provide an underdetermined model of language performance. The importance of the meld of theory and methodology is critical here and, if our goal is one of understanding language use, then we must look at that use appropriately—as it occurs in time.[3]

The second way in which models may be considered 'static' is when temporally independent analyses of language form the inferential base for positing mental processes. Traditional formal linguistic description, for example, represents structures independently of temporal constraints, a fact that is likely to void the relevance of at least some linguistic descriptions to the enterprise of psychological explanation. Further, the data on which such linguistic models are based—linguistic intuitions and observation of distributional characteristics—are clearly post-perceptual; they are simply not direct reflections of the processes involved in real-time language comprehension. At the very least, such intuitional data is filtered through whatever cognitive process is involved in bringing unconscious knowledge to consciousness; and, it has been amply demonstrated that

---

[3]Note that on-line processes in language are assumed here to provide a segmental analysis that yields a roughly literal interpretation and structural description of a given proposition/sentence, whereas post-perceptual processes involve evaluation or other manipulation of the basic structural description in light of world knowledge, processing demands, or other higher order information sources.

most unconscious events are opaque to conscious access. Overall, if we are to achieve any substantive understanding of language as a psychological process it will be necessary to examine the microstructure of this process as it occurs *in real time;* an adequate psychological description of language must incorporate specification of the availability, structure, and content of the various levels of representation throughout the temporal course of language performance.

2. *Issues of modularity, autonomy, and interactivity.*   The above considerations lead to a related and equally critical issue in language (and other cognitive) processing—the degree to which the system is decomposable into independent, internally consistent, specifiable subprocesses. Because our perceptual models require that some basic invariant unit of analysis be posited, the concept of modularity exists to some degree in most models of language. However, we have neither an account of: (1) what distinct processing modules (sources of information) are functional in language understanding; nor (2) an account of the degree to which these operate independently of each other.

With respect to the first issue, whereas systematic observation in linguistics has allowed for the characterization of a number of sources of information in formal linguistic analysis—phonetic, morphemic, syntactic, and so on—such descriptions are based on the rules and constraints of linguistic description and few have yet to be shown to have anything but an indirect (or indeterminant) relationship to the types of information that are *functional* in the ongoing process of language comprehension. Such observations do, however, provide a strong prima facie case for the 'functional reality' of these information types, and as such are important stepping stones in the search for functional properties of the system. Similarly, evidence from language pathology has given us hints as to functional modules in the language system. However, establishing psychologically relevant characterizations of the sources of information involved in language processing can only be accomplished by empirical examination.

The second issue—hotly contested in the past few years—concerns the basic nature of the integration of information that occurs during language comprehension. This issue contrasts two models of language processing: a maximally interactive system (in which any type of contextual information can affect processing of any other piece of information) and a highly modular system (a system comprised of autonomous subroutines in which contextual information does not affect processing internal to any subroutine). At the outset, it can be noted that language comprehension, like most other cognitive processes, is clearly "integrative" in nature. To understand an utterance we must, in some fashion, retrieve information about the words in that utterance, discover the structural relationships and semantic properties of those words, and interpret these in the light of the various pragmatic and discourse constraints operating at the time. Further, we know that all of this takes place at a remarkably rapid pace, a fact (along with much post-perceptual data) that has led some theorists to view the comprehension system in general as being a maximally interactive, 'top down' process. The modularist view of such integration, however, is that the subcomponents comprising the comprehension process constitute autonomous levels of processing, each of which may operate with its own unique temporal and distributional characteristics. It is only the outputs of these processes that interact (by providing inputs to other subprocesses); the internal operation of any one process is not changed by information derived from another (see, e.g., Garrett, 1978).

Resolution of the issue of whether the system in general is a modular or an interactive one is perhaps the most important key to establishing a viable psychological model of perceptual processing. Certainly, it is clear that a system that supports the nearly infinite variability in processing that is inherent in any *maximally* interactive model requires a significantly more powerful underlying mechanism to be posited than one which consists of relatively autonomous, context free, internally consistent routines. Failure to provide clear definition of (and detail of constraints on) the nature of information integration almost guarantees that we will posit either far too powerful or far too weak an underlying mechanism as the basis for language behavior, a failure that may well be a critical one in our attempts to understand the process.

It should be noted that the question of autonomy of processing levels, that has been presented here as a means for delineating the micro-structure of the language faculty, has a macro-level counterpart. It is, in fact, important that we determine the extent to which mental operation in general is comprised of independent, domain-specific systems (e.g.,—language, imagery) as opposed to domain-general functions. For example, to what degree do general memory functions or automaticity factors penetrate our structural and functional descriptions of the language system? And (taking another view) to what extent are the models of language processing that we develop distorted by various 'output' functions that recruit both world knowledge, situational constraints and (possibly) domain-general processing strategies in their operation? In both micro- and macro-level cases, as will be demonstrated in part below, the weight of the evidence appears to favor the modular approach. However, the critical link in our ability to detail the relevant independent structures and processes, and the nature of their interaction, is our ability to bring appropriate empirical methodology to bear on the issue. This ability bears discussion.

3. *Methodological Considerations.* Let us begin with an obvious but nonetheless important observation: how we characterize information sources and (their) procedures is intimately tied to the assumptions we make about the nature of the experimental techniques used to reveal them. There is simply no passive window that allows for the examination of mental processes without affecting those processes to some extent. Thus, the representational characterization of these information types, the processes involved in their interaction, and the

processes by which we attempt to examine these interactions must all be modeled simultaneously. Each of these constrains our ability to examine and model the other. It is, therefore, essential to stress that adequate models of our empirical tasks are a critically important part of our attempt to understand mental representation. Although it is encouraging to note that some experimental techniques have undergone extensive critical examination (witness the study of phoneme monitoring over the past 6 years), far too little is known of even the most basic characteristics of most of these techniques.

There are at least two corollary issues that are of importance here. The first relates to the critical need for discovering and testing new experimental techniques. Given that much of the problem in understanding language comprehension is best resolved by empirical examination of the real-time characteristics of language processing, we are in need of experimental techniques that are sufficiently flexible to examine such processing on-line and, simultaneously, that are as non-intrusive into the process under study as possible. Further, we are in need of tasks that are differentially sensitive to different *levels* of analysis in language processing (e.g., acoustic/phonetic, lexical, syntactic, inferential, etc.). It is important that the examination of mental processes be seen to be a multi-leveled enterprise. We need to discover the nature of each of the several putative levels of processing as they are computed during the on-going perceptual analysis of language. To do this, it will be necessary to develop batteries of 'on-line' tasks, each of which has known properties that reflect different aspects of the mental process under study.

The second corollary issue concerns the need for determining the relationship that holds between conscious and unconscious processing as well as the role of automaticity in such processing. Because language is a highly overlearned, automatized system, until we understand the manner in which automaticity affects our ability to interrupt, examine, and have intuitions about language, we will have little idea of the value of the information we gather from such experimental tasks. Further, as it is generally only the final result of any perceptual process that can be brought to the conscious awareness of a listener, any task involving conscious decisions by the listener as to the nature of the material he has heard is likely not to give a direct reflection of the series of unconscious, automatic stages involved in that perceptual process.

It has been argued (appropriately, I believe) that support for nearly any theoretical position can be obtained if the 'right' experimental task is used (Forster, 1979). One can, for example use tasks which will 'demonstrate' that context either does or does not have an effect on subsequent processing. The problem is that not only does any one task bring with it its own particular set of required strategies, each task also has a different degree of sensitivity to different mental processes. Although it is clear that we cannot have the ideal situation— one in which the task would add none of its characteristics to the process under study and would also be maximally sensitive to underlying processes—some of the tasks that have been used appear to approach that ideal more closely than others. One key factor appears to be the *point of application* of the experimental tasks during the comprehension process. As suggested previously, tasks that attempt to examine a process only after it is completed (post-perceptually) reflect quite different aspects of processing than do tasks that occur with more temporal immediacy to the process under study. Minute differences in the degree of temporal separation of the experimental task from the occurrence of the particular process one wishes to study, can cause major differences in the results obtained, as will be seen in the next example.

A final methodological issue relates more to the experimental situation than to the tasks used. In general, it is the case that the environment in which a particular operation is examined will influence the characteristics discovered. If, for example, one is interested in lexical processing in fluent language comprehension situations, then the process must be examined in a fluent language environment, and not in 'isolated' lexical tests (for the obvious reason that the cognitive/ linguistic operations involved in these different task situations will, themselves, differ). Similarly, if the target of inquiry is *speech* comprehension, we must be wary of drawing conclusions from the processing of written material. Although some underlying operations are undoubtedly shared across performance situations, it is not a trivial chore to determine which of the inferences that we make about cognitive processes are best attributable to 'underlying' processes and which are best attributable to superficial or task-related operations. Results from some task situations undoubtedly reflect more of how subjects *can* perform given constraining (and often complex) experimental situations than of what subjects normally do during language comprehension. (Obviously, when sufficient evidence from different 'task' sources coincide, we have the basis for arguments about underlying representations; however, that situation is not a common one.) Thus, because the task a subject sees him or herself as having—the *query* he or she poses to his or her cognitive system—may determine the cognitive strategies employed in that situation, a conservative position may well be the best one to adopt in research. In doing so, some relevant evidence will undoubtedly be overlooked, but the model we sketch will not be forced to account for much of what may be irrelevant or inappropriate data.

Information Interaction during Lexical Processing—An Illustrative Example

The point of the argumentation thus far has been to delineate a number of issues that are critical to understanding the psychological processes underlying language. The issues that have been elaborated constitute concerns that, in some cases, form the more unique contributions of psycholinguistics to this enterprise and, in others, represent concerns shared by a larger subset of cognitive psychologists (including questions of modularity, automaticity, autonomy, functional information sources, the role of consciousness in experimental investigation, and the penetration of domain-general processes into the examination of putatively independent cognitive faculties). In some cases an obvious stance has been taken on these issues, but in others the necessary information is simply not available for such a posture. It is the case, however, that the only viable approach to resolving most of these issues is the careful empirical examination of the micro-structure of the language process through the appropriate match of experimental technique to the questions of interest.

In what follows an extended example of methodological approaches to some of the problems raised earlier is presented. The example is intended to provide both an illustration of the problems involved in evaluating experimental approaches to these issues and (ultimately) to provide evidence for a particular characterization of the cognitive system in man.

The general issue motivating the research reported here is the question of the degree to which we can characterize comprehension as being a highly interactive (as opposed to an autonomous, modular) system. The approach my colleagues and I have taken in this enterprise has been to examine the nature of the influence of prior contextual information upon the processing of subsequent material during sentence/discourse comprehension. In this, we have attempted to map the temporal characteristics of such contextual effects. As argued previously, although we know that 'context' exerts some influence upon the interpretation of language, the important question is where and how such influence takes place. The work discussed here takes the domain of lexical processing as the testing ground for examination of this issue. This choice has been made for a number of reasons. First, words are perhaps the most likely candidates for being truly functional sources of information during language processing (word recognition is acknowledged to play a role in nearly every psychological and linguistic account of language). In addition, lexical representation and processing are commonly considered to be (the) major points of intersection of acoustic-phonetic, syntactic, semantic, and discourse information in language. Thus, it is a logical realm within which to examine information interaction.

The work to be examined focuses on the effects of prior-occurring 'higher' order contexts upon lexical access and interpretation. In this, the interactionist/modularist issue corresponds to a distinction between what have been labeled the Contextually Predictive and the Contextually Independent models of lexical processing. Under the Contextually Predictive theory, contextual information constrains and directs all of the various stages of lexical processing, whereas under the Contextually Independent model, higher order context will only act on the output of the lexical access process. Thus, the problem can be characterized as one concerning whether or not lexical *access* is independent of contextual constraint.

The overwhelming proportion of the evidence on this issue comes from studies that have examined the effects of contexts upon the interpretation of polysemous words. The reasons for this are straightforward. The meanings of lexical ambiguities can be differentiated rather easily, and thus the selective effects of higher order contextual constraints upon the functional activation of these meanings are relatively available to empirical examination. Further, not only are lexical ambiguities arguably as common as unambiguous words in language use, but nearly all words exhibit some type of indeterminacy in characterizations of their meanings. Thus, ambiguous words provide a useful and well founded vehicle for examining the effects of context upon lexical processing. Given a sufficiently sensitive task, one should be able to determine whether strongly biasing (higher order) contexts constrain the various stages of lexical processing by examining which meanings of the ambiguities are functionally activated in the presence of these contexts.

Most of the early studies on this issue utilized the so-called 'ambiguity effect' in their investigations. The principle behind this effect rested on evidence that the presence of a lexically ambiguous word caused increased processing difficulty (when compared to an unambiguous word) in sentence comprehension. Thus, the absence of this 'effect' in the presence of a prior context was taken as an indication that the context had simplified the access process. Holmes, Arwas, and Garrett (1977), for example, demonstrated that time to classify a sentence as meaningful is increased by the presence of a lexical ambiguity and, also, that the number of words recalled in a rapid serial visual presentation task is fewer in the presence of an ambiguity. MacKay (1966) demonstrated that the presence of ambiguous words in sentence fragments causes sentence completion times to be slower. And, in an approach that we will examine in slightly more detail, Foss and his colleagues (Cutler & Foss, 1974; Foss, 1970; Foss & Jenkins, 1973) as well as Cairns and Kamerman (1975) and Cairns and Hsu (1979) demonstrated that monitoring for a phoneme takes longer following an ambiguous word (compared to an unambiguous word). Further, this effect was shown not to diminish even in the presence of a prior biasing context.

The phoneme monitoring task operates on the following principle: in any finite capacity processing device in which two competing tasks are being performed simultaneously (e.g., listening for a prespecified target phoneme (monitoring) and comprehending the sentence), changes in difficulty in one task should be reflected in speed or accuracy of performance changes in the other. Thus, the experimental finding that phoneme monitoring took longer following an ambiguous word than an unambiguous control word in these sentential materials, even in the presence of a prior biasing context, was interpreted as demonstrating that such prior context did *not* affect lexical access. Unfortunately, this interpretation constituted support for the Contextual Independence hypothesis solely by virtue of *failing* to find support for the Contextual Interaction hypothesis (i.e., failure to find a reduction in monitoring latency). The problem with this experimental inference became apparent when the work of Swinney and Hakes (1976) demonstrated a reduction in the phoneme monitoring latency following ambiguities in the presence of *very* strong biasing contexts. However, the Swinney and Hakes research (and its apparent support for the Contextually Interactive Hypothesis) was also doomed to reinterpretation, but for an entirely different set of reasons, many of which have been suggested earlier. In all of the work using phoneme monitoring, the phoneme target, of necessity, occurs "downstream" from the ambiguity that is being examined (it usually begins the word following the ambiguity). The temporal gap between occurrence of the ambiguity and detection of the phoneme target in a following word is, thus, fairly extensive relative to the magnitude of the latency effects reported with this task. Thus, whereas claims deriving from monitoring data all contain the key assumption that the task actually measures lexical access and not some process that occurs following access, the phoneme monitoring task may well only reflect some type of post-access decision process. If so, in situations where a prior biasing context was not very strong this post-access decision process that incorporates context might take a relatively long time to complete, long enough so that the phoneme-monitoring decision is engaged while this process is still at work, thus producing no evidence of contextual interaction. However, in the presence of a very strong biasing context (the Swinney & Hakes study) this post-access decision process could occur rapidly enough so as to reduce the processing load caused by the ambiguity prior to the time when the phoneme monitoring task comes into play. In short, it seems likely that the phoneme monitoring task was not actually reflecting the real-time access of information for ambiguous (or other) words preceding the phoneme target but, rather, was reflecting post-access processing (see Cairns & Hsu, 1979; Swinney, 1982, for related arguments). Thus, despite appearing to be an 'on-line' measure, the monitoring task may not be capable of yielding data appropriate to the examination of the hypotheses under question.

In order to be able to provide more relevant evidence for the hypotheses under question, an experimental task was needed that was capable of directly reflecting the access of each of the several meanings of an ambiguous word as well as being capable of measuring lexical access immediately (with no temporal delay) during on-going sentence comprehension. To these ends, a task—the cross modal lexical priming technique (CMLP)—was devised which coupled the auditory presentation of an ambiguous sentence with a visual, lexical decision task. This technique utilizes the existence of automatic semantic priming (response facilitation) to detect the activation of word meanings during sentence comprehension. It has been demonstrated that lexical decisions (word/nonword judgments) are facilitated when the target word is semantically related to a previously presented word (see, e.g., Meyer, Schvaneveldt, & Ruddy, 1975; Neeley, 1977). This effect has been demonstrated to hold cross-modally, where the lexical decision is made to a visually presented letter string and the related word is presented auditorily as one of the words in a sentence (Swinney, Onifer, Prather, & Hirshkowitz, 1979). It is the fact that priming can be driven from words in sentences that makes the task of interest here. Not only is the task flexible (the 'primed' visual word can be presented at any point during which a subject is listening to a sentence), but it has also been demonstrated to reflect the relative degree of activation of the target word in a sentence as a function of the amount of facilitation obtained for the lexical decision. (The amount of priming for an experimental word is determined by comparison to an unrelated, but otherwise equivalent, control word presented at the same point in the sentence). There are three additional aspects of this task that are worth noting. First, processing of the sentence is relatively natural, at least until the appearance of the visual stimulus (that occurs immediately *after* the subjects hear the critical word in the sentence). Secondly, the task does *not* require subjects to try to manipulate the sentence or consciously relate the visual material to the sentence in any way, and thus the task is considerably less intrusive into the natural process than several other on-line tasks seem to be. Finally, when there is a sufficiently low ratio of materials in which the visually presented word is (as compared to 'is not') related to some other word in the sentence, subjects rarely report noting this critical relationship and, in fact, rarely appear to try to do so after the first few practice trials.

Thus, given a task that is an index of the degree of activation of word meanings in sentences, the rationale for its use in the examination of the effects of context upon the access of ambiguous words is no doubt apparent. If prior contexts constrain or predetermine lexical access, then one should only expect to find priming for visual words related to the contextually relevant meaning of an ambiguity; if context prevents the inappropriate meaning from being accessed and activated there can be no basis for a priming response from that meaning. On the other hand, if access is an autonomous (and exhaustive) process one would expect *all* meanings of a particular acoustic/phonetic word form to be accessed, and thus visually presented words related to all meanings, (contextually appropriate or not) should be primed, at least momentarily.

The initial experiments (Swinney, 1979) examined the effects of local lexical-semantic contexts upon the access of meanings for unsystematic ambiguities, utilizing a set of materials from which subjects listened to sentence pairs such as the following (where the material in parentheses constitutes the context manipulation):

> Rumor had it that, for years, the government building had been plagued with problems. The man was not surprised when he found several (spiders, roaches, and other) bugs in the corner of his room.

Visual lexical decisions were required for words related to the contextually appropriate meaning of the ambiguity [ANT], for words related to the contextually inappropriate meaning of the ambiguity [SPY], and for unrelated control words [SEW]. These materials were presented both at a point immediately following occurrence of the ambiguity △ and at a point three syllables later △.

The results clearly supported the Contextual Independence model of lexical access; significant facilitation occurred for lexical decisions to visual words related to *each* meaning of the ambiguity, even in the presence of a strong, constraining lexical-semantic context, when tested immediately following the ambiguity △. Importantly, at the delayed test point △, significant facilitation was obtained *only* for lexical decisions made to words related to the contextually relevant meaning and *not* the contextually inappropriate meaning. This result strongly supports the concept that an independent post-access decision process takes place that utilizes context to determine the appropriate meaning for the word. The result of this process is that activation is maintained for the appropriate meaning but the other, inappropriate, meanings of a word are allowed to rapidly decay (or, perhaps, be suppressed). These results also provide strong support for the assumption that this task is sensitive to the processing of words throughout the time-course of sentence comprehension.

The experimental work aimed at examining the autonomy and interactionist hypotheses of lexical processing has been presented in a fair amount of detail in order to demonstrate several of the methodological issues inherent in attempts to examine the nature of representation and processing involved in real-time cognitive operations. As demonstrated, with sufficiently fine-grain analysis techniques one can hope to find reasonably definitive answers to questions about autonomy and modularity in cognitive processing. Obviously, however, convincing evidence for modularity is not supplied by a single demonstration. While the data presented thus far support the argument that lexical access takes place independently of any biasing effects of lexical/semantic context in sentential environments, claims of true autonomy require demonstration that other types of context also fail to affect lexical access. A number of relevant investigations have been undertaken in the past 2–3 years in our laboratory and in other research facilities. It has been shown, for example, that the presence of various types of local and globel cues as to the syntactic (structural) role that must be played by subsequent lexical material also do not constrain lexical access. In a series of experiments employing categorical ambiguities (e.g., cross, watch) in sentential contexts with prior occurring cues as to their grammatical role (e.g., ''. . . *the battered cross* . . .''; ''. . . *to quickly* cross . . .''), it has been shown using CMLP that *both* noun and verb interpretations of the ambiguity are momentarily accessed (Prather & Swinney, 1977; see also Tanenhaus, Leiman, & Seidenberg, 1979).

Similarly, again using CMLP, preliminary evidence supports the position that lexical access is independent of effects from higher-order discourse constraints (including context provided by 'theme' and 'given' information in discourse). In all, lexical access appears to be an exhaustive procedure that operates independently of existing contextual information; such information only has its effects after the interpretations for a word are computed.

Separate evidence about the process that incorporates contextual information independently of (and following) lexical access has been obtained using real time processing techniques (CMLP) on a specific diagnostic population of subjects— chronic schizophrenics. The motivation for this work rested in the fact that the schizophrenic population is known for a tendency to ignore contextual information in processing (contextually independent, bizarre, associations are a major diagnostic for this disorder). It was reasoned that if contextual effects on word interpretation are independent of lexical access, and if the cognitive disruption involved in schizophrenia affects the use of context but *not* the underlying perceptual access process, then data from schizophrenic patients should be able to provide evidence confirming (or refuting) the modularity claims made above. Using sentences containing polarized lexical ambiguities (words with one very likely and one very infrequent meaning) it can be predicted that if the perceptual access process is intact for schizophrenics both the high and low frequency meanings should be activated immediately △ after occurrence of the ambiguity, regardless of the contextual bias of the sentence. (This is, of course, the result already reported for 'normal' subjects.) However, when the test point is delayed △ thus allowing sufficient time for the posited post-access decision process to have taken place, it can be predicted that a departure from the pattern for normal subjects should be found (reflecting a disruption in the post-access contextual integration module). This was precisely the result obtained. In the delayed text condition △, only the *most* frequent meaning for the ambiguity was activated, even when the contextual information in the sentence dictated that the *less* frequent interpretation was the correct one. (As expected, for normal control subjects only the contextually appropriate, *less* frequent, meaning was activated in this condition.)

Thus, whereas the research has certainly not exhausted the inventory of possible higher order constraints on access, the range of data that does exist strongly suggests that access of lexical information is unconstrained by higher order contexts during sentence comprehension. Lexical access appears to be a contextually independent process operating on a bottom-up principle of analysis, a fact that is not, however, represented in a number of influential theories of lexical processing (e.g., Forster, 1976; Morton, 1969) that were developed based on data taken from post-perceptual or non-sentential examinations of lexical processing.

It is interesting to note that this modular characterization of the language system has been somewhat mirrored in an entirely different domain—the processing of visual figures. Prather (1980), using a speeded decision technique *during* the processing of complex visual figures, has found evidence that even for very young children more than one possible organization for the lines comprising the figure are elaborated during the course of visual perception (even if the figure has one obviously correct and 'recognizable' organization). These data, taken from on-line perceptual analyses of visual processing, stand in contrast to the standard arguments about visual perception (i.e., that only one perceptual organization is perceived at a time) which, again, were obtained from post-perceptual report data (see e.g., Elkind, Anagnostopoulou, & Malone, 1970).

Just as distinguishing between on-line and post-perceptual analysis sheds light on issues critical to modeling cognitive processes and representations, so too can the contrastive examination of data from differing experimental environments (as was discussed earlier). For some time the most popular models of lexical processing have characterized the access procedure as a *terminating* ordered search (e.g., Forster 1976; Hogaboam & Perfetti, 1975). It has been argued that candidate word forms are compared against internal representations in order of their frequency, in a search that terminates once a contextually appropriate entry is encountered. Much of the data for this position, however came from isolated word processing techniques (lexical decisions made on isolated words or word pairs). When this claim was examined with an on-line technique (CMLP) *during* the course of sentence processing (Onifer & Swinney, 1981), it was found that even when the most highly frequent interpretation of an ambiguity was required by the sentential context, *all* interpretations of the ambiguity were momentarily accessed. These results argue against any type of terminating search in access *during sentential processing*—but point up the potential difference between processes operating in isolated word processing situations and sentential processing.

Finally, examination of the temporal characteristics of language processing leads to (sometimes surprising) evidence about the organization of the system, evidence that could never be discovered by simple intuitions. For example, it has been argued (Bobrow & Bell, 1973) that there is a special processing system for idiomatic interpretation. Evidence for this position came in the form of post-perceptual 'report' data in which it was found that only the literal *or* the idiomatic interpretation of a grammatical idiom (e.g., 'break the ice') was 'heard' by subjects, depending on the context. Again, however, analysis with more fine-grain processing measures (Swinney, 1981; Swinney & Cutler, 1979) has demonstrated that *both* idiomatic and literal interpretations are elaborated simultaneously for such materials, and there is simply no evidence to support a separate 'idiomatic mode' of processing. However, there *is* evidence from semantic priming studies that suggests that different ''rise times'' (a measure reflecting availability of information from lexical entries) exist for information that is: (1) a part of the literal (segmental) interpretation of a sentence; in contrast to that which is (2) derived from various inferential processes operating on this basic interpretation, such as might be found in metaphor comprehension. In general, information derived from such inferential analysis appears not to ''arise'' until a point considerably later than when it could first logically become available. Such evidence, although tentative, does suggest that we should consider these different types of informational analysis to constitute separate components of our language processing system.

GENERAL CONSTRAINTS ON USING PSYCHOLOGICAL DATA TO SUPPORT THEORETICAL POSITIONS IN THE COGNITIVE SCIENCES

The theory and research presented has been focused on demonstrating some of the underlying assumptions that psycholinguistic inquiry brings to cognitive science. One of the goals in this enterprise has been to provide evidence for a particular view of the nature of the psychological processes underlying language comprehension—namely, that the appropriate characterization is one of a modular, largely bottom-up system, that is composed of autonomous subprocesses each operating in an automatized and routinized fashion. Such a view is certainly not unique to psychological theorizing, although the real-time evidence presented here (and elsewhere) may provide somewhat stronger support for this characterization than has previously been possible.

There is a further consideration, however. It is largely (but not solely) in terms of its focus on language as a temporally distributed perceptual process, that the issues and assumptions detailed earlier differ from approaches to the question in a number of other fields. I have attempted here to demonstrate the value of tailoring both the theoretical constructs and the experimental techniques used to examine these constructs to the real-time facts of mental processing. Any model that fails to provide an account of the micro-structure of the operations involved in comprehending information in real time will fail to capture much of the essence of the mental structure underlying human language. However, it is precisely on this point that some further consideration of the role of psychological evidence in interdisciplinary studies in cognitive science is necessary. It is apparent in the work documented earlier that seemingly small changes in experimental technique can produce evidence supporting diametrically opposed theoretical positions: on-line phoneme monitoring results supported an interactive model of lexical processing, but on-line priming studies supported an autonomous modular view of that same process. To the theorist outside of psycholinguistics who is not familiar with the assumptions underlying the need for gathering real-time processing evidence (or about the sensitivity of various techniques to real-time processes) there is no basis for choice between the two sets of results. Thus, for example, it would be an obvious move for someone holding a ''frames'' model of language processing to assume that operations within a given ''frame'' are totally interactive (or, even, that the concept of time-dependent interaction is irrelevant to the theory) and to point to the psychological data from phoneme monitoring as support for such an interactive view. Were it simply that the data are wrong, the problem would be trivial. However, that is obviously not the case. The data simply reflect a level of processing that may be different from the level targeted for analysis with the technique. Thus, the ''frames'' theorist may be misleading himself by virtue of not recognizing that the data are not descriptive of the level he believes is being addressed, or, worse, by failing to recognize that the issue of levels is relevant here. Unfortunately, unless those using such data to support inferences are informed as to the nature of the issues motivating real-time analysis, the use of such data will always be inappropriate and, except by pure chance, incorrect. One can find such situations all too often in the literature, often by researchers who have the best of scholarly motives in attempting to examine their own work in the light of evidence from another field. As a simple example from a field outside of language, one can easily find general models of memory in AI that refer to experimental work in psychology as providing support for their position, where that data reflects only a single stage of memorial processing, one that may not be relevant to the level of description at which the AI model is operating. The point is that unless researchers in fields outside of cognitive psychology specify the level of mental processing that constitutes their target of inquiry, and understand the assumptions behind the generation of psychological data they assume supports their position, there cannot be a fruitful interdisciplinary exchange of information involving such evidence.

One might take the pessimistic view here that until we have a performance model detailed in its entirety we cannot accurately make such decisions about levels of analysis in our work—be it interdisciplinary or otherwise. However, I believe that this is an unnecessary constraint to apply. The approaches within the cognitive sciences can fruitfully borrow from each other even without a complete performance model, as long as the assumptions underlying work in each constitute mutually held conditions.

It seems that one of the most useful goals to be shared by the fields comprising cognitive science will be that of attempting to determine the structural and functional (independent) modules underlying mental ability, and the nature of their operation and interaction. In any such enterprise, however, it will be necessary to recognize that mental processes are not undifferentiated with respect to levels of analysis or operation over time. Thus, sensitivity to real-time processing issues—both theoretically and methodologically—seems critical to the growth of cognitive science, and, it seems likely that ''language'' will continue to provide a major forum for this work.

REFERENCES

Bobrow, S., & Bell, S. On catching onto idiomatic expressions. *Memory & Cognition,* 1973, *1,* 343–346.

Cairns, H. S., & Hsu, J. R. Effect of prior context upon lexical access suring sentence comprehension: A replication and reinterpretation. *Journal of Psycholinguistic Research,* 1979.

Cairns, H. S., & Kamerman, J. Lexical information-processing during sentence comprehension. *Journal of Verbal Learning and Verbal Behavior,* 1975, *14,* 170–179.

Cutler, A., & Foss, D. J. *Comprehension of ambiguous sentences. The locus of context effects.* Paper presented at the Midwestern Psychological Association, Chicago, May 1974.

Elkind, D., Anagnostopoulou, J., & Malone, S. Determinants of part-whole perception. *Child Development*, 1970, *41*, 391–397.

Forster, K. I. Accessing the mental lexicon. In R. Wales & E. C. T. Walker (Eds.), *New Approaches to language mechanisms*. Amsterdam: North Holland, 1976.

Forster, K. I. Levels of processing and the structure of the language processor. In W. E. Cooper & E. Walker (Eds.), *Sentence processing: Psycholinguistics studies presented to Merrill Garrett*. Hillsdale, N.J.: Lawrence Erlbaum Associates, 1979.

Foss, D. J. Some effects of ambiguity upon sentence comprehension: *Journal of Verbal Learning and Verbal Behavior*, 1970, *9*, 699–706.

Foss, D. J., & Jenkins, C. Some effects of context on the comrpehension of ambiguous sentences. *Journal of Verbal Learning and Verbal Behavior*, 1973, *12*, 577–589.

Garrett, M. F. Word and sentence perception. In R. Held, H. W. Liebowitz, & H. L. Teuber (Eds.), *Handbook of sensory physiology*. **Vol. VIII:** *Perception*. (Berlin: Springer-Verlag, 1978.

Hogaboam, T. W., & Perfetti, C. A. Lexical ambiguity and sentence comprehension. *Journal of Verbal Learning and Verbal Behavior*, 1975, *14*, 265–274.

Holmes, V. M., Arwas, R., & Garrett, M. F. Prior context and the perception of lexically ambiguous sentences. *Memory & Cognition*, 1977, *5*, 103–110.

Mackay, D. To end ambiguous sentences. *Perception of Psychophysics*, 1966, *1*, 426–436.

Meyer, D. E., Schvaneveldt, R. W., & Ruddy, M. G. Loci of contextual effects on visual word recognition. In P. M. A. Rabbit & S. Dornic (Eds.), *Attention and performance V*. London/New York: Academic Press, 1975.

Morton, J. The interaction of information in word recognition. *Psychological Review*, 1969, *60*, 329–346.

Neeley, J. Semantic priming and retrieval from lexical memory; Roles of inhibitionless spreading activation and limited capacity attention. *Journal of Experimental Psychology: General*, 1977, *106*, 226–254.

Onifer, W., & Swinney, D. Accessing logical ambiguities during sentence comprehension: Effects of frequence-of-meaning and contextual bias. *Memory & Cognition*, 1981,

Prather, P. *The development of perceptual processing: Organization and representation*. Doctoral dissertation, Tufts University, 1980.

Prather, P., & Swinney, D. *Some effects of syntactic context upon lexical access*. Presented at a meeting of the American Psychological Association, San Francisco, Calif., August 1977.

Swinney, D. Lexical access during sentence comprehension: (Re) Consideration of context effects. *Journal of Verbal Learning and Verbal Behavior*, 1979, *18*, 645–659.

Swinney, D. The structure and time-course of information interaction during speech comprehension: Lexical segmentation, access, and interpretation. In J. Mehler, S. Franck, E. C. T. Walker, & M. Garrett (Eds.), *Perspectives on Mental Representation*. Hillsdale, N.J.: Lawrence Erlbaum Associates, 1982.

Swinney, D. Understanding non-literal language: The temporal course of literal and non-literal interpretations. Paper presented at Psychomanic Society, 1981, Philadelphia, Pa.

Swinney, D., & Cutler, A. The access and processing of idiomatic expressions. *Journal of Verbal Learning and Verbal Behavior*, 1979, *18*, 523–534.

Swinney, D., & Hakes, D. Effects of prior context upon lexical access during sentence comprehension. *Journal of Verbal Learning and Verbal Behavior*, 1976, *15*, 681–689.

Swinney, D., Onifer, W., Prather, P., & Hirshkowitz, M. Semantic facilitation across sensory modalities in the processing of individual words and sentences. *Memory & Cognition*, 1979, *7*, (3), 159–165.

Tanenhaus, M., Leiman, J., & Seidenberg, M. Evidence for multiple stages in the processing of ambiguous words in syntactic contexts. *Journal of Verbal Learning and Verbal Behavior*, 1979, *18*, 427–440.

9 Competitive Argumentation in Computational Theories of Cognition

Kurt VanLehn
John Seely Brown
Xerox Palo Alto Research Center

James Greeno
University of Pittsburgh

During the past two decades, artificial intelligence and linguistics have had a major impact on the form of theories in cognitive psychology. Prior to about 1960, most theories in cognitive psychology considered information in relatively abstract terms, such as features, items, and chunks. Starting in the 1960s, and increasing during the 1970s, an additional theme in psychological theory has been to take into account the specific information that is present in the tasks that provide the material for theoretical analyses. The difference can be seen, for instance, in psychological analyses of problem solving that were developed in the 1950s, compared with analyses that have been worked out in the 1970s. In the earlier analyses, problem solving was considered as *selection* of a response (solution) that initially had low probability. Factors in the situation were examined for their facilitating or inhibiting effect on selection of the needed response. In computational models, specific task situations are represented, and programs are written that use the information in those representations to simulate processes of actually *constructing* solutions to specific problems.

The capability of analyzing the details of processing specific information is clearly an advance. For example, it enables psychological analyses of human behaviors that one would label ''understanding'' that are much more detailed than those provided previously. However, there is a well known danger to such an approach. Analyses can become mired in their increased precision and detail, with the result that it is extremely difficult to separate fundamental principles from their supporting detail. Yet explicit principles are needed in order to define or at least constrain the classes of processes and structures that are postulated.

235

In particular, when such detailed analyses aspire to be empirical theories, they face difficulties in achieving an adequate treatment of individual differences. In most analyses, there has been considerable obscurity in the boundary between what is meant to be true of all subjects, and what is meant to be true of a particular subject. To modify the knowledge base, rules, or other structures in order to fit a model to an individual subject is often easy enough, but how to place well defined limitations on such changes is an open problem. Identifying the universal components of a model and the general principles that constrain the processes is a start toward assessing the "degrees of freedom" of theories that tailor their predictions with non-numeric parameters (such as sets of rules). However, determining the tailorability of such models is at best vaguely understood. Yet it is crucial. A model may have so much tailorability that it can be tuned to match almost any data, rendering it vacuous.

Our concerns are similar to a number of recent contentions that the methodology of artificial intelligence (AI) and related fields is not productive for formulating and defending *theories* of mind (Dresher & Hornstein, 1976; Fodor, 1978/1981; Pylyshyn, 1980, forthcoming; Winograd, 1977). Unlike early criticism of AI (e.g., Dreyfus, 1972), which centered on whether computers could ever be intelligent, these critics concentrate on AI's supposed contributions to psychology. They note that despite its potential, well supported cognitive theories based on AI technology have not been forthcoming. We tend to agree with their conclusion, and offer a brief analysis of why this is so.

Approaches: top down, bottom up, neurological

AI has demonstrated that computer programs can behave with a certain degree of intelligence. However, no consensus has emerged concerning necessary or sufficient design principles for creation of intelligent programs, or even on the limitations imposed by the computational medium on intelligence. It currently appears that the goal of manifesting intelligence *per se* is not in itself constraining enough to force particular architectures or principles to be used.

The failure of AI to demonstrate the existence of "top down" constraints on cognitive theories, principles inherent in the nature of cognition independent of the medium of its implementation, suggests searching "bottom up," starting with concrete instances of intelligent human behavior. That is, given that the goal is to find out how human cognition works, it currently seems advisable to ground the models/analyses in human data. In this chapter, we consider only empirical theories, or rather, the problems of obtaining empirical theories, that maintain the advantages of an AI-like treatment of task information while striving to meet scientific criteria for empirical theories.

The technology of AI adapts readily to a bottom up approach. There are programs that simulate a subject's behavior quite faithfully and in considerable detail (e.g., even eye movements during problem solving can be predicted. Cf.,

Newell & Simon, 1972). However, despite the addition of empirical responsibility, the current methodology of employing simulation models is still weak in several respects. There has been virtually no attention to the tailorability (degrees of freedom) of simulation models. There has been little argumentation for the individual principles and components of the model. Because the entailments of each principle have not been separated from the performance of the model's simulation as a whole, one is asked to accept the model *in toto* with no explanation as to why it has the principles it does. Although questions of observational adequacy have been treated, questions of descriptive and explanatory adequacy have been almost universally ignored. Indeed, without the additional consideration of tailorability, measures of descriptive and observational adequacy have little meaning.

Some have asserted that it is necessary to add a third kind of constraint by moving to the periphery, where neurological data can be brought to bear on information processing theories. Although this has yielded some exciting results (Marr, 1976; Marr & Nishihara, 1978), the findings obtained thus far seem not to provide strong constraints on the hypotheses about processes such as language comprehension or problem solving.

The missing ingredient for scientific progress is not behavioral or neurological data, we believe, but scientific reasoning that explicates the principles underlying successful models of cognition and connects them with the data. The emphasis must be on the connection; explication alone is not sufficient. Efforts at explicating programs have increased in response to critics (e.g., Kaplan, 1981) who point out that a typical AI/Simulation "explanation" of intelligent behavior is to substitute one black box, a complex computer program, for another, namely the human mind. Extracting the principles behind the design of the computer program is a necessary first step. But many other questions remain to be addressed: What is the relationship between the principles and the behavior? Could the given cognition be simulated if the principles were violated or replaced by somewhat different ones? Would such a change produce inconsistency, or a plausible but as yet unobserved human behavior, or merely a minor perturbation in the predictions? Which alternatives if any can now be rejected in favor of the chosen principles? This connection of explicit principles to the data seem to be critical to progress in computational theories of cognition.

Nature and Importance of Arguments

Computer science has given psychology a new way of expressing models of cognition that is much more detailed and precise than its predecessors. But unfortunately, the increased detail and precision in *stating* models has not been accompanied by correspondingly detailed and precise arguments *analyzing and supporting* them. Consequently, the new, richly detailed models of cognitive science often fail to meet the traditional criteria of scientific theories. By *sup-*

port, we refer to various traditional forms of scientific reasoning such as showing that specified empirical phenomena provide positive or negative evidence regarding hypotheses, showing that an assumption is needed to maintain empirical content and falsifiability, or showing that an assumption has consequences that are contradictory or at least implausible. It is of course not new to desire that cognitive theories have this kind of support. Nor is it a new contribution to point out that current theorizing based on computational models of cognition has been lax in providing such support (Pylyshyn, 1980; Fodor, 1981). Perhaps what is new is discussing how such supporting argumentation could be developed for computational models of the mind.

The focus of this discussion is on the kinds of arguments that are applied to specified theoretical principles. We're not going to advocate building a particular kind of theory, nor do we wish to dispute over criteria for theories (e.g., falsifiability, tailorability). Instead, we discuss what kinds of tools are available or can be fashioned that will help one build computational theories of cognition that will meet some widely accepted standards that have so far proved difficult for such theories to meet. The prime tool of this discussion, actually a class of tools, is the *competitive argument.* Unlike famous technological tools of scientific advance, such as microscopes or Golgi stains, which offered clearer views of tissue structure and other factual material, competitive argument seems to be a tool for analyzing and clarifying the theoretical issues implicit in a computational model of a cognitive faculty.

There often is a great need for such a clarifying instrument when the model under development employs computations. The relationship between a principle and data must often be an indirect one and can take several forms. For instance, a principle might serve as a constraint on a class of processes, perhaps by defining a processing language and an interpreter; the processes in turn might have a mapping to observable behavior defined, for example, by some grain-size assumptions. The indirectness of the relationship between principles and data provides additional reasons for providing clear and adequate supporting argumentation. With appropriate argumentation, it is possible to show, among other things, that it is the principles that are responsible for the computations' empirical coverage and not some obscure or accidental details of the particular computer implementation of the theory. Moreover, the arguments can show that the principles have some force in that they are refutable.

Of course, the idea of argumentation related to specific theoretical principles is not new, and many examples could be listed showing how psychologists have considered principles in relation to their supporting evidence, their testability, and the plausibility of their consequences. However, argumentation regarding specific principles has been relatively rare in computational theories. This is partly because theorists have spent their major effort coming to grips with the precision and subsequent detail of the computational medium and of course with the details of fine-grained, conceptual task analyses.

We view argumentation regarding specific principles as a part of a natural progression. The progression includes stages of task analysis, articulation of principles, and competitive argumentation. In the third section of the chapter, we give an example of the evolution of one particular theory through these stages.

At the beginning of the study of a task domain, a great deal of analysis is necessary before even a crude simulation of behavior is possible. One must learn what details of the task should be included and what can be suppressed. Early examples such as Newell, Shaw, and Simon's (1963) model of proving theorems in logic, Bobrow's (1968) model of solving algebra word problems, and Evans' (1968) model of solving analogy problems (to name just a few examples) were valuable contributions partly because they showed that their mechanisms were *sufficient* for producing correct solutions for an interesting variety of problems in their respective domains. In psychological studies such as Newell and Simon's (1972) models of performance in cryptarithmetic, logic exercises, and chess, and Simon and Kotovsky's (1963) model of solving series completion problems, the sufficiency criterion has been extended to require general similarity to performance by human problem solvers. By and large, the first venture into a task domain does not yield a precise articulation of the principles that structure competence in it. But it is important to emphasize just how difficult it is to forge these first formalizations, and how much is learned from them. Furthermore, the understanding of processes involved in performance provides an essential resource for subsequent investigation of the task domain.

Such a subsequent investigation often aims at clarifying the general principles or components that mastery of the task involves. Often there is a separation of highly specific task information from more general information. Ernst and Newell's (1969) discussion of the General Problem Solver is a classic example. The general procedures of means-ends analysis were implemented as a distinct program that was run in conjunction with representations of several different problem domains. Similar remarks apply to natural language understanding systems, such as Winograd's (1972) and Woods' (1972), that separated a general syntactic parser from a task specific lexicon and grammar, and a semantic representation language from some task-specific information written in it. Hypotheses that identify some of the components of a theory as being relatively more general than others provide a step in the direction of making a theory principled, in the sense that we propose.

In our view, a significant strengthening of computational theories can be achieved by explicating their principles and the *entailments* of the principles. We believe that a substantial advance is achieved if a theory can be developed beyond being a black box with a certain measured adequacy. This involves laying bare the theory's principles and their entailments, showing how each principle, *in the context of its interactions with the others,* increases empirical coverage, or reduces tailorability, or improves the adequacy of the theory in some other way. With such developments, the theorist provides explanations of *why* the particular principles were chosen. The support structure for each principle is laid bare.

The internal structure of the theory—the way the principles interact to entail empirical coverage and tailorability—comes out best when the theory is compared with other theories and with alternative versions of itself. That is, the key to supporting theories appears to be *competitive argumentation*. This style of support has succeeded in certain deep, non-computational theories. We suggest that it can be adapted to the increased rigor and detail of computational theories.

In practice, most competitive arguments have a certain "king of the mountain" form. One shows that a principle accounts for certain facts, and that certain variations or alternatives to the principle, while not without empirical merit, are flawed in some way. That is, the argument shows that its principle stands at the top of a mountain of evidence, then proceeds to knock the competitors down. The second section of this chapter illustrates the notion of such an argument with an example.

Competitive arguments hold promise for establishing which principles are crucial for analyzing cognition. To show that some constraint is crucial is to show that it is *necessary* in order for the theory to meet some criteria of adequacy. To show that it is *sufficient* is not enough. Indeed, any successful theory that uses some principle is a sufficiency argument for that principle. But when there are two theories, one claiming that principle X is sufficient and another claiming that a different, incompatible principle Y, is sufficient, sufficiency itself is no longer persuasive. One must somehow show that X is better than Y. Indeed, this sort of competitive argumentation is the only realistic alternative to necessity arguments. *Competitive arguments form a sort of successive approximation to necessity.*

Argumentation in non-computational fields

Well reasoned, competitive argumentation occurs in non-computational fields. Linguistics is a particularly good example. Throughout its history, linguistics has had a strong empirical tradition. Prior to Chomsky, syntactic theories were rather shallow and almost taxonomic in character. The central concern was to tune a grammar to cover all the sentences in a given corpus. Arguments between alternative grammars could be evaluated by determining which sentences in the corpus could be analyzed by each. When Chomsky reshaped syntax by postulating abstract remote structures, namely a base grammar and transformations, argumentation had to become much more subtle. Because transformations interacted with each other and the base grammar in complex ways, it was difficult to evaluate the empirical impact of alternative formulations of rules.

As Moravcsik has pointed out (Moravcsik, 1980), Chomskyan linguistics is virtually alone among the social sciences in employing deep theories. Moravcsik (1980) labels theories "deep" (without implying any depth in the normative sense) if they "refer to many layers of unobservables in their explanations. . . . 'Shallow' theories are those that try to stick as close to the observables as possible, [and] aim mostly at correlations between observables. . . . The history of the natural sciences like physics, chemistry, and biology is a clear record of the success story of 'deep' theories. . . . When we come to the social sciences, we encounter a strange anomaly. For while there is a lot of talk about aiming to be 'scientific,' one finds in the social sciences a widespread and unargued-for predilection for 'shallow' theories of the mind [p. 28]."

Computational theories of cognition are "deep" theories because much of the mechanism and representation that they postulate is quite unobservable. This depth is another reason that argumentation has been rare in computational theories. The principles and components that structure the theory's computation are remote from the data. The derivation of predictions from them is often so lengthy and convoluted that only by executing the computation on the computer can the theorist tell what the current version of the theory predicts. To assign empirical responsibility to a *component* of the remote structure is possible only in rare cases. The depth of computational theories makes establishing the empirical necessity of their principles or architecture extremely difficult, even given that their sufficiency has been demonstrated.

When theories are "shallow," then argumentation is easy. In a sense, the data do the arguing for you. Most experimental psychology is like this. The arguments are so direct that the only place they can be criticized is at the bottom, where the raw data is interpreted as findings. Experimental design and data analysis techniques are therefore of paramount importance. The reasoning from finding to theory is often short and impeccable. On the other hand, when theories are deep in that the derivation of predictions from remote structures is long and complex, argumentation becomes lengthy and intricate. However, the effort spent in forging them is often repaid when the arguments last longer than the theory. Indeed, each argument is almost a micro-theory. An argument's utility may often last far longer than the utility of the theory it supports. This utility may take the form of a crucial fact, as discussed later.

The transition from studying shallow theories to studying deep ones is sometimes accompanied by something like culture shock. The heightened concern with intricate argumentation from apparently casual empirical observations strikes some as totally unwarranted. Yet this concern can be highly significant, since it can show where the theoretical connections are weakest. It is here that theories draw fire from their critics, and rightfully so, for there are many trails between surface findings and remote structures, and the first one traversed is not always the best.

There is another orientation that can block acceptance of argumentation. Cognitive science often equates the merit of a theory with the complexity of the task domain that it addresses. This is entirely appropriate when the theory consists solely of a task analysis. If the task domain is trivial, an analysis of its information structure is often trivial, or at least less interesting than an analysis of a complex task. As we suggested earlier, task analyses are and will continue to be an important first step in cognitive studies. They have dominated the field since the 1950s and distinguish cognitive psychology from its more task-information-free predecessors. Their dominance has made it almost inevitable that the complexity of the theory's task domain has been strongly associated with the theory's perceived merit, even if that theory goes well beyond an analysis of its domain. Often, the complexity of argumentation forces the research to be conducted in simple task domains. Since the task analysis is only a first step and a relatively easy one compared to eliciting principles and constructing supporting arguments, simple task domains are a good choice for such research. Moreover, the simplicity of the domain may allow discovery and sharpening of research tools, the cognitive equivalent of the stains of neuroanatomy or the restriction enzymes of microbiology. Such tools, developed in a simple domain, can be used to illuminate a whole area. Yet, choosing a simple domain makes it difficult to attract the attention of the cognitive science community, which often equates simple tasks with trivial theories. Yet this attention is sorely needed, not only to encourage the investigators, but because competitive argumentation, even more so than other forms of support, thrives on challenge. The entirely appropriate habits of the early years of cognitive science seem to be dampening the emergence of principled, well-argued theories.

Crucial facts

Ultimately, every theory is abandoned. In the long run, it might seem that the effort spent on careful argumentation is wasted since argumentation is so thoroughly embedded in the theory's framework. We do not believe this is the case. A long term effect of an argument is to raise the prominence of a certain set of observations, making them "crucial facts," in the linguistic jargon. For example, a previously obscure class of sentences (the "promise-persuade" sentences) became important as supporting data for two transformations of early transformational grammars (Raising and Equi-NP Deletion). A typical pair from this class is

> Ronald promised Margaret to pay the bill.
> Ronald persuaded Margaret to pay the bill.

Promise-persuade data have shaped *every* successor of transformational grammars. They have become, like active/passive sentence pairs, crucial facts that any serious grammar must explain.

Another example comes from behavioral theory in the domain of instrumental conditioning. Hull (1943) formulated a hypothesis in which the strength of a tendency to respond depends on a global motivational factor, drive, and a factor depending on the organism's experience, habit strength. The habit strength of a response was assumed to depend both on the number of times the response had occurred and been followed by reinforcement, and on the amount of reinforcement received. In an ingenious experiment, Tolman and Honzik (1930) placed rats in a maze and permitted them to move about but provided no apparent reinforcement. Then food was placed in a specific location, and the rats were given learning trials. The rats with exploratory experience learned to go to the food more quickly than rats that had not received the experience. This phenomenon of *latent learning* played an important role in the further development of behavior theory.

Commonly, crucial facts play the role of deciding among two important hypotheses. They could almost be called "decisive" facts for this reason. The two examples above were decisive. Promise-persuade sentences demonstrate that there must be two distinct transformations operating in roughly the same syntactic domain. Although the surface syntax is similar, advocating a single-transformation approach, the apparent differences in which noun phrase is the implicit subject of the subordinate clause (i.e., who pays the bill) shows that the more complex hypothesis of two transformations is necessary. Similarly, the fact of latent learning played a decisive role. It showed that the effects of experience may not be reflected directly in performance. The distinction between learning and performance had to be made more complex. In Hull's theory, latent learning was accommodated by changing the assumption that habit strength grows as a function of the amount of reinforcement. In later versions of the theory (e.g., 1952) Hull assumed that habit strength depends only on the *number* of occurrences of the response that have been followed by reinforcement, and not on the *amount* of reinforcement. To account for the affect of amount of reinforcement, Hull assumed it acts as a motivational factor, called incentive, which influences performance rather than learning. Hull was then able to account for latent learning by arguing that some small amount of reinforcement probably was provided in the exploratory trials—for example, the rat likes going home to its cage, so it takes removal from the maze as reinforcement for its last few moves. Since there was reinforcement, there was learning. But the amount of reinforcement was small, so the incentive to exhibit this learning was small. Hence, the rat's learning was not reflected in performance until food was introduced.

Few, if any, crucial facts have emerged as yet in relation to computational theories of cognition. For example, SHRDLU's famous sentence "Pick up the big red block," is not considered a necessary part of a parser's repertoire, nor must a learner learn to recognize an arch made of blocks, despite the centrality of these examples in Winograd's and Winston's work. The reason no old examples seem worth accounting for in new theories is that no theoretical issues hinged on them in the old theories. They were used for illustration, and not to argue principles.

Arguments and crucial facts often survive longer than theories.   Theorists repeatedly appeal to crucial facts, we believe, because they already know quite a few ways of accounting for these facts, most of which are somehow flawed. This makes them convenient tests for a new theory. Although the counter-arguments will most often not lift directly over to the new theory, they will at least hint at where to look for trouble.

We suggest that long term progress may be exactly the accretion of crucial facts, arguments, and possibly even techniques or types of arguments. Theories come and go. Wherein lies our accumulated knowledge? Perhaps the knowledge that separates a "mature" field from a young one is the empirical ramifications and other entailments of its ideas—the arguments connecting them to the facts and to each other.

Competence theories and star data

Recently, efforts have been increasingly directed at formulating theories of what people *can* do rather than what they *did* do in the given situation. In part, this follows from the realization that the performance of subjects when confronted with a task is strongly determined by the requirements of the task. To use Simon's metaphor (Simon, 1981), studying the path of an ant across the beach tells one more about the beach than the ant. To learn something about general principles of cognition (the ant), it is necessary to abstract detailed observations (the path). This introduces a level of inference to that usually encountered when performance is used for inferences about underlying cognitive processes and structures. Founding this new level, of course, involves argumentation.

Another reason for emphasizing competence (*can* do) over performance (*does* do) is that it allows side stepping, to a certain degree, the difficult issue of allowing some tuning of performance parameters around individual differences without introducing unlimited degrees of tailorability into the theory.

Assessing the underlying competence is no easy empirical task. The conclusions one draws can be colored by subtle variations in the task. An illustrative case is the controversy over the development of competence in counting. Gelman's classic "magic" experiments showed that the elements of number competence are present much earlier than strict Piagetian theory would predict (Gelman & Gallistel, 1978).

A successful style of argumentation has emerged in service of competence theories. If one defines a performance theory as a theory of what the subject *does* do, and a competence theory as a theory of what the subject *can* do, then clearly evidence about what the subject *can't* do will be very important for narrowing the range of behavior allowed by a competence theory. Such evidence is called *star data.*

A star datum is a simulated behavior that no human's behavior would ever match. It is named after the lingustic convention of starring sentences that are not in the language. How one ascertains the non-existence of a behavior may vary in different domains. By definition, star data are behaviors that are not found naturally. Thus, star data can not be observed in the same (potentially) objective ways that ordinary data are observed. In particular, in many domains, a subject can easily perform the behavior when asked to (e.g., utter "Furiously sleep ideas green colorless") even though they would never do so naturally. In current linguistic practice, sentences are starred according to the judgment of native speakers. For Repair Theory (the theory that the forthcoming example is drawn from), expert diagnosticians were the judges of whether the model's behavior represented a star datum or plausible human behavior that just hadn't been observed yet.

Star data can be particularly useful in supporting claims about mental representations. One of the techniques that a theorist can use to structure knowledge is to stipulate that it must be represented in a certain formalism, often called the representation language. A narrowly defined language reduces the ways that a piece of knowledge can be decomposed by forcing its decomposition to fit into the forms allowed by the representation language. This technique raises formalisms from the status of mere notations to bearers of important theoretical claims. Star data can be useful in supporting such claims: If it can be shown that the theory would generate certain star data if it were not constrained by the given representation language, then one has a strong argument for the utility, if not the actual psychological reality of that representation language. If the representation language is extremely successful in constraining knowledge structures, one might even be inclined to propose it as a mental representation (Fodor, 1975).

AN EXAMPLE OF AN ARGUMENT

In this section, we give an example of an argument. However, arguing is impossible in vaccuo, so several paragraphs must be spent in describing the theory the argument is taken from and the data that supports it. Like most arguments, this one depends strongly on assumptions of the theory which themselves have supporting arguments. This leads to rather involved, convoluted reasoning—a characteristic of competitive argumentation exacerbated by the complexities of protocol data.

The argument presented here is taken from VanLehn (1983). It concerns Repair Theory (Brown & VanLehn, 1980). Repair Theory examines certain cognitive aspects of procedural skills. The basic idea of Repair Theory is that while following a procedure students will barge through any trouble that they encounter by doing only a small amount of superficial "patching." That is, they make a minimal change to the procedure's execution state in order to circumvent the trouble and get back on the track. Sometimes they perform different repairs for the same trouble, other times they use the same repair for periods of time, and sometimes they seem to abstract the patch and make it a part of their procedure. Repair Theory is concerned with formalizing this overall impression so that it can be tested. In doing so, it aims to articulate a formal representation for knowledge about procedures—a *mentalese* for procedures (Fodor, 1975)—and an architecture for interpreting that knowledge and for managing trouble during its interpretation.

The procedural skill taken for study is ordinary multi-column subtraction. The subjects are elementary school students who are in the process of learning that procedure. The main advantage of these choices, from a psychological point of view, is that for these subjects, subtraction is a virtually meaningless procedure. Most elementary school students have only a dim conception of the underlying semantics of subtraction, which are rooted in the base-ten representation of numbers. When compared to the procedures they use to operate vending machines or play games, subtraction is as dry, formal, and disconnected from everyday interests as the nonsense syllables used in early psychological investigations were different from real words. This isolation is the bane of teachers but a boon to the cognitive theorist. It allows one to study a skill formally without bringing in a whole world's worth of associations. This isolation provides an elegant opportunity for building a microscope into the mentalese of procedural knowledge and the architecture for interpreting it.

The data supporting the theories comes from the Buggy studies (Brown & Burton, 1978; VanLehn, 1981). From the errors of thousands of students taking ordinary pencil and paper subtraction tests, a hundred primitive *bugs* were inferred. Bugs are a formal device for notating systematic errors in a compact way. They are designed so that any of the observed systematic errors can be expressed by a set of one or more bugs. (Actually, bugs often do not combine linearly. The co-occurrence of two or more bugs is called a "compound" bug.) This notation is precise in that it describes not only what problems are answered incorrectly, but what the contents of those answers were and what steps were followed in producing them. The technicalities of this form of data have been discussed in other papers (Brown & Burton, 1978; Burton, 1981). The behavioral data that support the arguments will be presented here as ideal protocols of the subject's local problem solving. This simplifies the exposition considerably.

A critical distinction in Repair Theory is between regular execution and *local problem solving*. Regular execution depends of course on what the representation of procedures is. If procedures are represented as pushdown automata, for instance, regular execution is just following arcs, pushing and popping the stack, testing arc conditions, executing primitive arc actions, and so forth. If the procedure is somehow flawed, perhaps because the student mislearned it or forgot part of it, then regular execution may get stuck. When a student executing a procedure reaches an impasse, the student is unlikely to just halt, as a pushdown automaton does when it can't execute any arcs leaving its current state. Instead, the student will do a small amount of problem solving, just enough to get unstuck and resume regular execution. These local problem solving strategies are called *repairs* despite the fact that they rarely succeed in rectifying the broken procedure, although they do succeed in getting past the impasse. Repairs are quite simple tactics, such as skipping an operation that can't be performed or backing up in the procedure in order to take a different path. Repairs do not in general result in a correct solution to the exercise the procedure is being applied to, but instead result in a buggy solution. The theory explains the large variety of observed subtraction bugs as the result of a few flawed underlying subtraction procedures being subjected to local problem solving involving a surprisingly small set of repairs. Local problem solving is twofold: detecting impasses, called *criticism*, and getting around them, called repair. Some bugs result from several instances of impasse/repair during the application of an underlying flawed procedure to a problem.

There are strong constraints on criticism and repair. Both criticism and repair are very simple and local. Two main types of criticism are detecting when an action's precondition is violated (e.g., trying to decrement a zero) and detecting when regular execution halts because none of its methods are applicable to the current situation. Repairs are also simple and local: for example, skipping an operation or backing up in the procedure. A basic principle of Repair Theory is that *any repair can be applied at any impasse,* subject only to the condition that it succeed in getting the procedure past the impasse. The empirical impact of this principle is that the theory predicts that *the set of all possible bugs is exactly the set of all possible repairs applied to all possible impasses.* To summarize: the student's observed behavior is a combination of regular execution and local problem solving, where local problem solving consists of detecting an impasse and applying one of a set of repairs to it.

Assumptions

Although a great deal more can be said about Repair Theory and bugs, it is time to turn to the illustrative argument. It concerns the architecture/representation that the theory uses to model human conceptions of procedures, and in particular, the component of the architecture that is called the short-term or working memory in production systems. This argument is part of a longer argument that the representation should be an applicative language. The principle at issue here is how the model should represent focus of attention. As students solve a subtraction problem, they focus their attention on various digits or columns of digits at various times. This can be inferred from the information that students read from the paper, or perhaps from eye-tracking studies. For this chapter, it will be assumed that focus and focus shifting exist. The issue to discuss is how to represent them. A leading contender will be a "you are here" register that stores the focus, that is, where on the paper the attention is being focused. This register, in tandem with a "currently active goal" register, puts a clean division between focus of attention and control state. However, register-based architectures will be shown to make focus *too* independent of control state. A different architecture, one that unites focus and control, will be shown to fit the facts more closely. It's named schema/instance for reasons that will become clear in a moment.

We would like to speak in an informal way of goals and subgoals, intending that these be taken as referring to the procedural knowledge of subtraction itself, rather than expressions in some particular representation (e.g., production systems, and-or graphs, etc.). In particular, we'll assume that borrowing is a subgoal of the goal of processing a column, and that borrowing has two subgoals, namely borrowing-from and borrowing-into. Borrowing-into is performed by simply adding ten to a certain digit in the top row. The borrowing-from subgoal is realized either by decrementing a certain digit, or by invoking yet another subgoal, borrowing-across-zero. (This way of borrowing is not the only one, but it was the one taught to all our subjects.) These assumptions, or at least some assumptions, are necessary to begin the discussions. They are, we believe, some of the mildest assumptions one can make and still have some ground to launch from. We've assumed that focus exists and that a goal-structured control regime exists. One other assumption is needed before the main argument can be presented. We'll assume that a repair called *Backup* exists. This repair is most easily described with an example of its operation.

The Backup repair in action

Figure 9.1 gives an idealized protocol. It illustrates a moderately common bug (Smaller From Larger Instead of Borrow From Zero). In a sample of 417 students with bugs, five students had this bug (VanLehn, 1981). The (idealized) subject of Fig. 9.1 does not know all of the subtraction procedure. In particular, he does not know about borrowing from zero. When he tackles the problem 305–167, he begins by invoking a process-column goal. Since 5 is less than 7, he invokes a borrow subgoal (episode *a,* see Fig. 9.1), and immediately the first of borrowing's two subgoals, namely borrowing-from (episode *b*). At this point, he gets stuck because the digit to be borrowed from is a zero, which can not be decremented in the natural number system. In Repair Theoretic terms, he has reached an *impasse.*

The theory describes several *repair strategies* that can be used at impasses to get unstuck. The one that interests us here is the Backup repair. It gets past the decrement-zero impasse by "backing up," in the problem solving sense, to the last goal that has some open alternatives. In this case, there are four goals active:

borrowing-from
borrowing
processing a column
solving the subtraction problem

a. **305**   In the units column, I can't take 7 from 5, so I'll have to
   −167   borrow.

b. **305**   To borrow, I first have to decrement the next column's top
   −167   digit. But I can't take 1 from 0!

c. **305**   So I'll go back to doing the units column. I still can't take 7
   −167   from 5, so I'll take 5 from 7 instead.
   2

   2 1
d. **305**   In the tens column, I can't take 6 from 0, so I'll have to
   −167   borrow. I decrement 3 to 2 and add 10 to 0. That's no
   2   problem.

   2 1
e. **305**   Six from 10 is 4. That finishes the tens. The hundreds is easy,
   −167   there's no need to borrow, and 1 from 2 is 1.
   142

FIG. 9.1   An idealized protocol of a student with the bug Smaller From Larger Instead of Borrow From Zero.

The borrowing-from goal has failed. The borrow goal has no alternatives: one always borrows-from then borrows-into. The next most distant goal, namely column-processing, has alternatives: one for columns that need a borrow, one for columns that do not need a borrow. So Backup returns control to the column-processing goal. Evidence for backing up occurs in episode c, where the subject says "So I'll go back to doing the units column." In the units column he hits a second impasse, saying "I still can't take 7 from 5," which he repairs ("so I'll take 5 from 7 instead"). He finishes up the rest of the problem without difficulty.

The crucial feature of the analysis above, for this argument, is that Backup caused a transition from a goal (borrowing-from) located at the top digit in the *tens* column to a prior goal (processing a column) located at the *units* column. Backup caused a shift in the focus of attention from one location to another as well as from one goal to another. Moreover, it happens that the location it shifted back to was the one that the process-column goal was originally invoked on, even though that column turned out to cause problems in that further processing of it led to a second impasse. So, it seems no accident that Backup shifted the location back to the goal's original site of invocation. It is because Backup shifts both *focus* and *control* that it is the preeminent tool to be used in the argument that follows.

Overview

The issue is how to represent focus. To keep the argument short, just two alternative architectures are discussed: register-based and schema/instance. As it turns out, just two facts are needed. One is the bug just described. The other will be introduced in a moment. The argument is organized around these two facts. Figure 9.2 is an outline of the argument.

It will be shown that the schema/instance architecture generates the first bug (I.A in the outline), but the simplest version of the register-based architecture, a single focus register, generates a star bug instead (I.B.1). Patching its difficulties by making the Backup repair more complex leads to problems with retaining the falsifiability of the theory (I.B.2). However, using several registers instead of just one register allows the bug to be generated simply (I.B.3). So the conclusion to be drawn from the first fact will be that the register approach will be adequate only if there is more than one focus register.

The second part of the argument introduces a new bug involving the Backup repair. Once again, the schema/instance architecture predicts the fact correctly (II.A). Two different implementations of multiple registers fail (II.B.1 and II.B.2) by generating star bugs. Smart Backup would fix the problem but remains methodologically undesirable (II.B.3). Postulating various complications to the goal structure of the procedure (II.B.4 and II.B.5) allow the correct

I. Fact 1: Smaller From Larger Instead of Borrow From Zero
   A. The schema/instance architecture generates it
   B. Registers
      1. A single register architecture generates a star bug
      2. "Smart" Backup is irrefutable, hence rejected on methodological grounds
      3. Multiple register architecture allows generation of SFLIBFZ

II. Fact 2: Borrow Across Zero
   A. The schema/instance architecture generates it
   B. Registers
      1. One register per goal: can't generate the bug
      2. One register per object: can't generate the bug
      3. "Smart" Backup is irrefutable, hence rejected on methodological grounds
      4. Duplicate borrow-from goals: entails infinite procedure
      5. Duplicate borrow goals: equivalent to schema/instance

FIG. 9.2   Outline of the argument.

predictions to be generated, but they have problems of their own. So the second part concludes that the register-based alternative is inadequate even when various complex versions of it are introduced.

In overview, the argument is a nested argument-by-cases where all the cases except one are eliminated. To aid in following it, the cases are labelled as they are in the outline of Fig. 9.2.

I.A Schema/instance generates the bug

The basic idea of a schema/instance architecture is that the location of a goal is very strongly associated with the goal at the time it is first set. That is, when a goal like borrowing is invoked, it is invoked at a certain column, or more generally at a certain physical location in the visual display of the subtraction problem. In a schema/instance architecture, this association between goal and location, which is formed at invocation time, persists as long as the goal remains relevant. That is, the goal is a *schema,* which is *instantiated* by substituting specific locations, numbers, or other data into it. Most modern computer languages, such as LISP, have a schema/instance architecture: a function is instantiated by binding its arguments when it is called, and its arguments retain their bindings as long as the function is on the stack.

The schema/instance architecture allows the bug of Fig. 9.1 to be generated quite naturally. Suppose the process-column goal were strongly associated with its location, namely the units column, in some short-term memory. Then Backup causes the resumption of the goal at the stored location. Another way to think of this is that the interpreter is maintaining a short-term "history list" that temporarily stores the various invocations of goals *with their locations.* In regular execution, when the borrowing goal finishes, the process-column goal is resumed *at the same place as it started.* That is, in the long-term representation of the procedure, the process-column goal is a schema with its location abstracted out. It is bound to a location (instantiated) when it is invoked. It is the instantiated goal that Backup returns to, not the schematic one.

This schema/instance distinction, which is at the heart of almost all modern programming languages, entails the existence of some kind of short-term memory to store the instantiations of goals, and thus motivates this way of implementing the Backup repair. But there are of course other ways to account for focus shifting during Backup. Several will be examined and shown to have fewer advantages than the schema/instance one.

I.B.1 A single register architecture generates a star bug

Suppose that instead of using the schema instances to implement focus storage, the architecture used a single register, a "you are here" pointer to some place in the problem array. There would be no problem representing the subtraction procedure in such an architecture. In order to shift focus left, for example, an explicit action in the procedure's representation would change the contents of this register as the various goals were invoked.

Suppose first off that Backup is kept as simple as possible, and in particular, that it doesn't go rooting around in the procedure in order to find out how to reset the register. Under this parsimonious account of Backup, the single register architecture causes trouble. Because the "you are here" register is simply left alone during backup, a star bug is generated. It is illustrated in Fig. 9.3. At episode *b*, Backup resumes the process-column goal, but the "you are here" register is not restored to the units column. Instead, the tens column is processed. The units column is left with no answer despite the fact that its top digit has been incremented. In the judgment of several expert diagnosticians, this behavior would never be observed among subtraction students. It is a *star bug*. The theory should not predict its occurrence.

I.B.2 Smart Backup repair makes the theory too tailorable

To avoid the star bug, the Backup repair would have to employ an explicit action to restore the register to the units column in episode *c* of the protocol of Fig. 9.1. But how would it know to do this? Backup would have to determine that the focus of attention should be shifted rightwards by doing an analysis of the goal structure contained in the stored knowledge about the procedure. It would see that in normal execution, a locative focus shifting function was ex-

a.
$$\overset{1}{4}02$$
$$-106$$
In the units column, I can't take 6 from 2, so I'll have to borrow. First I'll add ten to the 2.

b.
$$\overset{1}{4}02$$
$$-106$$
I'm supposed to decrement the top zero, but I can't! So I guess I'll back up to processing the column.

c.
$$\overset{1}{4}02$$
$$-106$$
$$\overline{0}$$
Processing it is easy: 0−0 is 0.

d.
$$\overset{1}{4}02$$
$$-106$$
$$\overline{30}$$
The hundreds is also easy. I'm done!

FIG. 9.3   An idealized "protocol" for a star bug.

ecuted between the borrow-from goal and the process-column goal. For some reason, it decides to execute the *inverse* of this shift as it transfers control between the two goals.

Not only does this implementation make unmotivated assumptions, but it grants Backup *the power to do static analyses of control structure*. This would give it significantly more power than the other repair heuristics, which do simple, local things like skipping an operation, or executing it on slightly different locations. It gives the local problem solver so much power that one could "explain" virtually any behavior by cramming the explanation into the black boxes that are repair heuristics. That is, allowing repairs to do static analyses gives the theory too much tailorability. It is much better to make the heuristics as simple as possible by embedding them in just the right architecture.

I.B.3 Multiple register allow generation of the bug

Another way to implement Backup involves using a *set* of registers. The registers have some designated semantics, such as "most recently referenced column" or "most recently referenced digit." That is, the registers could be associated with the type or visual shape of the locations referenced (e.g., as Smalltalk's class variables are). Alternatively, they could be associated with the schematic goals. (Some Fortran compilers implement a subroutine's local variables this way by allocating their storage in the compiled code, generally right before the subroutine's entry point.) Process-column would have a register, borrow would have a different register, and so on.

Given this architecture, Backup is quite simple. Returning to process-column requires no locative focus shifting on its part. Since the process-column register (or the column register, if that's the semantics) was not changed by the call to borrow, it is still pointing at the units column when Backup causes control to return to process-column. This multi-register implementation is competitive with the schema/instantiation one as far as its explanatory power (i.e., Backup is simple and local, and the architecture has motivation independent of the Backup repair in that is used during normal interpretation). However, it fails to account for certain empirical facts that will now be exposed.

II.A Another bug, and schema/instance can generate it

The argument in this section is similar to the one in the previous section. It takes advantage of subtraction's recursive borrowing to exhibit Backup occurring in a context where there are two invocations of the borrow goal active at the same time. This means there are two potential destinations for Backup. It will be shown that the schema/instance mechanism is necessary to make the empirically correct prediction.

A common bug is one that forgets to change the zero when borrowing across zero. This leads to answers like:

$$
\begin{array}{r}
2 \\
3 \\
4^1 0^1 2 \\
-1\ 3\ 9 \\
\hline
1\ 7\ 3
\end{array}
$$

(The small numbers stand for student's scratch marks.) The 4 was decremented once due to the borrow originating in the units column, and then again due to a borrow originating from the tens column because the tens column was not changed during the first borrow as it should have been. This bug is called Borrow Across Zero. It is a common bug. Of 417 students with bugs, 51 had this bug (VanLehn, 1981).

An important fact is seen in Fig. 9.4. The bug decrements the 1 to zero during the first borrow. Thus, when it comes to borrow a second time, it finds a zero where the one was, and performs a recursive invocation of the borrow goal. This causes an attempt to decrement in the thousands columns, which is blank. An impasse occurs. The answer shown in the figure is generated by assuming the impasse is repaired with Backup. This sends control back to the most recently invoked goal that has alternatives. At this point the active goals are:

borrow-from   (the recursive invocation located at the thousands column)
borrow-from-zero   (at the hundreds column)
borrow-from   (at the hundreds column)
borrow   (at the tens column)
process-column   (at the tens column)

In this procedure, the borrow-from-zero goal has no alternatives (it should always both write a nine over the zero and borrow-from the next column, although here the write-nine step has been forgotten). The borrow-from goal has alternatives because it has to choose between ordinary, non-zero borrowing and borrowing from zeros. Since borrow-from was the most recently invoked goal that has alternatives left, Backup returns to it. Execution resumes by taking its other alternative, the one that was not taken the first time. Hence, an attempt is made to do an ordinary borrow-from, namely a decrement. Crucially, this happens in the hundreds column, which has a zero in the top, which causes a new impasse. We see that it is the hundreds column that was returned to because the impasse was repaired by substituting an increment for the blocked decrement, causing the zero in the hundreds column to be changed to a one.

The crucial fact is that the Backup repair shifted the focus from the thousands column to the hundreds column, even though both the source and the destination of the backing up were borrow-from goals. This shift is predicted by the schema/instance architecture. However, the empirical adequacy of the register architecture is not as high.

II.B.1 One register per goal can't generate the bug

Suppose each schematic goal has its own register. Borrow-from would have a register, and it would be set to the top digit of the thousands column at the first impasse (episode *b* in Fig. 9.4). Hence, if Backup returns to the first invocation of borrow-from, the register will remain set at the thousands column. Hence, Backup doesn't generate the observed bug of Fig. 9.4. In fact, it can't generate it at all: the only register focused on the hundreds column is the one belonging to borrow-from-zero. That goal has no open alternatives, so Backup can't return to it. Even if it did, it wouldn't generate the bug of Fig. 9.4. So one register per goal is an architecture that is not observationally adequate.

II.B.2 One register per object doesn't generate the bug

Assuming the registers are associated with object types fails for similar reasons. Both the impasses (episodes *b* and *d*) involve the same type of visual

a.
$$\begin{array}{r} \overset{0\ \ 1}{102} \\ -\ \ 39 \\ \hline 3 \end{array}$$
Since I can't take 9 from 12, I'll borrow. The next column is 0, so I'll decrement the 1, then add 10 to the 2. Now I've got 12 take away 9, which is 3.

b.
$$\begin{array}{r} \overset{0\ \ 1}{102} \\ -\ \ 39 \\ \hline 3 \end{array}$$
Since I can't take 3 from 0, I'll borrow. The next digit is 0, but there isn't a digit after that!

c.
$$\begin{array}{r} \overset{0\ \ 1}{102} \\ -\ \ 39 \\ \hline 3 \end{array}$$
I guess I could quit, but I'll go back to see if I can fix things up. Maybe I made a mistake in skipping over that 0, so I'll go back there.

d.
$$\begin{array}{r} \overset{1}{\underset{}{\overset{0\ \ 1}{102}}} \\ -\ \ 39 \\ \hline 3 \end{array}$$
When I go back there, I'm still stuck because I can't take 1 from 0. I'll just add instead.

e.
$$\begin{array}{r} \overset{1}{\underset{}{\overset{0\ \ 1}{102}}} \\ -\ \ 39 \\ \hline 173 \end{array}$$
Now I'm okay. I'll finish the borrow by adding 10 to the ten's column, and 3 from 10 is 7. The hundreds is easy, I just bring down the 1. Done!

FIG. 9.4   An idealized protocol of a student with aversion of the bug Borrow Across Zero.

object, a digit, and hence the corresponding register would have to be reset explicitly by Backup in order to cause the observed focus shift.

II.B.3 Smart Backup makes the theory too tailorable

But providing Backup with an ability to explicitly reset registers would once again require it to do static analysis of control structure—an increase in power that should not be granted to repairs.

II.B.4 Duplicate borrow-from goals

One could object that we have made a tacit assumption that it is the same (schematic) borrow-from that is called both times. If there were two schematic borrow-froms, one for an adjacent borrow, and one for a borrow two columns away from the column originating the borrow, then they could have separate registers. This would allow Backup to be trivial once more. However, this argument entails either that one have a subtraction procedure of infinite size, or that there be some limit on the number of columns away from the originating column that the procedure can handle during borrowing. Both conclusions are implausible.

II.B.5 Duplicate borrow goals

But one could object that there is another way to salvage the multiple register architecture. Suppose that the schematic procedure is extended by duplicating borrow goals (plus registers) as needed. The bug could be generated, but this amounts either to a disguised version of schemata and instantiations, or an appeal to some powerful problem solver (which then has to be explained lest the theory lapse into infinite tailorability). So, this alternative is not really tenable either.

This rather lengthy argument concludes with the schema/instance architecture the only one left standing. What this means is that representations that do not employ the schemata and instances, such as finite state machines with registers or flow charts, can be dropped from consideration. This puts us, roughly speaking, on the familiar ground of "modern" representation languages for procedures, such as stack-based languages, certain varieties of production systems, certain message passing languages, and so on.

The example illustrates the main points

The preceding argument illustrates several of the main points of this chapter. The structure of the argument was to first establish a need, in this case for a data flow scheme, then to examine several alternative architectures that meet the need. This pattern of establishing a need and examining alternatives is characteristic of competitive argumentation.

The argument introduced a crucial fact: Whenever problem solving backs up to a previously invoked goal, the goal is resumed with the same instantiating information that it had during its original invocation. This was illustrated with two bugs (Fig. 9.1 and Fig. 9.4). We can vouch for the truth of this observation in the local problem solving that accompanies subtraction performances, and we expect it to remain uncontradicted by evidence from other domains. In Newell and Simon's classic study of eye movements during the solution of cryptarithmetic puzzles, for example, there is ample evidence that backing up restores not only the goal, but the focus of visual attention that was current when the goal was last active (Newell & Simon, 1972, pp. 323–325).

The arguments turned mostly on limiting the tailorability of the theory and on avoiding star bugs. Since all the competing explanations for the crucial fact were able to account for it one way or another, it was their entailments that decided their relative merits. Sometimes the machinations necessary to account for the facts would introduce so much power into the theory that it could trivially account for any data; in these cases, the hypothesis was rejected as introducing so much tailorability as to make the theory irrefutable. In other cases, the hypothesis entailed the generation of certain absurd behaviors: star bugs. Such generations were treated like the generation of false predictions despite the fact that empirical claims about existence can never be proven false.

The emphasis on what people *can* and *can't* do as opposed to what they *do* do is apparent in the use of bugs rather than raw protocols as the data supporting the argument earlier. Bugs are an idealization of human behavior. They describe systematic errors and leave aside unsystematic errors (i.e., *slips* in the sense of Norman, 1981) such as $7 - 5 = 3$. Also, bugs isolate distinct behaviors: The bug Smaller From Larger Instead of Borrow From Zero occurred five times in our sample but always in combination with various other bugs (VanLehn, 1981). It never occurred alone. There is a layer of inference involved in this idealization of the raw data that has not been presented here (Burton, 1981; VanLehn, 1981). In addition to these difficult sorts of idealization, there are simple ones such as choosing not to collect timing data or self-report data from subjects as they solved problems. Although the objective of Repair Theory is in part to understand the processes involved in applying procedures to problems, it can be successfully approached, we believe, within a competence theory framework.

Lastly, the argument illustrates what we mean for a theory to have "depth." This attribute correlates with the length and complexity of the theory's arguments. In Repair Theory there are multiple layers—protocols, bugs, interpreter state, and knowledge structures. There are precise relationships between these layers. The format of the structures at each layer as well as the nature of the relationships between them require supporting argumentation to show not only that the proposed architectures are sufficient to account for certain crucial facts, but also that they are the leading edge of a convergence upon empirical necessity in that a careful drawing out of the entailments of competing proposals reveals that each of the competitors is flawed.

AN EXAMPLE OF THE PROGRESSION TOWARD
COMPETITIVE ARGUMENTATION

It was suggested earlier that cognitive science research is following a natural progression that is entailed by its emphasis on precise and detailed use of task-specific information. That emphasis necessitates an early phase where information latent in a new task domain is uncovered, often by creating a rough computer simulation of the task behavior. These early formulations become refined as important principles and components are separated from the more task-idiosyncratic information. These are put forward as a sufficient formulation of domain knowledge and skill. From these first articulations of principles, attention naturally turns to supporting the principles and/or revising them. From competition among various analyses of the task domain, a successive approximation of what principles and components are necessary emerges. It was toward this later stage that the bulk of this chapter was addressed. Yet, it seems appropriate to end by showing, again with an example, how this phase of competitive argumentation arises naturally from those that must precede it.

A domain that illustrates this natural progression is young children's understanding of principles of number and quantity. Until recently, based largely on Piaget's seminal investigations, most developmental psychologists accepted the conclusion that significant conceptual competence (in the sense of general ability) regarding number is not achieved until children are about seven or eight years old. Yet, children are able to count sets of objects well before they enter school at four or five years old. On the standard view, children's ability to count objects reflects a procedural knowledge of a rote, mechanical nature, rather than understanding. Changing the definition of "number competence" to exclude counting, which develops too early and hence offers a counterexample to classical stage theory, is exactly what Lakatos (1976) would call "monster barring." Just as Lakatos would predict, this caused considerable attention to be focused on counting.

Observations by Gelman and Gallistell (1978), and others, provided evidence that significant conceptual competence underlies preschool children's performance in counting tasks. One example of such evidence is the observation that although many preschoolers use idiosyncratic lists when they count, these lists are used consistently. For example, a child might count with the list, "one, two, three, six, ten." The consistency of use of such lists is taken as evidence that children appreciate the need for a set of symbols with stable order, and that while they can sometimes acquire the wrong set, the principle of stable order is part of their conceptual competence.

Another example involves performance in tasks where children count objects with an additional constraint superimposed on the counting task. In a typical experiment, the child is asked to count five objects arranged in a straight line, then the experimenter adds a constraint by saying, "Now count them again, but make this the *one*," pointing to the second object in the line. Most five-year-old children make completely appropriate adjustments of their counting procedures in order to accommodate these constraints. Even children whose modified counting is not completely appropriate still perform in ways that preserve some of the constraints of counting. Their performance provides quite strong evidence that counting reflects a good understanding of number, and is not a rote, mechanical procedure, because they generate novel procedures that conform to some, but not all, of the principles of counting.

Given these results, it became untenable to bar counting from theories of the development of cognitive competencies. One response has been to construct a computational model for the development of number competency. This is where our example of the natural progression of computational analyses of cognition begins.

An analysis of children's counting was conducted by Greeno, Mary Riley, and Rochel Gelman (forthcoming). First, a process model was formulated that simulates salient aspects of children's performance on a variety of counting tasks. This was necessary just to come to grips with the latent information that the tasks required. Next, an analysis was developed in which the procedures in the process model were derived from premises that correspond to certain principles of counting. These premises formalized and extended a particular decomposition of counting competence proposed by Gelman and Gallistel (1978). The steps of the derivation involved use of planning rules for including procedural components in the counting procedure. The outcome of the analysis was a planning net (VanLehn & Brown, 1980) that showed how the various components of the counting procedure are formally related to the counting principles. The result of this phase of the research was to articulate clearly the counting principles, indeed to formalize them as operable rules, and to show that they were sufficient to generate the kinds of counting performances observed by Gelman and others.

From a clear articulation and a first demonstration of sufficiency, the research has begun to focus on supporting these particular principles in various ways. To argue that the particular decomposition of competence chosen is correct, we have been investigating how removing certain principles from the set while adding constraints corresponding to the experimenter's request to "Make that one be *one*," results in the derivation of procedures corresponding to the various counting performances observed under the stressed conditions. If successful, this demonstration could support the particular set of counting principles, and rule out others that are (hypothetically) equally successful at generating counting procedure in unconstrained situations. Thus begins a phase of competitive argumentation, following on naturally after a first formulation of a principled, detailed and precise analysis of counting competence.

This example of natural progression also illustrates how experimental facts can become crucial. The fact that performances of children's counting degrade along the lines predicted by a certain set of principles is used to decide between that set of principles rather than some others. It also appears to preclude an explanation of counting as some indecomposable, "rote" procedure. Because these experiments have been used to decide important theoretical issues, we expect them to remain crucial facts.

CONCLUSIONS

The development of models that simulate the processing of specific information in detail has required large investments of time in developing tools for model building as well as obtaining a working understanding of the power and limitations of computational concepts and theoretical methodology. Hence, many computational theories have lacked explicit principles, many have not used data, and virtually none have the argumentative support that we have discussed in this chapter. During the period of this early development, it has been inevitable and perhaps even desirable that computational theories should have been relatively unprincipled and unsupported. However, the rapid initial growth in computational experience, understanding and tool building seems to be leveling out now. We suggest that it is time to use that investment to initiate a parallel growth in our understanding of what constitutes principled computational theories of mind and what tools would facilitate their construction and especially their support.

Some kind of defense of individual theoretical principles, whether competitive or not, seems necessary to expedite scientific progress. If the support for a theory is not analyzed so that one can see how the evidence bears on each part, then the theory must be accepted or rejected as a whole. In contrast, argumentation allows the theory to be revised incrementally. Indeed, perhaps it is because of the infrequent use of argumentation in cognitive science that its theories "have stood on the toes of their predecessors, rather than their shoulders" (Bobrow, 1973).

Competitive argumentation can have several advantages. In addition to its function of showing the lack of support for some theoretical principles while favoring other principles, it adds information at a more general level, enriching the understanding of the connection between facts and abstractions. A second potential advantage is that by making explicit the reasons for rejecting a principle, when future development of the theory brings the chosen principle into conflict with others or into conflict with new facts, one can sometimes dust off one of the fallen competitors and patch its flaws, rather than searching for a replacement from scratch. In this respect, an argument functions like the "support" assertions that must be saved in order to do dependency-directed backtracking (c.f., de Kleer & Doyle, 1981; Stallman & Sussman, 1977). In short, the criticism of alternative explanations that a typical argument provides is value added to the demonstration of empirical support for the chosen principle. Third, contrasting two alternative explanations sharpens both, making it easier to understand the positions involved by explicating the considerations that mutually support them as well as those that distinguish them. Fourth, argumentation guards against reinvention of the wheel, for if no argument can be found to split two proposals, one begins to suspect that they are equivalent in all but name. Finally, there is the possibility that arguments will outlive the theory they were crafted to support. They might survive not only as crucial facts, but also perhaps as argumentative techniques. Some of the most significant contributions to mathematics have been innovative proof techniques, techniques that have far outshown the theorems they supported. Perhaps cognitive science will also evolve a repertoire of argumentative techniques.

ACKNOWLEDGMENTS

We would like to thank those whose have helped us with their criticisms of early drafts: Danny Bobrow, Johan de Kleer, Bob Lindsay, Mark Stefik, Bonny Webber, and the editors. This research was supported in part by grants N00014-82-C-0067 and N00014-79-C-0215 under contract authority identification numbers NR667-477 and NR667-430 from the Office of Naval Research, Personnel and Training Research program.

REFERENCES

Bobrow, D. G. Natural language input for a computer problem-solving system. In M. L. Minsky (Ed.), *Semantic information processing.* Cambridge, Mass.: MIT Press, 1968.

Bobrow, D. G. Address given at Third Annual International Joint Conference in Artificial Intelligence, Stanford University, Stanford, Calif., 1973.

Brown, J. S., & Burton, R. B. Diagnostic models for procedural bugs in basic mathematical skills. *Cognitive Science,* 1978, *2,* 155–192.

Brown, J. S., & VanLehn, K. Repair Theory: A generative theory of bugs in procedural skills. *Cognitive Science,* 1980, *4,* 379–426.

Burton, R. B. DEBUGGY: Diagnosis of errors in basic mathematical skills. In D. H. Sleeman & J. S. Brown (Eds.), *Intelligent tutoring systems.* London: Academic Press, 1981.

de Kleer, J., & Doyle, J. Dependencies and assumptions. In A. Barr & E. Feigenbaum (Eds.), *The AI handbook,* Los Altos, Calif., Kaufmann, 1981.

Dresher, B. E., & Hornstein, N. On some supposed contributions of artificial intelligence to the scientific study of language. *Cognition,* 1976, *4,* 321–398.

Dreyfus, H. L. *What computers can't do.* New York: Harper & Row, 1972.

Ernst, G. W., & Newell, A. *GPS: A case study in generality and problem solving.* New York: Academic Press, 1969.

Evans, T. G. A heuristic program to solve geometric analogy problems. In M. Minsky (Ed.), *Semantic information processing.* Cambridge, Mass.: MIT Press, 1968.

Fodor, J. A. *The language of thought.* New York: Crowell, 1975.

Fodor, J. A. *Representations: Philosophical essays on the foundations of cognitive science.* Cambridge, Mass.: MIT Press, 1981.

Gelman, R., & Gallistel, C. R. *The child's understanding of number.* Cambridge, Mass.: Harvard University Press, 1978.

Greeno, J. G., Riley, M. S., & Gelman, R. *Conceptual competence and young children's counting.* Unpublished manuscript, forthcoming.

Hull, C. H. *Principles of behavior.* New York: Appleton-Century, 1943.

Hull, C. H. *A behavior system.* New Haven: Yale University Press, 1952.

Kaplan, R. M. *A competence based theory of psycholinguistic performance.* Psychology colloquium at Stanford University, May 1981.

Lakatos, I. *Proofs and refutations: The logic of mathematical discovery.* Cambridge, England: Cambridge University Press, 1976.

Marr, D. Early processing of visual information. *Philosophical Transactions of the Royal Society, B,* 1976, *275,* 483–524.

Marr, D., & Nishihara, H. K. Visual information processing: Artificial intelligence and the sensorium of sight. *Technology Review,* 1978, *8,* 2–23.

Moravcsik, J. M. Chomsky's radical break with modern tradition (A commentary on Chomsky's Rules and Representations). In N. Chomsky, Rules and representations. *The Behavioral and Brain Sciences,* 1980, *3,* 1–63.

Newell, A., & Simon, H. A. *Human problem solving.* Englewood Cliffs, N.J.: Prentice-Hall, 1972.

Newell, A., Shaw, J. C., & Simon, H. A. Empirical explorations with the logic theory machine: A case study in heuristics. In E. A. Feigenbaum & J. A. Feldman (Eds.), *Computers and thought.* New York: McGraw-Hill, 1963.

Norman, D. A. Categorization of action slips. *Psychological Review,* 1981, *88,* 1–15.

Pylyshyn, Z. W. Computation and cognition: Issues in the foundations of cognitive science. *The Behavioral and Brain Sciences,* 1980, *3,* 111–168.

Simon, H. A. *Sciences of the artificial.* Second edition. Cambridge, Mass.: MIT Press, 1981.

Simon, H. A., & Kotovsky, K. Human acquisition of concepts for sequential patterns. *Psychological Review,* 1963, *70,* 534–546.

Stallman, R. M., & Sussman, G. J. Forward reasoning and dependency-direct backtracking in a system for computer aided circuit analysis. *Artificial Intelligence,* 1977, *9,* 135–196.

Tolman, E. C., & Honzik, C. H. Introduction and removal of reward and maze performance in rats. *University of California Publications in Psychology,* 1930, *4,* 257–275.

VanLehn, K. *Bugs are not enough: Empirical studies of bugs, impasses and repairs in procedural skills* (Tech. Rep. CIS-11). Palo Alto, Calif.: Xerox Palo Alto Research Centers, 1981.

VanLehn, K. On the representation of procedures in Repair Theory. In H. Ginsberg (Ed.), *The development of mathematical thinking.* New York: Academic Press, 1983.

VanLehn, K., & Brown, J. S. Planning Nets: A representation for formalizing analogies and semantic models of procedural skills. In R. E. Snow, P. A. Federico, & W. E. Montague (Eds.), *Aptitude, learning and instruction: Cognitive process analyses of learning and problem solving.* Hillsdale, N.J.: Lawrence Erlbaum Associates, 1980.

Vere, S. A. Induction of concepts in the predicate calculus. Proceedings of the Fourth International Joint Conference on Artificial Intelligence, Tbilisi, USSR, 1975, 281–287.

Vere, S. A. Inductive learning of relational productions. In D. A. Waterman & F. Hayes-Roth (Eds.), *Pattern directed inference systems.* New York: Academic Press, 1978.

Winograd, T. *Understanding natural language.* New York: Academic Press, 1972.

Winograd, T. On some contested suppositions of generative linguistics about the scientific study of language. *Cognition,* 1977, *5,* 151–179.

Winston, P. H. Learning structural descriptions for examples. In P. H. Winston (Ed.), *The psychology of computer vision.* New York: McGraw-Hill, 1975.

Woods, W. A., Kaplan, R. M., & Nash-Webber, B. *The lunar sciences natural language information system: Final report* (BBN report number 2378). Bolt Beranek and Newman Inc., Cambridge, Mass. 1972.

10 Cognitive Science is Methodologically Fine

Eugene Charniak
Brown University

Introduction

I remember a conversation with Terry Winograd when we were both graduate students at M.I.T. in the Artificial Intelligence Laboratory. Terry was explaining to me how the transformational linguists refused to consider all of the various phenomena that illustrate the need to take world knowledge into account when discussing language. I was amazed by this (having by then assimilated the MIT Artificial Intelligence Lab's view of the world) and I asked Terry several times, "but what do they say when you point out X," where X was some case where real world knowledge came into play. His reply was always something like:

> They will say, "That's interesting. Someday we will have think about that." Of course, someday may be quite a ways off.

I doubt if Terry remembers this conversation. There was nothing particularly novel about its content. But it was the first time I came to grips with the idea that people could have very different ideas about what makes for good science, and furthermore, that they could defend these ideas in ways that seemed impervious to my arguments. If it seems odd to you that I could have reached graduate school without realizing this, I can only plead that my undergraduate degree was in physics, and if any discipline can hide this fact from its undergraduates, it is physics.

I mention this bit of personal history because it was my first brush with problems of methodology. I also mention it because in my innocence I had an idea that may affect others in cognitive science, that there is a "right" methodology that everyone should use. I no longer hold this view. Thus I will be arguing that cognitive science does not have methodological problems, and with certain minor changes, I think things should go on as they are. This chapter explains why.

METHODOLOGIES IN COGNITIVE SCIENCE

Roughly speaking there are two sorts of methodological problems. The first, and most serious, is that the methodology being used is incapable of producing the desired results. An extreme example would be someone proposing to contribute to cognitive science by studying the growth of toenails.

Now, what we find in cognitive science is not this, I would claim, but rather methodological babble. That is, cognitive science consists of several previously established disciplines, each with its own methodology. Each of these methodologies is a serious one, and, as I argue, any of them is capable of producing the results we need for cognitive science. Most, if not all, of us would agree with this, and hence we can turn to a somewhat more subtle methodological point. Even if we agree that any of the various methods being used in cognitive science *might* produce the theory we are all looking for, in practice it is bound to turn out that some of these will ultimately prove to be more useful than others. If it is possible to decide which one this is, then obviously we could speed things up by all adopting it.

In this chapter I assume that all of the methodological arguments are of this second sort. In most cases this is correct, although sometimes the debates get sufficiently heated that one cannot be sure. Nevertheless, because I argue that this second, and weaker argument, is still wrong, then anything I say will, perforce, apply to arguments that purport to show that alternate methodologies are not simply inefficient, but *evil*.

METHODOLOGICAL ARGUMENTS

The Argument from Common Sense

Roughly speaking there are two kinds of arguments put forward to support one methodology over another. One, calling for support from historical precedent, I discuss in the next section. The other essentially is no argument at all, but rather an appeal to common sense. An example of this is found in the chapter by Swinney. I might add, before I start, that I like the research done by all of the authors I am commenting on. However, methodological arguments are another matter, and Swinney in particular seems guilty of arguments which, when examined closely, do not really hold up. For example, consider the following quote from Swinney's chapter:

> *Static* models of putative language structure are of questionable (and, at best, indeterminate) importance in the enterprise of understanding language performance. Models can be classified as static on two grounds: the first of these can be seen in the distinction drawn between those mental processes involved in real-time language perception (the *process* of understanding a sentence) and those involved in various post-perceptual language functions (recall, recognition, paraphrase, or some other such manipulation of already perceived sentential/propositional material). Not only do quite different mechanisms likely underly perceptual and post-perceptual processes, but data taken from one of these two sources may not legitimately reflect processes in the other [p. 219].

There is, of course, a ring of common sense about this. If I want to know what happens to a horse's legs while it is running, the obvious thing to do is to look at them when they are running. Conceivably one might get some idea by looking at, say, the anatomy of the horse's legs, but it would nevertheless seem a rather roundabout way of doing things. Of course, in terms of history, it was not until the photographic camera that one could stop the otherwise too rapid motion so that one could actually observe what happened, but once the camera was available, a thousand pictures were worth all the theories you could muster.

But common sense or no, this argument will not go through. The clue to its fall is already apparent when we noted that we needed a camera to do the experiment. In other words, if there is equipment around that will show what is happening, then one should simply *look*. But what if the equipment is not available? Or, to put this slightly differently, if we have the experimental techniques that will reliably tell us what is going on during all phases of language comprehension, then the obvious thing to do is to use them and find out. But we do not have such techniques. We have some techniques that seem to tell us a little, but little of what we know is certain, and, as Swinney points out himself, very small changes in the experiment can radically change its outcome. In his own work he found that changing the method of measuring cognitive load changed the results of his word sense ambiguity experiments.

In cases when the experimental techniques are not all there, then one may legitimately ask, will one learn most from pushing the experiments, or, might easier tasks, such as looking at comparatively static situations (like judgments, of grammaticality, or ability to remember stories) be the way to go. In some cases it may well pay to go the static route. One such case concerned theories of how the genetic material replicated itself in the cell. Biology today is still trying to tie down the details of this process. Nevertheless, we have had a general idea of what must be going on for many years now, ever since Watson and Crick produced a model for the three dimensional structure of DNA—a model that wore its replication process on its sleeve, as it were. In this case, at least, the study of static structure said a great deal about process.

The Argument from History

The other major way of arguing for a particular methodology is on the basis of past performance. That is, on the basis of history.

Now history it seems to me is probably the best grounds for such arguments, because in the absence of controlled experiments, about all we can do is look to see what worked in the past. The trouble is, that when one turns to the historical record, it does not support a single methodology. That is, sure, you can support your methodology with selected examples from the history of science, *but so could everyone else in cognitive science!* To illustrate this, let me give several historical arguments each supporting a different methodology. An amusing feature of my arguments is that I can use the same historical event to argue for different positions. History by itself says nothing about current situations. One needs a historical event, plus an analogy from the historical event to the present. If we change the analogy, we change the argument.

Perhaps like most transformational linguists you would like to ignore details of process in order to concentrate on basic laws. Well, you are in good company. Newton's work can be viewed this way, and Kepler fits this mold very well. Kepler developed his laws of planetary motion by looking at the data and just trying to fit simple geometric ideas to the numbers. The rules he eventually found were of this sort and Kepler had no idea of how his laws might result from some physical process. Indeed, it is not even clear that he considered this necessary for a scientific explanation.

Another good example to support a competence methodology would be Darwin. For evolution to make sense Darwin had to assume on one hand that successful traits would be passed on unchanged to descendants. On the other hand, somehow other traits were modified, so that there would be some differences to select amongst. How did either of these processes work? Neither he nor his contemporaries had the foggiest idea, except for some obscure monk who nobody ever heard of.

Suppose, however, that you are of an atheoretical bent. Hence, not only do you dislike competence theories, but you tend to distrust all current theorizing. Instead you believe that what we need is more detailed studies of individual cases, even if we have no theory for how they connect up to anything else. Well, you can use Darwin for your purposes as well. The pre-Darwinian study of the classification of animals and plants involved a lot of bright people spending a lot of time looking at the existing animals and plants, noting the similarities and differences between them. In doing this they developed taxonomic biology, but as to what this taxonomy meant they had not a clue. Nevertheless, without their work Darwin would probably have been just another victorian gentlemen naturalist. The point is (according to this argument), before we can get a theory we need to get our facts right.

A slight variation of this argument, particularly suited to a neuroscience type, would be to argue that the best way to understand the brain is by working up from the small to the large. After all, isn't that the way biology conquered the cell. Nobody every proposed developing an abstract "competence" theory of the cell. What they had to do was a lot of slogging experimental work to find out how all of the parts fit together.

Yet another historical example that argues against a competence method of research deals with post Watson-Crick genetics. After the discovery of the structure of DNA, an obvious problem was determining the code by which four bases found in DNA could specify many more amino acids. Early in the game, Watson and Crick suggested that there were 20 amino acids, and that the few others that were found were formed by modifications of the amino acid after protein synthesis. The number 20 is important. In the course of work on the coding problem, a sub problem came up—how could one tell where the code for one amino acid stopped and the next began. To make this more specific, because the minimum number of nucleotydes needed to specify 20 amino acids is three, let us assume that the code is a code of triplets. The question becomes, how do we decide on the boundaries of the triplets. So, if we saw the sequence Adenine Adenine Guanine Thymine Guanine, should this be divided up AAG TGx or perhaps xxA AGT Gxx, or xAA GTG? Where do the commas go, to use the terminology of the time. One possibility, thought of by Crick, was that because three nucleotydes can code many more than 20 possibilities ($4 \times 4 \times 4 = 64$) perhaps one could design a code such that, at any point only one combination would be legal, and all of the others illegal. So in the above example, only one of AAG, AGT, or GTG would be legal. This scheme was called the *comma-less code*. The fantastic thing about this code is that when you calculate how many amino acids could be used given three nucleotydes per amino acid, the number comes out *20*. Well, when people saw this, they flipped. Although noone would have put it this way, here was a real competence theory in biology. *Unfortunately, the comma-less code was completely wrong.* When, some years later, the code was finally broken, the reality was much less elegant. The way the cell knows which three to use is by laboriously "counting" from a fixed point on the DNA. Moral, competence theories are bad for your health.

I could, if pressed, go on like this for some time. Perhaps we could even make it into a parlor game. But there is a serious point. None of the methodologies used in cognitive science today are a priori, unsound. One can find example after example where such a technique worked in the past. This is not to say, however, that absolutely "anything" can work. Obviously some techniques are beyond the bound of legitimate scientific inquiry. The most obvious modern example is the use of politics to decide scientific questions, as with Lysenko. To take a less drastic example, I have some doubts about the seriousness of the Semiotics I have come across. But we are not discussing things at this level of the bizarre. We are talking about serious scientists doing serious work. Any of the techniques being used in cognitive science today *might* produce the "big breakthrough," the question is, which one will actually do it. There is no way, arguing from first principles, to decide this.

ACHIEVING UNITY

So far I have been arguing that the reasons proposed for preferring one methodology over another simply do not stand up. Thus, my argument has been basically a negative one. There is no way to achieve unity through methodological debates. Does this mean that we in cognitive science are doomed to live with perpetual confusion. No, this does not follow. As Kuhn points out, scientific groups do reach consensus, and they do so because someone makes a discovery, or thinks up a theory that suddenly puts everything into place, creating the agreement that previously was lacking. One such case, the situation in the study of heredity in the early 1950s, has many parallels with current cognitive science, so it would pay to look at it more closely.

The period is the few years just preceding the Watson and Crick model of DNA. What makes this period and topic of interest to cognitive science, is that just as language is currently studied by several disciplines, each with its own methodology, the same was true of heredity in the 1950s. The biochemists were primarily concerned with the chemical properties of DNA. However, those who thought it might prove to be the genetic material surely hoped that chemical properties would say something about heredity. We also had the biologists, and within this group, at least two major subgroups. The traditional geneticists had a host of techniques developed for determining the abstract structure of the genetic material, whatever it was. Many years of this sort of research had been slowly pushing the level of resolution down further and further. Then we had the phage group, who wanted to attack genetics by seeing how reproduction occurred in bacteria phage, the simplest reproductively active organism known, and as such, presumably the easiest to understand. Finally, we had the molecular biologists. They were interested in establishing the exact molecular configuration of biologically interesting molecules. They had a bright new tool, namely X-ray diffraction, that promised to illuminate the three dimensional structure of biological molecules.

We know from history that it was the last group who produced the big breakthrough, and as a result of this breakthrough, the entire genetics community was unified insofar as they then had a common theory between them. However, even in retrospect it is hard to say that one should have "known" that it would be molecular biology that would grab the prize. For one thing, the structure of DNA might have been too complicated to fall to the rather unsophisticated techniques available at the time. For another, even if Watson and Crick had found the structure, finding *any old* structure would not have been enough to create the breakthrough. The structure they found, of course, had other properties that made it important. For one thing, it immediately suggested a method of DNA replication, and for another, it immediately suggested how the information used to create the rest of the cell was encoded, although, as we noted earlier, the exact code was not obvious. It is only because the structure had obvious lessons for everyone, that it became the all important unifying fact in genetics.

Of course, genetics was not unified in the sense that everybody gave up what they were doing and started working on molecular biology. This did not happen. Rather, each group took the DNA structural hypothesis, and extracted from it the information relevant to their concerns. Furthermore, it was fortunate that they did so, for the next breakthroughs in genetics did *not* come directly from molecular biology. Indeed, the obvious next things to do in molecular biology, such as finding the structure of RNA, which people believed was an intermediate in the chain from DNA to protein, these things did not work out at all. (Thus reenforcing the point that doing molecular biology of DNA was not an ''obvious'' right thing to do.) RNA was too complicated, and it was biological approaches that lead to the next set of important discoveries, such as the messenger. Finally, it was biochemistry of a rather slogging sort that finally decoded DNA.

This last point is a rather important one. Even in the face of a unifying theory, the response was not to turn en mass to the methodology that produced this theory. Thus, one day cognitive science may be unified in terms of theory, but I doubt that it will ever be unified in terms of methodology. However, as I have been trying to argue, that is fine.

WHAT TO DO WHILE WAITING FOR THE REVOLUTION

My argument thus far is relatively simple. There are no a priori reasons for believing that any of our current methodologies is more likely to produce an adequate theory of, say, language comprehension than any of the others. Thus, we should accept our multiple methodologies, and wait for the day that one or other of them will produce a theory around which we can all rally. However, even the existence of such a theory will still not eliminate the methodological babble. We will all continue on, but now exploring the same theory. Of course, in saying this, what I seem to be saying is that we should all sit back and wait. Then one day some brilliant person will come along and propose a revolutionary new theory of intelligence—one which will make everybody happy. This may be correct, but it isn't very useful. The trouble is that it gives us no guidance in the meantime.

Unfortunately, I do not believe there is really much guidance to be had for the meantime. The best I can do is one very small point, and for this, let us return again to the example of the Watson-Crick structure for DNA.

As I noted earlier, in the early 1950s there were at least four methodologies around, all concerned with genetics: biochemistry, classical genetics, phage biology, and molecular biology. Although it was the latter group that "solved" DNA, it is interesting to look more closely at the information that went into this solution.

Before one can solve the DNA problem one has to know that DNA is important. This way by no means obvious in the first half of this century. Most people who thought about it believed that protein would prove to be the physical realization of genes. Protein is found everywhere within the cell, and the wide range of forms it can take make it a plausible carrier of genetic information. DNA, on the other hand, was considered a dumb molecule, probably a structural filler for the cell.

The first suggestions that DNA might be the genetic material after all came from two sources. A microbiologist, Avery, tried to isolate the material that was responsible for the genetic modification of pneumococci. His tests said DNA. Secondly, two biologists from the phage group, Hershey and Chase, did an experiment to see what was transferred from the bacteria phage to the cell at the point when the phage infected the cell. It was known that most of the phage remains outside the cell, with only a small fraction entering the cell. After the infection, the cell starts producing more phage, so it is probably the genetic information that is transferred. In a famous experiment (the "Waring Blender" experiment) they too found that it was DNA that seemed responsible.

Results like these suggested to people like Watson and Crick that DNA was worth studying. They did so using the methodology of molecular biology, but a large number of important facts were imported from biochemistry. There were several facts about how the various components of DNA tend to combine with one another, and, at the very end of the search, the missing piece was finally supplied by a physical chemist who told Watson that the structures given in books for the bases of DNA were probably in an incorrect form. It also is possible that Chargaff's rules played a role in the solution, although Watson and Crick have different memories on this point. Chargaff, a biochemist, found that the amount of adenine in a cell's DNA was always equal to the amount of thymine, whereas guanine equalled cytosine.

My point in this is that for this one case at least, it was the interaction of several different fields that allowed people in one of the fields to put together a coherent theory of heredity at the molecular level. The lesson for cognitive science should be obvious. We may have different methodologies, but keeping track of each others results is very important.

WORD SENSE DISAMBIGUATION IN COGNITIVE SCIENCE

That we should talk to each other may seem a rather trivial point, but its importance was brought home to me by Swinney's chapter. As an editor of

Cognitive Science, I am probably as aware of work in psychology as anyone in artificial intelligence (AI), but I must confess that until I read Swinney's chapter I had not known of his work at all. What I would like to do in the rest of this section is explain why his results are important to me.

It should surprise no one that word sense disambiguation is an important problem in AI. The amount of work on the problem is actually somewhat less than would be expected from the importance of the issue, but this can probably be explained by the difficulty of the problem. There is, however, one very interesting idea that has been put forward by several people in the field. To understand it, I will have to assume the reader has some familiarity with the "frame" (or "script") hypothesis of language comprehension. Roughly speaking, this theory suggests that our knowledge of the world is organized topically, so that it forms clusters, one cluster per topic. These clusters are called frames (or scripts), and it is assumed that during language comprehension we are, in some sense, moving clusters from "long-term storage" into some more active memory, aprocess typically called "activation." Naturally, it will only be the topics that have come up in the conversation or story that will be activated.

Now, in the most straightforward theory of word sense disambiguation, a list of all possible word senses is attached to each word. Actually, we could invent more complicated theories where the senses are organized into more complicated structures than a simple list, but such distinctions are unimportant for our current discussion. So, for example, the word "call" would have attached to it at least 10 such senses, as illustrated by the following:

[1] Jack picked up the phone and called Fred.
[2] Jack called to Fred to come upstairs.
[3] Jack called Fred a dirty name.
[4] Fred raised fifty cents and Jack called him.
[5] Jack called for an investigation.
[6] Jack called the meeting for 10 a.m.
[7] Jack called the debate a draw.
[8] The banker called the loan.
[9] Jack called the dance.
[10] The game was called due to rain.

A radically different approach to word sense disambiguation depends heavily on the use of scripts. In this approach rather than listing all possible senses in a large dictionary (or lexicon), we would have a "default" lexicon that would give only the most standard definitions. For "call" it might just be restricted to shouting and telephoning. However, we would then have associated with each script a lexicon, that gives for each notion used in the script the word associated with it and vice versa. Thus, in the script describing the card game **poker** we would have an entry for "call" pointing directly into the part of the script that deals with "calling a bet." Naturally, the word sense would point not to another

English phrase such as "call a bet" but rather to some yet unspecified semantic representation. So, if we were reading a story about people playing poker on a train and saw the word "call" we would have the following situation:

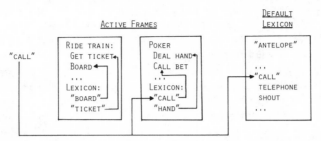

So, in looking for a word sense, we would first look to see if the word appeared in the lexicon of an active script, in which case we would automatically use the contextually suggested meaning.

Now there are many potential advantages to such an approach. From a purely algorithmic point of view, when the number of word senses gets up into the teens, working your way linearly through them does not seem like the epitome of efficiency. In the scriptal lexicon approach we circumvent this problem because most word senses will not even be around at the time the choice has to be made.

Secondly, this approach will automatically guarantee that the contextually relevant approach will be the first one tried. There is even some examples that suggest that this is exactly what people do:

The astronomer married a star.

People tend to have trouble with examples like this, seemingly because they initially get the wrong sense for "star." Given scriptal lexicons we can easily understand why.

Finally, this approach would account for our introspective feelings that we "immediately" know the correct sense in most cases, and do so without considering the alternatives. It would also account for the obvious fact that when asked for meanings for "call" for example, people have a hard time coming up with more than two or three. If we had a pointer from the word to all of its senses, then why do we not get them all. If this pointer is not accessible to our conscious mind, why do we get any?

Despite all of these advantages, I have always been rather luke warm to this theory. My reasons are purely efficiency ones, because, despite the claim that this is an efficient theory, it will only be so in very special circumstances. In particular, to be efficient, this theory requires that at any point in reading or speaking there must only be a small number of active frames (or scripts). Although this is a frequent claim of frame-based comprehension theories, I doubt that it will hold up. For one thing, most concepts, like **poker,** do not correspond to a single frame. Rather, most of us in AI believe that what we know about poker is divided up into several frames, each at a different level of generality. This is called the "isa hierarchy" and the idea is that by placing information about, say, card playing, in a **card-playing** frame, then we can share this information between poker, bridge, canasta, and so on. If we did not do this, then this information would have to be repeated in each frame, which seems wasteful at best. So, for, say, five card stud, we would have frames corresponding to **five-card-stud, poker, card-playing,** and possibly others at still higher levels of generality.

Naturally, each of these frames would have its own lexicon. So, whereas "call" would be found in **poker,** "hand" would be found in **card-playing.** Hence, we have just gone from one frame to at least three. Furthermore, typical stories do not just talk about, say, people playing poker. People may be playing poker with pornographic playing cards on a train on the way to a birthday party while one player eats a hamburger and another talks about yesterday's baseball game. It does not seem unreasonable that we could have 50 or 60 active frames at a given time. If this is so, then we would have to go through all 50 or so, looking in each to see if it has a contextually relevant sense for the word at hand. Better by far to look through 10 or 20 word senses.

Now this is hardly the last word on the subject. For example, it may be possible to come up with solutions to this search problem. But now comes Swinney with his rather interesting results. Admittedly, they may not stand the test of time, but if they do, they would seem to rule out the scriptal lexicon approach as a possible theory of human language processing. The connection is clear. If people do, as Swinney claims, access *all* of the senses of a word at the time the word occurs in the sentence, then the scriptal lexicon approach is impossible, because it would predict that many of the word senses, namely those stranded in lexicons associated with non-active scripts or frames, would be unavailable for inspection.

But Swinney's results create real problems when we try to accommodate them with examples like "The astronomer married a star." If we do access all of the senses initially, why do we seem to pick the wrong one first in these examples? We can simply state that when we choose between senses we always use context first, hence going astray, but without any logical reason for this, it makes an unsatisfactory theory.

Quite frankly, I do not know if I should be hoping that Swinney's results stand up or not. On one hand, he promises to dispose of a theory that bothers me a lot, but in many respects it is a very attractive theory. If his results do stand up, however, they will be important.

CONCLUSION

My conclusions are easily stated. There are no a priori arguments that would lead one to prefer one of the cognitive science methodologies over another. Furthermore, even when we have managed to come together on a theory of language comprehension, or any other aspect of cognitive science, we will still have different methodologies, because we will still be examining different aspects of the problem.

With this all said and done, however, I must confess that in my heart of hearts I do not quite believe this. I know which is the best methodology—mine. But then the objective observer in me raises his head and objects. We all have our opinions, he tells me, but they are nothing more than guesses. Science is a betting sport. We bet on methodology.

ACKNOWLEDGMENTS

This research was supported in part by the Office of Naval Research under contract N00014-79-C-0592, and in part by the National Science Foundation under contract IST-8013689.

IV TOWARDS COGNITIVE SCIENCE

11 Methods and Tactics Reconsidered

P. G. Polson
J. R. Miller
W. Kintsch
University of Colorado

Instead of an abstract discussion of methodology, the strategy adopted in this book was to work from concrete examples. By examining some actual research projects in detail, we proposed to investigate how researchers work and think in various branches of cognitive science, and what the implications of these different modes of operation are for the emergence of a mature field of cognitive science. In this section of the book we attempt to draw some conclusions. What have we learned from these examples? Are we now better equipped to avoid some of the pitfalls that, as we have argued, in Chapter I, threaten the young discipline of cognitive science? What indeed is or can be the nature of that discipline?

In the first chapter in this section, the editors provide a brief summary of the discussion so far. No attempt is made to discuss the data or theories reported here in any detail, instead we concentrate on their implications with respect to problems of methodology in cognitive science. It has always been apparent that cognitive science must live with diversity. What these discussions show is that this is not such a bad fate. There are difficulties, such as communication problems, that arise in the face of such diversity, but there is also great opportunity for building a richer, sounder science of cognition. Everyone seems to have his or her own favorite method, and is quick to see the flaws in other people's favorites, but important results that are significant irrespective of the methods used to obtain them have come from every quarter. Perhaps the best cognitive scientists can do is to make sure that their own methods are used as effectively as possible. Good research (or bad) cannot be tied uniquely to any particular methodology.

277

IS AI JUST COGNITIVE PSYCHOLOGY IN A DIFFERENT GUISE?

This is a time of flux for psychology, especially with regard to its relation with artificial intelligence. As more and more psychologists turn to the construction of simulation models as a means of testing their theories, and as more and more researchers in artificial intelligence turn to experiments as a means of identifying useful components for their systems, the distinction between these two fields has become quite unclear. Depending on one's feelings toward psychology, it might appear either that the two fields have happily merged into one, or that psychology has retreated from studying the issues with which it ought to be concerned, ceding the true study of cognition to AI.

As the chapters by Lehnert, Clancey, and Haugeland demonstrate, the truth is somewhere between these extremes. As we all probably knew before opening this book, there are important areas of overlap between these domains; there are also notable areas of disagreement. What these chapters illustrate is that the sources of these agreements and disagreements lie in the roots of these domains; it will not be surprising if they remain with us for some time to come.

Cognitive science as a process of gradual refinement.   One source of similarity of these domains is the use of the hypothesize-test-revise research strategy; both Clancey and Lehnert pointed out that experiments as well as AI systems can be fit into this technique of gradual refinement. Lehnert's exception to this parallel, that experiments that fail to confirm the beginning hypothesis are less interesting than faulty AI programs, is questionable. It is certainly true that experiments fail as a result of faulty design (i.e., improper counterbalancing of experimental conditions), but these failures correspond to the failure of an AI program as a result of an atheoretical programming error. However, experiments can—and do—fail in theoretically interesting ways. It is not necessarily the case that an experiment is limited to either confirming a hypothesis or saying nothing about the question being studied: many, if not most, experiments are designed so that something is learned whichever way the data come out.

In any event, it is true that there are some important ways in which the hypothesize-test-revise strategy differs in its application to program construction and experimentation. First, the relation between a theory and an experiment is rather different than that between a theory and its implementation as an AI program. When an experiment is constructed to test a theory, it is usually constructed in relative isolation from the theory; the only direct contact between the two is that the experiment must accurately depict those aspects of the theory that it intends to test. It is quite rare for a theory to be changed as a result of the *construction* of an experiment; modifications more typically occur at the next stage in the process, should the experiment produce data that cannot be explained by the theory.

In contrast, theories are very likely to be modified as a result of the implementation process itself: most verbally-expressed theories inevitably gloss over often important details, all of which must be specified if the theory is to be implementable. This specification, and the rigor it enforces on the theory construction process, is one of the most powerful arguments that can be cited in favor of computer simulation. Still more changes, of course, are likely when the resulting program reaches the validation stage, and flaws are detected in its performance. Although it is true that both theory instantiation and experimentation are iterative processes in search of an accurate representation of the true state of affairs, the parallel is not as clearcut as one might like.

Analyses of experiments and theory instantiations are also quite different. The power of a psychological experiment comes from the fact that, in principle, the interpretation of experimental data is a purely mechanical matter, converting the experiment's data into some numerical form and using straightforward statistical techniques to determine whether certain distributions of these numbers differ. The evaluation of an AI program is generally a much more subjective matter, relying upon the researcher's interpretation of the validity of the program's output (e.g., whether the summary generated by a prose comprehension program is a "correct" summary). This subjectivity is often inescapable—the whole point of much of AI is to explain cognitive processes that are individualistic and nonliteral. However, it also means that it can be difficult to determine when a ". . . theory is inadequate or [when it] executes but fails to stimulate the target phenomenon [Lehnert, p. 24].''

The sum of these points is that a psychological experiment is primarily vulnerable to the criticism that it does not realistically evaluate the theory it purports to address. The controls needed to achieve a well-defined experiment may oversimplify that experiment to the point where what is tested no longer bears any resemblance to the original question (cf., Lehnert, pp. 28–29. Clancey, p. 78). Errors in the design or analysis of an experiment are uninteresting from a theoretical or even analytical point of view: they are generally statistical in nature, and correspond to the technical flaws in a program that prevent it from running correctly. In contrast, AI research has two places where an experimenter's intuitions can be challenged. A program may not be an accurate implementation of its originating theory, and therefore fail to test it; the researcher may also misinterpret the program's results, leading him to claim success when he should not. This problem is only occasionally found in current work: at the level of analysis typical of most AI research, an experimenters intuitions about the appropriateness of his system's performance are usually acceptably accurate. However, when AI systems begin to be seriously evaluated as cognitive models, finding a rigorous way to analyze the products of a simulation will become an even more important and difficult question than it is now.

The role of experimentation in cognitive science.   How experiments are to be used as part of cognitive science is a question central to psychology's role in this multidisciplinary effort: this question is very nearly synonymous with "What does psychology have to offer cognitive science?'' The usual answer to this question (especially from the perspective of AI) is that psychology can provide guidance in the identification of cognitive structures and processes that can ultimately be implemented as part of an AI-based model: consider Clancey's list of topics that are relevant to his construction of a powerful tutoring system as well as to more basic research topics in psychology, such as the nature of working memory and various kinds of problem solving strategies.

Lehnert offers a somewhat more abstract discussion of the kinds of experiments that AI researchers find useful to their own efforts. Her discussion focuses on the importance of a *task orientation* to cognitive science experiments, and argues that experiments should relate directly to the input-output characteristics of programs that AI researchers might write. As an example of an experiment that would be of little use to AI, she presents the example of a psychologist studying the effects of eye color on the time required to execute some action: this task does not readily fit into an input-output task orientation, and, accordingly, could be of no help to an AI researcher.

This argument is open to objections. Much of the problem surrounds the example itself—it is hard to imagine how eye color could affect *any* cognitive process, which makes debating the example itself a questionable enterprise (see also Haugeland's comments on whether psychological experiments can, in fact be carried out without a task orientation: Haugeland, p. 88). However, the more glaring flaw here is a basic misunderstanding of what experiments that involve indirect measures—such as reaction time—are intended to do. Lehnert is correct in that they do not address input-output task orientations, but that is not the only question that an experiment can pose. Reaction time experiments can be an extremely useful tool for the study of the general architecture of the cognitive system, which is ultimately a critical question for researchers in all parts of cognitive science.

We do not have to look far for an example of architectural experiments that are relevant to cognitive science as a whole—Swinney's work (this volume) offers an excellent example. His experiments, which suggest that lexical access is unaffected by the semantic context of the sentence in which they appear, would seem to fall outside of Lehnert's criteria of ''relevant'' research, due to its use of indirect reaction time measures. This presumed irrelevancy is particularly surprising, because Charniak (this volume) finds Swinney's work quite interesting! Charniak points out that these results address one of the major problems in AI models of language comprehension, that of word sense disambiguation; in particular, they seem to rule out the schema-driven disambiguation strategy currently favored by many AI researchers.

Lehnert's prescription of relevant psychological research is too restricted—in general, experiments involving indirect methods such as reaction times address not picayune points of little interest to anyone outside a particular subarea of psychology, but rather very high level properties of the cognitive system, such as the characteristics of lexical access functions. Being able to interpret such an experiment's results from the perspective of this higher-level focus is what enables the researcher to avoid getting stuck at a local maximum (cf., Lehnert, p. 44). There are problems with this type of research—indirect methods are rarely as efficient as direct methods, and the jump from observing a difference in reaction time to postulating some aspect of cognitive architecture on the basis of that difference is not without risk—but these techniques can and do provide information that is of interest and importance to all of cognitive science.

What parts of the cognitive system must be studied?   One of the major contributions of cognitive science in general and artificial intelligence in particular is that it is very difficult to study a small, isolated piece of the cognitive system. Lehnert argues convincingly that researchers must cover a large set of processes: studying the limited domain of "prose comprehension" really means studying parsing and comprehension strategies, memory structures, belief systems, and a variety of other related structures and processes. Similarly, Clancey reminds us that we must be sure to look deeply enough into the processes we are studying: the production system (MYCIN) on which Clancey built his first tutoring system was sufficiently detailed to produce the "surface" behavior of medical diagnosis, but it lacked the deeper kinds of knowledge that would explain *why* its production rules were constructed as they were, which was precisely the kind of knowledge that is needed to build a powerful tutoring system.

Cognitive science only rarely studies a small, highly constrained domain; more often, it is concerned with the study of cognition as a whole. This is essentially the point made by Lehnert and Haugeland about local and global orientations (although some caution must be shown, as above, for Lehnert's tendency to identify architectural issues as "local," and not as worthy of study as global issues [cf., Haugeland's critique of Lehnert's distinction, p. 87). Needless to say, attempting to explain all of cognition is an overwhelming task. Inevitably, certain parts of a problem must be avoided or finessed, which leads to questions about a project's ad hoc or "kludgy" status: this is the AI equivalent to a psychologist studying a task and declaring that one of the variables that might be associated with this task in the real world is not critical to his experimental study of it.

There are two points to emphasize about these papers' discussions of kludges and ad hoc models. First, as Lehnert and Clancey note, it is difficult to determine which parts of a system can be "safely" kludged. Ideally, one kludges only those components that are ancillary to the processes being studied, but decisions about relevancy can be difficult. There was little danger in Feigenbaum's kludging the perceptual component of EPAM—at the level at which Feigenbaum was studying this question, perceptual processes would have only negligible effects on the learning of paired associates. This distinction is an instance of Simon's discussion of "nearly decomposable systems" (Simon, 1969): vision and learning are sufficiently distinct and non-interacting processes that they can be safely studied in isolation. However, Meehan's TALE-SPIN is somewhat different: the kludged parts of this system correspond to structures and processes that are admittedly relevant to the model's domain. Although implementing a vision system as a "front end" to EPAM would have added little to its explanation of paired associate learning, a version of TALE-SPIN that was knowledgeable of indirect speech acts, complex rhetorical structures, and more powerful inferencing strategies would have produced a very different set of stories as well as a different set of amusing stories caused by the unintentionally kludged gaps in these processes.

A second problem surrounding kludges is that the ability to detect an inappropriate or unintentional kludge is limited by the ability to detect incorrect performance by the program. When TALE-SPIN generated flawed stories, the flaws were generally so obvious that the corresponding errors in the program were relatively easy to identify and repair. However, continued improvement in this (or any) system would ultimately become difficult, as the errors and the flaws in the program that cause them become less obvious and less easily detected by the researcher's intuitions alone. At some point, more rigorous techniques, probably involving psychological experiments, would be required to identify the model's shortcomings. Of course, whether these final differences are sufficiently important to warrant exhaustive study is another question altogether; it may be more profitable to shift to a different problem where these remaining issues can be studied more directly.

AI is not cognitive psychology.   These chapters indicate that AI is not just a name for a more quantitative form of cognitive psychology. Both the methodologies and the ranges of research of these two disciplines differ in significant ways. Psychology has historically been concerned with smaller issues than has AI, if only because of the difficulty of kludging some part of an experiment without destroying its overall validity. In contrast, AI can tackle the larger issues, because the parts of these issues that are beyond the capabilities of the program can generally be finessed without rendering the program unexecutable and inference based on its performance invalid.

There can and should be a middle ground between these two areas: psychology needs to be more open to exploratory experiments that cover global issues, with many factors admittedly indeterminate and uncontrolled. Although traditional psychological paradigms prefer to compare several alternative hypotheses in search of the one that best explains the data, psychology can still benefit by getting a single idea of how some complex and relatively unstudied process might be done. Even though this idea will probably turn out to be wrong in the long run, it should point researchers from all disciplines if not in exactly the right direction, at least to a promising new area of research. This is not just a matter of psychology providing a service to AI—this kind of work will ultimately benefit psychology as well.

It is not clear that AI has to worry about the long run more than psychology—both AI and psychology have other people outside these disciplines looking toward them for answers, or at least advice. However, it is clear that in the short run, AI's primary interest lies more in the construction of running systems than in psychological validity, an interest that has both positive and negative effects. Building an AI system for some task without a detailed psychological explanation of how people perform this task is not an unreasonable thing to do, although the arguments by Schank & Abelson (1977) and Feigenbaum (1979) about the advantages of modeling a system after human strategies, should be recalled. AI need not wait for psychology to understand a task before beginning its own efforts. As noted by Haugeland, all disciplines have their own criteria by which to judge successful research, and, as Lehnert and Clancey remind us, psychology's criteria have not always led to research that has addressed the questions that AI has wanted answered.

In such cases, AI has little choice but to strike out on its own. In doing so, it may lead the way for psychology, but it may also reinvent the wheel, or even invent a square one. Not so long ago psychologists concentrated all their efforts on list learning experiments and nonsense syllables, but it was also not so long ago that AI believed that language translation could be done by word substitution, and problem solving by brute force search. What is needed here is a sound set of theoretical principles by which this research can be guided; as these historical examples indicate, simply adopting a global task orientation is no guarantee of successful research. It is the establishment of such a set of principles by both AI and psychology that will ultimately produce the best interaction between these disciplines, and insure the most productive study of those issues that are of interest to both groups of researchers.

LINGUISTIC ISSUES IN COGNITIVE SCIENCE

Relations between disciplines may sometimes be troubled and stormy, as we have just seen in our discussion of AI and psychology. But conflict is not necessarily interdisciplinary. Indeed, the views of the two linguists represented here appear quite incompatible. The gulf that separates Bresnan, Kaplan and Givón has no parallel elsewhere. It does, however, reflect the reality of present-day linguistics that is characterized by severe splits. Note that it is not even the case that these two linguists represent opposite poles within the field of linguistics proper. Although it is probably safe to say that Givón holds an extreme functionalist position, Bresnan and Kaplan are certainly not representatives of the extreme formalist wing. The true autonomous linguist would hardly concern himself with issues in cognitive science; for him, language is an object of study quite independent of any questions of language use. He is concerned with an abstract competence. The psychological reality of this ideal language, or the empirical testability or application of his theories are simply of no interest to linguists of that persuasion, as Bresnan and Kaplan have explained. Their own position is quite unlike that: it is not isolated neither from performance models nor from questions of psychological reality and computability by real-time processors, human or otherwise. Indeed, lexical functional grammar is a cognitive theory par excellence with clear implications for psychological experimentation on the one hand (Ford, Bresnan, & Kaplan, in press) and artificial intelligence programs on the other (Kaplan & Bresnan, in press).

Thus, if we look at linguistics as a whole, Bresnan and Kaplan's position lies somewhere in the middle of the field. In the context of the present chapters, however, their formal orientation and relatively restricted subject matter is unique and it contrasts most strongly with the work of their fellow linguist, Givón.

Their positions are so contrasting that they can't both be right, *if* they are talking about the same things. To some extent, we suggest, they may not be talking about the same things, and that their positions may be more compatible than they seem. It is a distinct merit of our contributors that they have provided us with clear accounts of their ways of doing linguistic research. Let us briefly look at these.

The pragmatic cline.   Givón's main point is quite simple. Natural language is essentially pragmatic, open-ended, inductive, and fuzzy. Logicians (and many linguists of similar persuasion) misrepresent this state of affairs by "rigorously" eliminating from the data base any evidence that does not fit their preconceived, discrete, logical categories. The metaphor that comes to mind is of a rich, varied plain whose four corners are occupied by small mounds. The logicians have built their temples on these mounds where they spend their days, occasionally walking over to a neighboring one along straight, narrow paths, oblivious to the world around them. That world, says Givón, is an enticing one, if one but looks at it, a bit of a jungle that is populated by all sorts of peasants and beasts. It is not that the logical corners of this world—truth and falsity, presupposition, definiteness, and definition—do not exist, but rather that by looking only at them a misleading picture of the world is constructed. There is a lot more to it than a few prominent markers.

Variations on this theme occur again and again in Givón's rich and stimulating chapter. He points out that discourse referents don't simply exist or not exist, but that speaker-dependent judgments are involved that are graded and probabilistic. Although there are frequent occasions when the speaker does or does not have a specific individual object in mind, these are just extremes in an inherently continuous pragmatic space. Thus, what he jokingly calls "semi-reference" muddles up the clean division between what is referential and what is nonreferential. Similarly, definite expressions are shown to involve graded judgments along several multidimensional scales.

What the speaker thinks the hearer will take for granted is at the heart of presuppositions, not truth relations between propositions (or even new versus old information). Examples for this claim are adduced from negation, restricted relative clauses, "if" and "because" clauses, and so on. The very nature of speech acts is scalar and they will not fit into the neat logical cubbyholes that might be constructed for them. Lexical meaning is context dependent, and the logic-derived definitional approach of many current lexical theories is fundamentally misguided (a similar argument was made with respect to psychological studies of semantic memory in Kintsch, 1980). Language change throughout history, as well as language acquisition in the individual are open-ended processes, depending on complex pragmatic inferences.

Thus, in case after case, Givón shows that natural language is much richer than the logician-linguists allow for, and more importantly, that a wrong interpretation of linguistic facts is obtained unless one takes all that richness into account.

It makes a great deal of difference to a cognitive scientist whether or not one accepts Givón's arguments. Consider, for example, definite descriptions. What kind of parser one writes, or what kind of experiment one designs, depends on whether one considers this to be a problem of assigning unique referential identities and of presuppositional relations in the logical sense, or whether one assumes instead that it involves Givón's cline with several continuous dimensions pertaining to the communication situation (example 18 in Givón's chapter). In a processing model that takes Givón's cline seriously the hearer/reader might be thought of as applying an open-ended set of comprehension strategies to reconstruct from a text a representation congruent with that of the speaker/writer as in van Dijk & Kintsch (1983) while in the more traditional approach the hearer/reader might be modeled as an automaton following a rule system, mostly of a syntactic nature (e.g., Wanner & Maratsos, 1977).

Linguistics, as practiced by Givón, is by its very nature open-ended. Multiple continuous dimensions are everywhere; though one might isolate this or that phenomenon, it is hard to be sure that one has captured every aspect and nuance of it. We might one day think that we had finally arrived at a complete description, only to have language itself play a trick on us and subtly slip away. It is hard to see progress in such a context. One might overlook the positive contributions of such work, that is, the actual descriptions provided of how language functions when people use it. But they are there, and it takes little imagination to see how psychologists could collect experimental data based on Givón's results, or how these results could be used in an AI program. What this kind of work does not do, however, is to satisfy our need for closure and neatness. At the end of it there stands no rule system, no formal structure to hold onto. That formality is exactly what Bresnan and Kaplan offer.

On mapping sentences into representations.   Bresnan and Kaplan's long-term scientific goals are probably very close to those of Givón. For both, the goal is a theory of language and language use. But Bresnan and Kaplan approach this goal in a very different way from Givón. They select certain language phenomena as a starting point and then develop an explicit formal structure to account for them. There is nothing fuzzy or open-ended about that: they are dealing with a rule system—a formalism—in much the same way others in linguistics and AI have done. It is, of course, the nature of the formalism itself that makes this work distinctive, along with the special goals Bresnan and Kaplan try to achieve with it.

Bresnan's and Kaplan's position is a functionalist one, too. They, too, think that cognitive constraints are crucial for understanding language. This is perhaps a minority position among linguists, but it is one that is easily appreciated by cognitive scientists. It is instructive to review the ground rules under which Bresnan and Kaplan operate. If we understand what they want to achieve, what they do and why become much clearer. The goal of all generative grammars is to devise a formal symbolic system to describe properties of natural language. Because there can be many descriptively equivalent formal systems, an additional constraint is introduced by Bresnan and Kaplan, namely, a concern with how cognitive processes use this system. The program implied by this goal is laid out clearly. First, the components of information-processing problems are defined—what is the initial state, what are the characteristics of the goal state, and what sort of processes can get us from one to the other. The problem is to find constraints for this process that guarantee a unique solution. Once these constraints are determined, an appropriate algorithm can be formulated and eventually be tested and evaluated.

Crucial assumptions are made right at the beginning of this research program that determine the flavor of the work. The problem for language acquisition is conceptualized as finding some learning function that maps an initial state (a universal, possibly innate knowledge base plus the primary language data available to the child) into a grammar, the mature knowledge of the language. Analogously, the language comprehension problem consists in finding a decoding function that maps this grammar plus the auditory-visual language data into a mental representation of meanings. More specifically, the syntactic mapping problem deals with the set of sentences and a mapping relation from these sentences to the set of representations of grammatical relations (that is, relations between the surface configurations of words and their semantic predicate-argument structure). This mapping has complex, many-to-many relations and is unique only for unambiguous sentences (lexically and syntactically). However, continuous clines of meaning a la Givón are ruled out.

Five constraints are proposed to make a solution to the mapping problem feasible. The first two have been used previously by Chomsky (1965): Both the domain and range of the mapping function are theoretically infinite, reflecting the creativity of language, and only a finite number of language elements must be used, the finite capacity assumption. The other three assumptions take us far beyond where Chomsky wanted to go. These are reliability or effective computability, order-free composition (sentences or phrases can be classified as grammatical or ungrammatical out of context), and universality. All of these are motivated by cognitive considerations (though Clark and Malt have questioned how "psychological" these "cognitive constraints" really are). On the basis of these assumptions, Bresnan and Kaplan proceed to build an algorithm—the lexical functional grammar—that generates structures that are meant to be isomorphic with the *mental* representations of the meanings of sentences. Because psychological considerations are so basic to this work, and because issues of computability are faced directly, there are numerous crosslinks between this theory and work in AI and psychology. Thus, this theory may turn out to be of pivotal importance in cognitive science.

Performance versus competence.   Is the gap between Givón and Bresnan and Kaplan really unbridgable? Could one not have something like Givón's context dependent processing strategies together with a representation of grammatical knowledge like lexical functional grammar?

Suppose we accept the lexical functional grammar as an adequate theory of grammatical competence. For cognitive science, the problem would not end here. We also would need a performance component to show how this competence is used by people or machines. The grammar is a knowledge source available to the language user, one among many. But knowledge use may be strategic. Knowing that R is the right grammatical representation for some sentences S, does not in itself tell us how we are to derive R from S. Algorithmic, context-free systems are conceivable, but perhaps not practical, given the amount of computation that would be involved. Strategies might help the parser to use its knowledge at the right place and at the right time. That is precisely what Kaplan's parser (Kaplan, 1972) attempts to do, and where some of Givón's observations could prove helpful.

There is nothing contradictory, in principle, about having a formal description of a knowledge structure (in this case, a grammar) and a strategic model for its use. Indeed, in other areas similar situations are accepted as normal. Consider models of arithmetic problem solving. We have at one level a highly developed formal mathematical system. That, however, does not correspond completely to the knowledge structure that children are actually employing in solving arithmetic problems. Hence, we need a second formal system, describing not arithmetic as the mathematician sees it, but as the children know it. Finally, we need a set of processing strategies, that show how this knowledge is activated and used in concrete contexts. The strategies are pragmatic, in the sense of Givón. But to make them work, we need the formal knowledge, too, that would not have to be context dependent. It is not obvious why an analogous situation could not exist in the area of language processing. Indeed, Lehnert's work on language (e.g., Dyer, 1982) exemplifies such a position. She is concerned with specifying rigorous formal processes for extracting meaning from a passage, but is doing so with comprehension strategies that are capable of handling the knowledge and context effects that are the concern of Givón.

Linguistics and cognitive science.   Sometimes, when faced with the diversity of approaches within cognitive science, we get a little nervous. Looking at the diversity within established academic disciplines—linguistics, in this case— ought to reassure us. If linguistics can live with such differences as we have seen in the two chapters in this volume, cognitive science can learn to expect and tolerate diversity, too. Eventually, of course, history will make its own choices, and the right way will be there for all to see. In the meantime, not being able to wait for that happy day, we must place our bets. It is of course dangerous to exclude certain observations from the domain of a science: What if the critical facts are dismissed? On the other hand, by no means does it follow that a science must deal with everything; it is too easy to become overwhelmed. The history of science has shown that it is not necessarily the case that the significant, crucial observation upon which a science is constructed is particularly salient or ecologically valid. A rather abtruse datum can turn out to be crucial, and there is no safe way to tell beforehand what is and what is not a productive observation. It is the old story: there is no recipe for doing good science.

It is interesting to note how the conflict between the approaches of Bresnan and Kaplan and Givón reflects the dual origins of cognitive science. Bresnan and Kaplan's work is best understood via the information transmission metaphor: a source and a receiver, with information flowing between them and language as the permanent record of that information transfer. That record is an object that can be studied for itself. Of course, even if one accepts such a framework, many additional decisions have to be made about exactly which aspects of language to study. In Bresnan and Kaplan's case, language is taken to be a set of sentences (or, more precisely, the grammatical form of these sentences). A crucial property of these sentences is that their length is unbounded. Note that the decision to focus on these aspects of language is quite arbitrary. One can ask, why sentences? How much of actual spoken language is in the form of sentences? Why not texts, for instance? Why grammatical form? Why unboundedness? Even moderately long sentences are quite rare in natural language. One can ask such questions, but one can ask equally well, why not? Only the eventual success, or nonsuccess, of the enterprise will provide an answer.

Givón represents the alternative approach of science: that from the inside. What is important for him are the means that one conscious mind uses to change the consciousness of another, through the use of language. Such a view predisposes one to ask questions about the function of language and to attend to the subtle nuances of language use. There is no corpus to be studied, but an ever shifting communication situation. The complexity of that situation plays a critical role, language being just one of many facets.

For cognitive science it is important to be clear about the linguistic issues and controversies that it is faced with. Intelligent choice presupposes a clear understanding of exactly what the issues are, and where the strengths and weaknesses lie on each side. There are, however, a number of points to be made about the role of linguistics in cognitive science that apply to all linguistic work, irrespective of the issues we have focused on thus far. Namely, what profit can linguistics derive from other sciences within cognitive science? Clark and Malt, in their chapter in this volume, have tackled this question. They are concerned with the question of how constraints from one science can affect the theorizing in another. Surely, this is a very basic problem in a cooperative discipline such as cognitive science. There are two kinds of constraints, those that are internal to a science and those that derive from related fields. It is that second type of constraint that interests us here. Clark and Malt are not trying to tell linguists what kind of constraints to use, but rather explore what kind of psychological constraint might be of use to a linguist. How can linguistics profit from psychology?

The use of linguistic constraints is a much more familiar affair in psychology. It is well known (e.g., see Fodor, Bever, & Garrett, 1974) that early psycholinguistics was almost exclusively constrained by linguistic theory. The derivational theory of complexity as discussed by Fodor et al. is a particularly clear case in this respect, though not a very happy one. Later, psychology learned to employ linguistic constraints with more sophistication. The question asked here is the reverse. Clark and Malt argue that linguists have not yet fully exploited the potential of psychological constraints. Many linguists today appear to be open to psychological arguments, but may not yet have taken full advantage of what psychology has to offer. That is a psychologist's view, but it is worth pointing out that none of the linguists were ready to accept it in the conference discussion.

Linguistic theorizing, of course, has a proud history of formalization, at least in some subareas. However, AI techniques offer possibilities that are not yet fully exploited or even appreciated. Initial attempts have been made—such as the cooperation between Bresnan and Kaplan (Kaplan & Bresnan, in press)—with exciting implications. As this sort of cooperation speeds throughout cognitive science, the face of linguistics may be permanently changed. It is important, however, that the possibilities of such theorizing not only are used by those linguists who have strong formalistic predispositions anyway. There is a real chance here for linguistic theory to find precise, objective formulations for some of the problems that resist formalization by other means. It is fine to collect examples demonstrating the open-endedness of communication via language, but to show that a successful language comprehension program can be based on such principles could give considerable weight to one's arguments. Language may by its very nature be fuzzy, but our theories need not be so. AI techniques may offer a solution: specific, precise models dealing with inherently open-ended problems.

If the trends towards cooperation with cognitive science that we see today continue, the linguistic landscape might be reshaped. As we pointed out in this volume's introduction, the development of cognitive science has caused the marketplace of ideas on linguistic issues are traded to change dramatically. Suddenly, there are a lot more people who care about language. It is no longer a peaceful (or not so peaceful) reservation for a few linguists with their special, historically conditioned interests, but is instead increasingly populated by cognitive scientists of many persuasions who are interested in what is being offered. Their decisions to buy or not to buy are a new factor within linguistics. It may even change the nature of the game.

THEORIZING IN PSYCHOLOGY

The chapters presented in the Psychology section and the commentary on them, dealt with a wide variety of issues involving theoretical and empirical methodologies with illustrations from the authors' own work. The Swinney and VanLehn, Brown, and Greeno chapters describe an interesting subset of the theoretical and empirical methods employed in cognitive psychology and characterize some of their limitations. Charniak, in his commentary, concludes that, at this point in the development of cognitive science, the large number of methodologies are both necessary and beneficial and that we can't a priori select "the correct set of theoretical and empirical methods." Charniak justifies his championing of the current methodological pluralism with an illustration from the recent history of the development of our understanding of the genetic code.

We agree with many of the points made by the authors, but we will discuss these chapters in a context provided by a paper by Newell (1973) entitled "You Can't Play Twenty Questions With Nature and Win: Projective Comments on the Papers of This Symposium." Newell's analysis was motivated by his ambivalence about the current state of cognitive psychology. He had high praise for the series of papers that had just been presented at the 1972 Carnegie-Mellon Symposium on Cognition. On the other hand, he had serious reservations about the overall direction of cognitive psychology. In particular, he could discern little progress towards understanding the basic principles of human cognition.

Newell's malaise was reinforced by his views on the current state of cognitive psychology. First, the field's empirical knowledge was a large collection of partially understood phenomena often defined by a particular experimental paradigm. Second, the field's theoretical knowledge was a large collection of binary contrasts concerning a wide range of psychological processes. It was difficult for him to discern an overall structure in the collection of empirical phenomena and theoretical contrasts. There were no clear goals guiding the research paradigms that generated this collection of theoretical and empirical results. Newell argued that the continued unguided aggregation of empirical phenomena and low level theoretical contrasts would not lead to a coherent theory of human cognition.

Newell speculated that this situation was due to two factors. First, cognitive psychology is almost totally reliant on hypothesis testing as a method of inference and on its associated sophisticated experimental methodology. The second was the field's preference for dealing with qualitative, incompletely formalized, low-level theories that are captured by various binary contrasts like serial versus parallel processing, and so on.

Newell had three proposed solutions for the problems uncovered in his analysis. The first solution was that the field should focus on the construction of detailed process models that include an explicit formalization of the control structures necessary to execute the psychological processes that the theorist hypothesized were responsible for behavior in a given experimental paradigm. His second was that we should develop detailed theoretical analyses of significant complex tasks like reading or playing chess. His third suggestion was that we should develop global theories that account for performance in a large number of different experimental paradigms.

Many of cognitive psychology's limitations and problems exist for cognitive science as a whole. In spite of Charniak's championing of our current methodological eclecticism, we have no assurances that the resulting collection of empirical and theoretical results will aggregate into a coherent theory of cognition and intelligent action. The list of empirical and theoretical results would be longer and even more heterogeneous for cognitive science, and it would still be difficult for a thoughtful reviewer to discover an underlying organization of this collection of results and to discern any progress towards a set of well defined goals.

Charniak's historical analogy may not be entirely conclusive. The events that Charniak described concerning the history of the development of our understanding of the genetic code support championing the methodological diversity that currently exists within cognitive science. However, the analogy between cognitive science and modern biology is less than exact, because the very diverse programs of research that led to our current understanding of the genetic code were driven by a common set of reasonably explicit goals. It was accepted that DNA was the genetic material and that the goal was to understand the structure and function of this complex macro-molecule. As we showed in Chapter 1, cognitive science is unified only by the most tenuous of high level goals, the understanding of intelligent action in demanding environments. What's missing is an understanding on how to decompose such a goal into an explicit set of research issues. One of Newell's criticisms of cognitive psychology was that we had not formulated such goals. The situation is even more disturbing for cognitive science because of the diversity both between and within the separate fields that make up cognitive science discussed in Chapter 1 and forcefully illustrated in the chapters in the Linguistic section. In addition, the interactions between the separate fields and their very disparate methodologies will not automatically lead to the evolution of the necessary high level goals.

As a first step in the development of cognitive science, we surely must accept the methodological diversity championed by Charniak. However, in order to bring about the cohesion necessary for cognitive science to become a serious substantive discipline, it may be necessary to generate a better understanding of the strengths and weaknesses of the methodologies within the separate disciplines. In fact, it is probably most important for us to understand the weaknesses of our own chosen methodologies and the strengths of the methodologies proposed by our colleagues in other disciplines. For this reason, analyses like Givón's of the formalist program in linguistics or Lehnert's chariacature of the experimental method in cognitive psychology may not be the most productive ones at this stage. Surely, these methodologies have important weaknesses and limitations, but researchers with first hand experience in these fields are often the ones best able to diagnose these problems, as Newell (1973) demonstrated, or Tulving and Madigan (1970) in their highly critical review of memory research. In short, we have to take Charniak's conclusion seriously, that all of the mainstream methodologies in the various disciplines that make up cognitive science have the capability of making important contributions to the solution of problems of shared interest.

How to play the game.   Turning now specifically to the discussion of the Swinny and the VanLehn, Brown, and Greeno chapters we note that these chapters provide us with additional insights beyond Newell's (1973) analyses. Swinney describes important results concerning the mechanisms of lexical access. Ironically, Swinney presents a textbook case of the kind of theoretical analysis that led to Newell's (1973) criticisms. And yet, Swinney's work does not appear to suffer from the weaknesses pointed out by Newell, in particular the lack of an overall theoretical rationale that plagues much of the current work on cognition. Newell objected strongly to a style of theoretical analysis that leads to an incomplete qualitative characterization of the processes involved in performing a task and the isolation of a binary theoretical contrast. Swinney has been attempting to distinguish between the assumption that the processes of lexical access are automatic and independent of other mechanisms involved in comprehension versus the hypothesis that the top-down processes have strong effects on lexical access only retrieving the contextually relevant meanings. There was general agreement by the conferees that Swinney's results were important and had numerous implications; Charniak presents an analysis of the impact of Swinney's results on his thinking.

What distinguishes Swinney's work from the research that led to Newell's critical analysis of the current state of cognitive psychology? The major difference is that Swinney's binary theoretical contrast is embedded in an informal but substantial theoretical context. Different theoretical analyses of comprehension distinguish lexical access as an important stage. In this context, showing that lexical access occurs independently of higher order processes has a profound effect on our conceptualization of the whole set of theoretical issues involving language comprehension. Newell's objection was that many of the binary theoretical distinctions that cognitive psychology explored with its elegant experimental technology had no adequate theoretical context. The impact of answering one of these questions in one way or another was obscured by the fact that there was no higher order context in which to embed the various theoretical and empirical results into a growing understanding of cognitive processes in general. We are not concerned enough with the development of the higher order theoretical structures necessary to generate a successful cumulative science.

Swinney's other methodological issue—that real time analyses of language are to be preferred over static analyses—is more controversial. We agree with Swinney's criticisms of earlier experiments that explored the same issue and obtained different results. In fact, Swinney's analysis is a specific case of a general issue that was explored in some detail by Newell (1973). Newell uses the term ''method'' to refer to the representation of the experimental situation and the processes that operate on those representations to produce the behavior observed by an investigator. Newell points out that the human information system is a powerful adaptive system. Thus, subjects may use very different methods to produce equivalent behavior, or minor changes in the situation or instructions may cause subjects to use different methods. Newell observes that experimental psychologists often do not carry out a sufficiently detailed analysis of the processes by which subjects perform an experimental task, and therefore can be seriously misled by their results. Swinney's criticism of earlier work implies that previous researchers had an inadequate understanding of the methods used to perform their experimental tasks and were led to incorrect conclusions about the nature of the lexical access process.

One of the most serious limitations of work in cognition is that it has not been based on detailed analyses of the methods used to perform experimental tasks that are the source of the empirical data that drive theoretical arguments. Cognitive psychologists can be peculiarly insensitive to the fact that the human information-processing system is adaptive and that minor changes in instructions or procedures can cause subjects to use different mechanisms to perform a seemingly very similar experimental task. The adaptive character of the human information-processing system is one of Newell's major reasons for believing that a simple minded type of ''20 questions'' theoretical methodology has no chance of success in cognitive psychology or any other domain of cognitive science. The adaptive character of the system makes cognitive psychology's willingness to explore endless variations of a once interesting experimental paradigm, the functional autonomy of methods (Tulving & Madigan, 1970), totally counter-productive. The process under investigation will not remain constant, ''sit still,'' under arbitrary changes of various aspects of the basic experimental procedure. It is all too easy to have minor changes lead to the use of very different processes by subjects. However, our current methodological paradigms do not take adequate account of this fact.

The trouble with complexity.   One of Newell's prescriptions for the future progress of cognitive psychology was that it should focus on detailed theoretical analysis of significant tasks: expert level chess, reading comprehension, and so on. Furthermore, these analyses should take the form of computational theories of performance in these various domains. The VanLehn, Brown, and Greeno chapter discusses the limitations of this strategy for advancing our understanding of cognition. They observe that significant segments of cognitive science have taken Newell's prescription seriously and that much of the work in the field involves the detailed theoretical analysis of increasingly complex task environments. In fact, they point out that the increasing complexity of the environment and the resulting computational model seem to be one of the primary goals. Their concern is that the increasing complexity may not lead to an increase in our understanding of the basic principles of intelligent action. Computational theories involve complex interactions between many processes and representations. However, successful models, in terms of accounting for performance in a complex domain, can give surprisingly little insight into basic principles. The major difficulty is that there is no way of disentangling the relative contributions of the components of a complex program. Our inability to decompose the separate effects of the various assumptions and principles incorporated into a computational theory severly limits our ability to uncover basic principles. It seems that one could write an article in the same vein as Newell's entitled, "My program is more complicated than your program, but do I know anything more about cognition?"

The limitations of computationally based theoretical methodologies derived from AI are important because of the central role that numerous individuals assign to such models in the further development of cognitive science. Mandler and Haugeland argue forcefully that computational techniques developed by workers in AI will play a central role in the formalization of theoretical work in all branches of cognitive science. In fact, both assign a first-among-equals status to the field of AI within cognitive science as a whole. As Mandler points out, AI provides us with conceptual tools necessary to formalize assumptions about representation and process that are basic to all of the cognitive sciences.

VanLehn et al. point out that the first step in developing a computational theory of a complex task is a task analysis. Initially, a program may incorporate processes that are not reasonable models of known cognitive mechanisms. The focus of the initial theoretical analysis is on simulating intelligent action in the new environment. The resulting computational theory may not be a well-reasoned combination of principles and mechanisms that give us a clean theoretical analysis of the cognitive mechanisms responsible for performance in this environment. Indeed, a collection of such programs will not by itself aggregate into a theory of intelligent action.

The programs themselves are often not well understood. The necessity or the entailments of individual mechanisms incorporated into the program are very difficult to disentangle because the overall performance of the program is some complex interaction of the separate mechanisms. VanLehn et al. argue that a critical component of our theoretical knowledge is an explicit understanding of the contributions of the individual components that make up a complicated model. These principles are what we can generalize across models and aggregate into a substantive theory of cognition.

VanLehn et al. propose that the entailments of the various mechanisms incorporated into computational theories can be understood more deeply by a process they call competitive argumentation. For example, we should understand the impact on overall performance of removing a given mechanism from the theory. However, it seems to us that VanLehn et al. are calling for a variety of theoretical analyses that psychologists have practiced for years in order to be able to employ their empirical methodology. The low level binary theoretical contrasts that drive so much of cognitive psychology's empirical work today are in fact the results of such analyses. One reduces a given theoretical issue to the presence or absence of a given mechanism or a comparison between complementary mechanisms. Swinney's work can thus be taken as a good example of the kind of theoretical program advocated by VanLehn et al. Swinney's work is driven by a very careful analysis of the implications of the assumption that the lexical access process is independent of other higher level mechanisms involved in speech and text comprehension. However, it is exactly this kind of analysis that enables cognitive psychologists to successfully employ their powerful empirical methodology.

A second and equally important issue dealt with by VanLehn et al. concerns the tailorability of computational theories. They argue forcefully that the most desirable kinds of theoretical explanations are strongly constrained in two senses. First, a process makes strong predictions about the kinds of behavior that can and cannot occur in a given task. Second, the principle has limited tailorability; by this they mean that the proposed mechanism cannot be tuned to match any data at will.

Tailorability was a central underlying theme in the interactions amongst the various conferees. It was agreed by all of the participants that the body of theoretical ideas under consideration today in the diverse fields of cognitive science are more than powerful enough to account for the data concerning cognition and intelligent action. In fact, our difficulties are quite the converse. The principles incorporated in many theories are so flexible and powerful they are not testable. In addition, various mechanisms don't have a clear set of empirical consequences and thus we have little theoretical understanding of precisely what the function is of a particular subprocess in a computational theory of a complex task. All conferees agreed that the issue of constraints was central for the further development of cognitive science, but they reached no consensus on what exactly an acceptable set of constraints ought to be.

REFERENCES

Chomsky, N. *Aspects of the theory of syntax.* Cambridge, Mass.: MIT Press, 1965.

Dyer, M. G. *In-depth understanding.* Technical report, Department of Computer Science, Yale University, 1982.

Feigenbaum, E. A. The art of artificial intelligence: I. Themes and case studies of knowledge engineering. *Proceedings of the Fifth International Joint Conference on Artificial Intelligence,* Cambridge, Mass., 1979.

Fodor, J. A., Bever, T. G., & Garrett, M. F. *The psychology of language.* New York: McGraw-Hill, 1974.

Ford, M. J., Bresnan, J., & Kaplan, R. A competence-based theory of syntactic closure. In J. Bresnan (Ed.), *The mental representation of grammatical relations.* Cambridge, Mass.: MIT Press, in press.

Kaplan, R. Augmented transition networks as psychological models of comprehension. *Artificial Intelligence,* 1972, *3,* 77–100.

Kaplan, R., & Bresnan, J. Lexical functional grammar: A formal system for grammatical representations. In J. Bresnan (Ed.), *The mental representation of grammatical relations.* Cambridge, Mass.: MIT Press, in press.

Kintsch, W. Semantic memory: A tutorial. In R. S. Nickerson (Ed.), *Attention and Performance VIII.* Hillsdale, N.J.: Lawrence Erlbaum Associates, 1980.

Newell, A. You can't play 20 questions with nature and win: Projective comments on the papers of this symposium. In W. G. Chase (Ed.), *Visual information processing.* New York: Academic Press, 1973.

Schank, R. C., & Abelson, R. P. *Scripts, plans, goals, and understanding: An inquiry into human knowledge structures.* Hillsdale, N.J.: Lawrence Erlbaum Associates, 1977.

Simon, H. A. *The sciences of the artificial.* Cambridge, Mass.: MIT Press, 1969.

Tulving, E., & Madigan, S. A. Memory and learning. *Annual Review of Psychology,* 1970, *21,* 437–484.

van Dijk, T. A., & Kintsch, W. *Strategies of discourse comprehension.* New York: Academic Press, 1983.

Wanner, E., & Maratsos, M. An ATN approach to comprehension. In J. Bresnan & M. Halle (Eds.), *Linguistic theory and psychological reality.* Cambridge, Mass.: MIT Press, 1977.

12

Some General Remarks on the Cognitive Sciences

Patrick Suppes
Stanford University

Now we are at the end of the conference, so you can relax. George Mandler and I will take you on a general tour of cognitive matters. In my younger days I used to be what some people call a 'nothing but-er'—devoted to nothing but the truth. A 'nothing but-er' is very clear there is only one way to do things, and his advice is to get cracking on doing things right. Well, with time I have developed the pluralistic attitude that I have been expressing here, but it is a market theory of pluralism. It does not mean that I believe all theories and all methodologies are equally good or equally worth buying. I am not going to try to appraise the products currently on the market, or say what I am willing to buy or sell. But it is an important part of the market that the market itself is pluralistic in another way. For different consumers there are different goods. Whatever your choices may be, you can probably buy them, and this is the way the market ought to be.

I want to start with four neglected issues that I was a little surprised we did not get into. The first is that of mental representation. In view of the extensive dialogue within psychology about mental representation, I am surprised we did not have more discussion of such issues that as knowledge, can always be expressed propositionally. The recent exchanges between John Anderson (1978, 1979) and Zenon Pylyshyn (1979, 1981) in the *Psychological Review* are fascinating.

Neglect of the second issue is less surprising here, but I think it is very important. It was alluded to, to some extent, in Gene's remarks about computational power. The issue is the nature of detailed front-end theories of listening and seeing, and the relevant data; for example, the very complex problem of speech recognition—whether we are interested in it from an AI standpoint or from a psychological standpoint of hearing. The way in which the sound-pressure wave gets converted into cognitively meaningful material is one of the intriguing complex and subtle stories of all time, surely. The competition between various approaches, beautifully reported in Allen Newell's ARPA report on speech recognition, was not talked about here. The competition between the brute-force computational power of the program called HARPY and the knowledge-based program called HEARSAY should have been, I think, of conceptual interest to this group because it presents in dire contrast two approaches to computational aspects of cognitive problems.

The third issue, only hinted at in Talmy Givón's chapter, was that of probabilistic aspects of cognitive processing. I was surprised that the complexity of the kind of data and processes, considered essential in cognition, did not generate more probabilistic questions. There are some natural probabilistic questions that arise in relation to David Swinney's examples, but here too I found that matters were not pursued in sufficient detail to raise the central questions of a stochastic approach. In particular, in attempting to give a causal analysis he did not proceed as far as looking at auto-correlations in time, that are at the heart of any stochastic analysis of real-time data. It seems likely to me that probabilistic questions were avoided for two reasons. On the one hand, there is the optimistic hope that the kind of theories considered will lead ultimately to determinate rather than probabilistic analyses of cognitive processing. On the other hand, the theories presented were in general not sufficiently developed from a mathematical and quantitative standpoint to warrant a full-blown probabilistic analysis of data relevant to the theories.

The fourth neglected issue was consideration of the methodology of protocols. I will not say more about it here. It was just very surprising to me that no speaker spent any serious time on the complicated methodological issues raised by the use of protocols for data.

Let me now turn to eight comments on matters that were discussed one way or another in the papers given at the conference.

My first point concerns the surprising diversity of methodology and theoretical viewpoint in the two linguistic papers. I suppose I am used to going to conferences at which linguists represented a solid phalanx of theory against opposing ideas of psychologists or philosophers. I am happy to welcome this diversity. It satisfies my marketplace model of ideas. Talmy Givón may feel that there has been a monopoly or at least an oligopoly in the linguistic marketplace in recent years, but surely that is a passing phase that will fade and perhaps already has.

From a broad historical standpoint, I like to think of the other linguist, Joan Bresnan, as Joan the Alexandrian. Here I am thinking of her in her concern with structural competence and not in her strong concern with empirical matters as expressed in parts of her paper. Much of what she wants to do would have been very comfortably received and appreciated by the ancient Alexandrian grammarians, who really started the serious work on grammar. I am sure that they would be into what she is doing after two weeks of up-to-date lectures by her or Noam Chomsky. I want to emphasize my great respect and admiration for this Alexandrian tradition. It is a conservative tradition but one that should be valued. Of course, it is easy to say that there is more than one Alexandrian tradition and Talmy Givón fits in with one that has been too much neglected. I have the tradition of the mechanicians in mind, of those who did all kinds of practical things. Heron is a good example of this school. Talmy Givón, like Heron, is in one sense interested in real-time processes.

But I want to emphasize that I am not trying to oppose these two viewpoints. We need them both; neither will solve all of the problems of interest and importance. The kind of conservative intellectual tradition that Joan represents will be with us, as far as I can see, forever and ever. The kind of thing that Talmy and many psycholinguists do, would not be something that the Alexandrian grammarians would have appreciated, but that in no sense denigrates the work. These grammarians simply would not have gone in for extensive transcription and recording of data and, above all, they would not have gone for looking at a lot of different languages. It was clear to them that Greek was the only serious language, just as it is clear to some that today English has the same status. There is in fact quite a good case to be made that only three languages have ever received really extensive linguistic attention, first Greek, then Latin, and now English. This concentration on the most important languages should continue but Talmy's work of a quite different character is also important. I want to emphasize the thoroughness of my pluralism here. There has been some mention by Marshall Farr and others of the struggle for unified theories in physics, especially unified field theories, but I think this example from physics is not a good one for the cognitive sciences. The unification of field theories in no sense annihilates the diversity and plurality of classical problems in physics. To mention just one simple example, consider the most important problem of nineteenth century classical particle mechanics, the three-body problem. This is the problem of predicting the paths of motion of three bodies interacting only by gravitational force. The three-body problem will be with us, I would predict, when all of the field theorists of modern physics have come and gone and long since been buried. What is even more surprising is that magnificent progress has been made recently on the three-body problem. It is easy enough to cite a variety of other examples from physics. Let us not make the mistake of thinking that in any sense physics is unified. Its pluralism is much more dazzling than that of linguistics, but so it should be because the physicists have been in serious scientific business for a much longer time.

The second point is that to some extent this Alexandrian tradition in linguistics contrasts with the AI tradition because the two disciplines are interested in different problems. Wendy Lehnert mentioned the fact that her grammar is messy and she hates to talk about it. She is not really engaged primarily in writing a grammar. She is interested in language in terms of different kinds of questions, and so different kinds of models arise. This does not mean there cannot be some overlap. I would expect those different traditions will remain separate and should remain separate—both in theoretical outcome and in methodology. Tensions will be evident, but pluralism will carry the day.

The same kinds of tensions exist between linguistics and psychology. Here, instead of saying "process," we might say, using Herb Clark's ideas, it is structure versus use. I was commenting to Joan Bresnan earlier on the analogue of the Helmholtz-Lie problem of space to her general problem of finding psychological constraints on language. Leaving all technical matters aside, the intuitive idea of the Helmholz-Lie problem is this. Helmholtz considered the fact that rigid bodies can be moved about without distortion in physical space one of the most important features of the space we live in. The problem has been to give an axiomatic characterization of space that led precisely to the conclusion that would be expected, namely, that only elementary spaces—what are sometimes called spaces with constant curvature—could satisfy such an axiom of non-distorting motion. The analogy to Joan Bresnan's problem is that very general axioms depending on only very limited aspects of experience have led to extensive mathematical investigations on the nature of space. I see her suggested enterprise as being in the same spirit. Such investigations have been the provenance of mathematicians, not physicists. In the same spirit, even though Joan Bresnan wants to begin with psychological constraints just as Helmholtz wanted to begin with physical constraints, it really would be mathematical linguists, not psychologists, who would continue to investigate the problem she poses. As should be apparent, I think the problem is important and deserving of a great deal of attention.

My fourth point concerns the new twist to the debates about competence versus performance. In the past I have been so brainwashed by linguists about what competence is that I was surprised and pleased when somebody made the very intelligent remark that there is, after all, an ordinary sense of competence. But I also like the linguistic controversy about the theory of competence. I think it is worthwhile and interesting. But it is not at all clear to me, once we leave the definite situation of having a grammar or a language, how we want to think about competence. I suggested in a discussion with John Seely Brown and Jim Greeno that we could have competence of procedures and competence of subject matter. There are many relevant puzzles from epistemology. For example, there has been much discussion in philosophy about this principle of competence for knowing: if you know p, you know any logical consequence of p. This is a requirement of closure under logical consequence. One of the things I suggested in the discussion with John and Jim is that we could look for closure conditions on elementary parts of a subject matter. Over the years I have had delightful conversations with Seymour Papert about elementary physics. I have felt there are parts of elementary physics that we could put in a very elegant way, where we have closure conditions. Closure conditions give us a closed miniature theory. I suspect my philosophical taste for the elegance of closure could easily drive me into a position that psychologists might not be happy with, but I think it is a topic worth understanding better, because I do not really understand, except for the case of grammar, what competence means. Here is an example I mentioned to John Seely Brown—I do not know whether he is entirely happy with it, but it is very much in the spirit of his work on subtraction. We could think of competence in subtraction in this theoretical way as the amount of arithmetic you can do with marks, or just Arabic symbols. Such a theory has been investigated and we have a very nice understanding of it—maybe that is the right sense of competence in arithmetic. This particular theory of the arithmetic of marks corresponds very nicely to what is taught in the schools through about the first four or five grades. I am looking at it as a subject matter, but I think there is another way that John talks about, and that is a procedural way of looking at competence. I am not quite sure what he has in mind. I will propose a task for him for the discussion: To give us one or two examples of competence in the sense of procedure. I also do not fully understand, as I told Jim Greeno, what he and Rochel Gelman have done to characterize competence. I think they are pointing out a very interesting direction to go, but I need more guidance.

Associated with the discussion of competence is a discussion that I am less sanguine about. It may be just prejudice on my part. I refer to the associated concept of understanding. What does it mean to understand something? We may be able to develop a competence notion of understanding. But I am leery of having a performance concept of understanding. I guess understanding as I use the concept is a good example of something without closure. It is very open ended. I do not know how much we can use it as a systematic concept. It is certainly important. I say, well, I understand what you say—and in fact, among philosophers it is a favorite phrase. When a philosopher says, "I do not understand," he is not speaking modestly. He is saying, "You idiot, why didn't you say something clear?" and it is used that way among philosophers as a kind of code. But I have some skepticism about having a good working systematic notion of understanding.

Now for a point that I felt we wasted time on. It seemed to me there was a tangled and not needed discussion about finite, recursive, and recursively enumerable objects. I think all the important notions for this conference are very independent of these logical refinements. So I am quite happy to keep everything finite and introduce the concept of being creative. You are creative if you have a relatively small apparatus and you have to deal with a very large finite setup. I won't try to pursue this idea technically but there is one new technical development that I think is of some interest to our discussions. This is the attempt to introduce in computer science and related disciplines a classification of the order of difficulty of problems that can in principle be decided in a fixed finite manner. A good example of something that has an effective finite procedure is Tarski's famous decision procedure for elementary algebra. If we write down any elementary algebraic formula, it is decidable in a fixed finite number of steps whether the formula is valid or not in elementary algebra. This decidability result contrasts sharply to the famous Gödel undecidable results for elementary number theory. On the other hand, from the standpoint of applications in computer science or in related cognitive sciences, the fact that such algebraic formulas are decidable is in another sense uninteresting because it is an unfeasible kind of decidability. Roughly speaking, a method of decidability is unfeasible if it is not polynomial in length where the polynomial is written in terms of the length of the formula itself. In Tarski's case there is an exponential explosion of the number of steps required to decide if a formula is valid or not. But even polynomial results are unrealistic from the standpoint of psychological processes. We usually seem to need something that is not even linear but sublinear, really fast and simple. My point is that what we need as a serious point of discussion in psychology is what kinds of proposals are psychologically feasible in terms of computation. Let us forget all about something being recursive but not primitive recursive, recursively enumerable but not recursive, and so on. Let us begin to worry now about some real questions and some new questions of computability, ones that might have some applicability to psychological theories of cognition.

There is a point I want to make about the tension in artificial intelligence between knowledge engineering, the Feigenbaum conception that what ought to be done is that which is most efficient in a computer framework, and the contrasting effort to model human performance. In general I am very sympathetic with the Feigenbaum concept of knowledge engineering. All of us accept it when it comes to computers doing arithmetic. No one seems to think that computers should do arithmetic in the fallible and very slow way that we humans do. A recent article in *Business Week* predicted that in the next two decades 45 million jobs in the United States will be changed because of the impact of automation in the factory and the office. On the surface, this seems to be a case of knowledge engineering, for we do not expect factory robots and word processors to model in any deep way human approaches to the relevant tasks.

On the other hand, I think that too much can be made of the tension. The reason is simple enough. Even if we want to emphasize the knowledge-engineering approach to AI, the problem of the human interface remains one of the most important aspects of the work. That human interface has got, in part at least, to model successfully human performance. In spite of the present tension between the two approaches to AI, it is my feeling that it will be dissolved by the increasing concentration on the interface problem. There is, of course, a long history of such interface problems, going back at least to the intensive work in World War II on human factors, especially in the instrumentation of airplanes. But I think we are just about to begin a new era, and perhaps the thriving and developing automation revolution in the office will be one of the most important driving forces behind a whole new principled approach in the cognitive sciences to man-computer interfaces.

My final point is to comment on the several discussions we have had about modularity versus integration in the cognitive sciences. There is a long history in psychology of placing bets on modularity. The idea is that you can study a small piece of a system very well and you have faith that results will contribute to the larger picture in a really meaningful way. Something like this attitude has certainly been a driving force behind the intensive and extensive developments of verbal learning theory and experiments. It is all too easy to poke fun at this tradition now in retrospect. There are many things about it of considerable scientific value but there is also no doubt that the sense of dissatisfaction is a real one and not simply due to a superficial change in fashion. Certainly it is very common in the cognitive sciences now to be skeptical of modularity and to feel that only integrated scientific approaches that deal with large parts of cognition can be really successful. The criticism of this view is that as yet it seems to be more a case of wishful thinking than of real achievements.

My pluralistic attitude is to praise both approaches and to counsel not getting stuck in either one as the only way to do serious science.

To make my point in a more extended way, let me superficially draw on a certain part of the history of physics. By the end of the nineteenth century it was clear that classical particle mechanics had reached a dead end in terms of its most important problem, the three-body problem, which I have already mentioned. Poincaré was given the Swedish royal prize not for solving the problem but for showing together with Bruns that you could not have closed form solutions. Fortunately, in the meantime all kinds of physics of a different sort was developing. It would certainly have been disastrous if some tightly integrated theory of physics as a whole that solved the classical problems first had been demanded.

Furthermore, retrogression of an unsuspected kind can occur. A good example is relativistic mechanics—I mean mechanics in the setting of special relativity, not anything so exotic as general relativity. In relativistic mechanics one can not only not solve the three-body problem, Newton's classical solution of the two-body problem no longer works, and no simple solutions exist. These problems of mechanics that I have mentioned are really hard. They are so hard that people have had to take up other lines of work in order to get any significant results. Once in a while, of course, somebody will have a really good idea, but such moments may be really few and far between.

I recommend in the cognitive sciences what we find in physics: the same pluralistic mixture of the old and the new, the integrated and the modular. It seems to me that this is the way we ought to think about the Alexandrian tradition in grammar. We are very fortunate to have Joan Bresnan and Noam Chomsky, as well as other people, continuing so well with this tradition. On the other hand, it would be a great mistake to think that where cognitive science touches on linguistic matters it must wait for progress of a linear sort. I mean by this, that first we will solve the Alexandrian problems of structure; then we can move on to Talmy Givón's problems of processing or Elizabeth Bates' problems of psycholinguistics. I predict it will not be that way at all. There will be a pluralistic, many-splendored development, just as in physics. It has, I think, been one of the mistakes in psycholinguistics methodologically to wait too often for the latest news to come down the corridor from structural linguistics. Just as Einstein did not wait for Minkowski, I do not think our psycholinguists or our process linguists should wait for the latest word from Alexandria. We have all seen the benefits of workers in AI charging off on their own without a care in the world for anything that was previously written or said. Such blitheness is not always in order, but it seems just right for the cognitive sciences in their current youthful state.

REFERENCES

Anderson, J. R. Arguments concerning representations for mental imagery. *Psychological Review,* 1978, *85,* 249–277.

Anderson, J. R. Further arguments concerning representations for mental imagery: A response to Hayes-Roth and Pylyshyn. *Psychological Review,* 1979, *86,* 395–406.

Pylyshyn, Z. W. Validating computational models: A critique of Anderson's indeterminacy of representation claim. *Psychological Review,* 1979, *86,* 383–394.

Pylyshyn, Z. W. The imagery debate: Analogue media versus tacit knowledge. *Psychological Review,* 1981, *88,* 16–45.

13 Cohabitation in the Cognitive Sciences

George Mandler
University of California, San Diego

COGNITIVE SCIENCE OR COGNITIVE SCIENCES?

It is not quite clear whether a single science is going to emerge from the current excitement and fussing that surrounds the cognitive sciences. What is clear is that the past several years have seen the emergence of an identifiable group of cognitive sciences. Among the practitioners of these sciences one may, to be sure, find many who are properly called cognitive scientist, *in conjunction with* the more traditional label of linguist, computer scientist, psychologist, anthropologist, and so forth. As a result I tend to view cognitive science more as an umbrella than as a hat. In order to do justice to the complexity and promise of the cognitive science enterprise, I shall abandon the umbrella metaphor, in part because it suggests inclemency. In the last section of this chapter, I present an alternative model (using a more comfortable metaphor) that describes how the cognitive sciences can live together and possibly merge into a single cognitive science. But I first consider the content of these sciences, their common history, and interfaces.

SOURCES AND DOMAINS OF THE COGNITIVE SCIENCES

It is a frequent and facile claim that one or another of the cognitive sciences became ''cognitive'' as a result of the motherly concerns of one or the other sciences. In fact there is a good argument to be made that the various disciplines became cognitive more or less simultaneously and in parallel, and that the AI field emerged—independently—at about the same time. The time was the second half of the sixth decade of the twentieth century. I cannot speculate here about the socio/historical reasons for that emergence, but it cannot be accidental that the same decade saw the first glimmerings of what powerful high speed computers can do. What is currently opaque are the social and cultural imperatives that determined this parallel (not serial) development. I cannot do justice to all the discoveries and rediscoveries of that half decade and the following ones, and will restrict myself to a brief description of the kinds of conceptual changes that marked the various cognitive sciences.

Common themes in the cognitive sciences

The central themes that emerged during those 5 years and that mark the cognitive sciences are the concepts of *representation* and *process*. They are the primary foci of all the relevant disciplines, and it is symptomatic of our acceptance and their importance that we rarely hear anybody question these two foundations. In fact, our acceptance of representation and process as basic to the cognitive enterprise is exemplified by the arguments that go on now about the nature of representations, about the utility of the distinction between the two concepts, and generally about the fine points of their use—not arguments about whether cognitive science should be *about* something like representation and process. We are more concerned about distinctions between analogic and propositional representations or between declarative and procedural knowledge.

For the *artificial intelligence* community the two concepts came with their baby's milk. From the beginning the AI business was about representation and process, essentially by definition. AI is concerned with representing structures and knowledge and with devising processes that can operate over them. In *cognitive psychology* the invention of useful theoretical devices, both in the representation and the processing of knowledge, was a hallmark of the renaissance of the 1950s. Dynamic systems were part of psychology's heritage from the French and German psychologies of the early twentieth century, but the models were crude and their influence faded during the behaviorist interlude. We have now rediscovered and have reestablished intellectual connections with the insights of Selz and the Gestaltists, with the French and Genevan schools, and with the pervasive contributions of the psychoanalytic enterprise. Mentalism once again became the modal approach (though in a new and more respectable form) and the old traditions of schematic representations and the analysis of consciousness (*Bewusstseinspsychologie*) became commonplace. In *anthropology* the late 1950s saw the emergence of representational models in an area focal to anthropological thought—the investigation of kinship systems. The development of transformational grammars in *linguistics* and the notion of deep structure similarly can be seen as the discovery of underlying representations and the processes that operate on them. By now, of course, linguists of all persuasions—

whether the messengers of universal truth or not—are concerned with the nature of the representations underlying language and the major theme of their internecine struggles are more concerned with representational matters than with matters of fact.

If these common concerns unite us, what distinguishes the various cognitive sciences from each other? The question raised in the context of this assembly is whether methodologies not only distinguish, but in fact divide our various disciplines. However, method (in the broadest sense) is inextricably bound up with subject matter and theory, and I believe that the best way to draw distinctions among the cognitive sciences is to look at their different domains. Such an examination will also lead to some insight into barriers to a real integration and a look at what an integration might imply.

The domains of the cognitive sciences

I start with *artificial intelligence* in part because it is the most difficult to circumscribe but also because it may well be the focal one of the group.[1] I see the potential (not yet quite realized) role of AI as extending toward all the other cognitive sciences. As keeper of the computational grail, the AI community may well turn out to be for the cognitive sciences what mathematics broadly has been for all the sciences. If mathematics is the queen of the sciences, AI could earn the mantle of the Prince of Wales of the cognitive sciences. The advent of the modern high speed computer has brought within reach the possibility of implementing and testing the kinds of complex dynamic systems that surely characterize the human mind both in its individual and social manifestations. From its beginnings the computational riches of the new technology suggested the possibility of implementation of social and psychological theories. Furthermore, the test of implementation promised to be a useful tool for keeping social and psychological theorists honest. If their theories are so vague that their assumptions, axioms, and postulates cannot even be properly stated for possible implementation, or if their theoretical statements, once implemented, lead to internal contradictions and lacunae of indeterminacy, then the most advisable watchword would be: Back to the drawingboard![2] Granted that we have not reached the stage in which this kind of enterprise is easily possible, or that many psychological and social theories even claim the degree of precision and prescription that is implied here, the goal is easily envisioned and—in principle—attainable.

I do not want to imply that the major role of AI is to test other people's

---

[1]My position is in general agreement with John Haugeland's paper at this conference and his description of the central role of artificial intelligence.

[2]The same theme was sounded in the discussion by Givón when he described AI as a "tool in testing" theoretical propositions. It is also useful in uncovering the implicit theories hidden in vague and incomplete formulations.

theories. Completely apart from developing the programs, algorithms, and formalisms that might be used in testing existing theoretical formulations, the AI practitioner has been and should be concerned with developing our vision of *possible* mechanisms and processes. In developing synthetic intelligences (and I adopt here Haugeland's happy phrase), AI can open up a wide arena of formalisms that can be tested, adopted, and adapted in the mimicry of intelligent organisms and social groups that is practiced by theoretical cognitive science.[3]

At the present time, we are moving somewhat slowly toward the goal of AI becoming the repository of possible intelligences (if in fact that is a proper goal). We see neither signs nor claims that the new Leibniz or Newton has emerged to present us with an innovative computational calculus. Actually, rather than narrowing its focus, the AI community has roamed far beyond the domain that I have outlined. As a relative newcomer to the business of scientific boundary building it certainly needs the fuzziness and vagueness that must precede the hardening of the arteries that comes with achieving a professional identity. One area of uncertainty that is of concern to all the cognitive sciences is the distinction between modeling synthetic or human intelligence. It is sometimes refreshingly difficult to tell the difference between an artificial intelligencer and a cognitive psychologist. In a sense that heralds the beginning of cognitive science in the singular. However, in their (understandable) desire to distance themselves from traditional psychology, some AI practitioners have fallen into the very traps that modern psychology has painfully learned to avoid.

Intuition, anecdotal examples, striking exemplars, and singular demonstrations of complex processes are poor substitutes for hard evidence. Free associations in response to interesting problems may be a good way of revealing underlying motives but it is not (and has not been in the past) a particular edifying way of doing science. There is no doubt that the lure of what I have dubbed *phenomenocentrism* is very powerful. Having abandoned anthropocentrism and ethnocentrism as the touchstones of human science, phenomenocentrism relies on the immediate phenomenal experience as revelatory of the structure of the world. Our phenomenal world is immediate, obvious, and thereby very convincing. Unfortunately, we often forget that it too is constructed, that it is a complex product of our culture and personal histories. Even apparent collegial consensus about one's phenomenal world might reveal nothing more than socio/cultural commonalities and norms. Although the process of creativity in science is dependent to a large extent on these intuitions and insights, that context of discovery should not be confused with the harsh reality of verification.

These are not proscriptive warnings. Much that is new, fascinating, and

---

[3]Again, this argument was presented by Brown when he asked for the development of robust systems that may tell us something about the human mind. An actual illustration was provided by Greeno when he described the use by Greeno, Riley, and Gelman of a formalism developed, in a different context, by VanLehn and Brown. A similar argument was made by Suppes when he called for the "engineering" use of AI.

important has been contributed by the AI work on human modeling, but it would be helpful if we made the distinction between AI models that claim theoretical relevance to human intelligence and those that do not. The distinction will obviously be a fuzzy one, but it is likely to help overcome some of the misunderstandings between various camps within the cognitive sciences. None of this denies the need for a continuing interplay between artificial intelligence and substantive issues. AI methodology depends to a large extent on the recognition of the kinds of problems that human and other intelligent systems face and present. Nothing human (or synthetic) should remain alien to the cognitive sciences.

The domains of the other cognitive sciences are more easily described. They have established more or less firm frontiers, often as a result of extensive armed skirmishes with neighboring tribes. Some of them are still involved in internecine conflicts, often trying to become dominant forces within their disciplines, while the other cognitive sciences (hopefully) cheer from the sidelines. The best example of this particular scenario can be found in psychology, where cognitive psychology is well on its way to occupying the vacuum left by a dying behaviorism.[4]

Cognitive psychology is often misnamed within psychology; it is frequently believed to be a psychology exclusively concerned with thought processes and knowledge representation. However, as cognitive psychologists reestablished a theory-rich discipline, their interests soon expanded into other areas, such as theories of skills and action and the analysis of emotional states. Their early concerns with the representation of knowledge and the complex processes of human thought earned them the cognitive label, but the domain engulfs all of psychology. Psychology has classically been considered to be the science of the isolated individual, whether acting and thinking alone or in groups of varying sizes. That particular distinction is breaking down as psychologists are becoming more sensitive to the fact that thought and knowledge are often best conceptualized as emerging from people/situation interactions (or better, symbioses). That particular trend has tended to blur psychological frontiers toward anthropology and sociology, while infusing new thought into the common areas shared with linguistics.

If one accepts the commonality between deep structure and underlying representations, *linguists* have been cognitive since the 1950s. All the various linguistic enterprises, whether formalist or functional, have addressed the problem of theory, of the representations and processes that generate human language. Until recently they have been marked with their preoccupation with verbal language, but the discovery that sign language is not opaque to linguistic analyses opens up new venues for linguists.

---

[4]I have discussed the prospects and problems of cognitive psychology at greater length elsewhere (Mandler, 1981).

Together with the unifying themes of representation and process, the cognitive sciences also share a pervasive constructivist approach to the behavior of organisms. The emphasis is to develop systems and structures that can be said to construct the observable, evidential aspects of human thought and action. It is evident in the analyses of emotion and consciousness in psychology, in the linguists' approach toward a system that constructs human language, in the joint AI/psychology models of perception, and—finally—in the acceptance of the dictum that even our phenomenal world is—in the last analysis—constructed. Philosophers of kindred persuasions hear echos of some of their own history in such talk.

TOWARD A SINGULAR COGNITIVE SCIENCE?

The signs of a true cognitive science can be seen in the projects that defy these domain designations. They occur in work on psycholinguistics between psychology and linguistics, in memory between psychology and AI, between linguistics and AI in studies of language processing, in the modeling of motor skills which combines psychology, AI, and neurosciences, to name just a few examples. There are now enough people around who have picked up expertise in more than one of the cognitive sciences to be called cognitive scientist without a parenthetical disciplinary designation. Cognitive science in the singular will be well established when there exists a critical mass of people who can easily move across disciplinary boundaries and who also will have the necessary expertise to use our powerful tool—the computer—in the exploration of complex dynamic systems. In the process we will hopefully establish new contacts with the cognitive sciences that have sometimes been seen as peripheral to the cognitive enterprise, such as the neurosciences (when they deal with intelligent systems), with cognitive anthropology, and with sociology.

There are remaining prejudices, fears, and assumptions that frequently interfere with effective interdisciplinary efforts. They are primarily found in the remnants of reductionism and in the intellectual imperialism of the individual disciplines. As long as one accepts a hierarchical notion of the sciences and an, in principle, inevitable reduction of the higher (softer) sciences to the more basic (harder) sciences, one tends to look over one's shoulder—if soft—to make sure that one's efforts are consistent with big sister, or—if hard—to make sure that little brother keeps within the boundaries of the permissible. For example, it is conceptually constricting and empirically unsound to use the current state of one's "harder" neighbor as deterministic constraints on one's own theorizing (e.g., psychology vs. linguistics of neurophysiology vs. psychology). Current fashions change in all fields and one would not want to be left high and dry because the assumptions adopted from one's neighbors have been dropped by them. Conversely, the more "basic" sciences frequently forget to look at the conceptual achievements of their "softer" neighbors. The obvious danger here is that the more basic is supposed to explain the less basic phenomena, but the latter are defined within another discipline. Reductionism is probably a mistake in principle, but it is certainly premature at the present state of the various arts. Collaboration and bridge building at the boundaries is a much more satisfying and—for the time being—more promising enterprise.

What next?

It may seem premature to project what the cognitive sciences should be doing next, given the rather fuzzy state of current affairs. But if we are going to be proper intellectual imperialists we should at least have long range battle plans. More important, I believe that there are problems central to the cognitive sciences that we have missed, problems whose solution would present significant advances for the various disciplines. I shall concentrate on a few topics that seem to have been missed (for good reason?).

The good and the beautiful.   The question of esthetic and ethical judgments (knowledge) has been approached gingerly by psychologists, embraced by philosophers, avoided by linguists, and ignored by artificial intelligencers. But if we truly pretend to have an understanding of all of human knowledge, we must surely also tackle these problems.[5] Solutions that see value-laden predicates stalking the world in search of appropriate objects, or that postulate valued objects lying around the landscape just waiting to be properly "appreciated" will not do. We must seriously worry about the proper and useful representation of esthetic and ethical knowledge.

Expressive Syntax.   Now that linguists have gone beyond vocal language, there is a large area of data and puzzles waiting for their skills—the domain of nonverbal communicative behavior. Current work on so-called expressive facial behavior as well as gestural languages has been descriptive at best, but in any case has never tried to approach the problem of the syntactic structures of these communicative acts.[6] As a result the field is theoretically relatively barren. Much is known about some of the details of these acts (their muscular "phonology," for example), a little has been attempted rather crudely about their semantics, but the syntax of these behaviors has remained virgin territory. At the same time, the modeling of these human actions will test the mettle of the AI community to the point of calling for significant advances in both software and hardware development.

---

[5]When I recently made a foray in the direction of the representation of value I was warned by a knowledgeable friend that here was a topic in which even the minefields have minefields.

[6]One exception is the promising approach toward the analysis of expressive interactions between mothers and infants.

Universals—Beyond language.   The search for language universals and the possible underlying universal grammar has become too narrow an enterprise. Is it possible that we may miss some important points about these universals because of our preoccupations with language? There are other human universalities that badly need further examination, universals in human toolmaking, food preparation, and eating habits (cuisines?), play and games, and others. Are we to look for the Universal do-it-yourself guide, the Universal recipe, the Universal game rules, or could we possibly approach the problem more generally and ask for rules for human universals that might reveal unanticipated gains for our study of language universals?

Perchance. . . .?   One of the oldest challenges to the human imagination has been the interpretation of dreams. Few advances have been made since the publication of *The interpretation of dreams*—many years ago. Dreams offer challenges to theories of semantic structure, of search and storage processes in memory, of visual representation, and to the wide range of problems posed by apparently independent parallel processors. The topic also might make it necessary for cognitive scientists to come to grips with the complex problem of the self, so badly in need of a firm hand and principled discussion. How, for example, would one want to structure a system in which it is possible for somebody to have a dream that is full of surprises? What part of the system is surprising (shocking, frightening, delighting) what other part?

MUDDLE: A PROCESS MODEL

Given that the preceding is a reasonable summary of the recent history and common themes of the cognitive sciences, I now turn to questions of cohabitation, of the rules that might make it more likely that we can live together constructively and possibly create a single cognitive science. I shall present—in the best tradition of the field—a model, acronymically designated as MUDDLE (Members of Undefined Distinct Disciplines Learning Eclectically). The basic notion underlying MUDDLE is that of a boarding house, in which the various cognitive disciplines are housed—each in its own room.[7] In the preceding pages I have described some of the origins of the present situation, how we all arrived at the boarding house because we do in fact have a common ancestry and a (very short) common history. These characteristics suggest how we have come to inhabit a single boarding house, what it is that goes on in the various rooms, and how we can better understand these sometimes mysterious activities. The boarding house metaphor is implemented in the MUDDLE model which assumes that the boarding house is inhabited by all the various cognitive disciplines. In

---

[7]It is possible that in the process of moving in the Artificial Intelligencers grabbed the best room, and John Haugeland even suggested that they hold the lease on the house.

order to define what may (or should) be the activities in that boarding house, we need a set of rules in the hallowed tradition of boarding houses. These rules contain both warnings and encouragements; they are designed to make the house a home, to create a family.

The rules of the house

We do not permit any discrimination on the basis of scientific origin

Rationale: No cognitive science is ''better'' or ''harder'' than any other science; none of them represents the final irreducible elements to which all the others must be reduced. Nor is it the case that it is better to have a lot of ''brilliant'' ideas than to do a good and clean experiment (or vice versa). Things are hard enough in the 1980s without the snobbishness of psychologists toward anthropologists, or neurophysiologists toward psychologists, or computer scientists toward experimentalists. If a cognitive science is eventually to emerge, we need to understand (and even respect!) each other's problems.

Dont ask management for another room; other rooms are not better

Rationale: The converse of snobbishness is a looking-over-your-shoulder sense of inferiority. It is unlikely that the answers to *your* problems will be found in somebody else's discipline. This is not to discourage interdisciplinary work (see later); rather the intention is to encourage work on tough problems that need to be solved within their own domains. Other disciplines may look neat and clean, but adopting their current state as ''constraints'' on your own work may be premature adulation.

Dont claim that other boarders are in the house only because you guaranteed the rent, or because they had to leave home

Rationale: Claims that some other cognitive science would not possibly be ''cognitive'' if it had not been for the influence of your particular scientific enterprise are gratuitous; they are also very likely to be untrue. The history of the past quarter of a century suggests that the reasons that bring us together come from common interests, usually uncommonly developed. Nor is it the case that your brothers and sisters would be helpless and uncomprehending against the flood of cognitive problems if it were not for the theories, methods, and data that you have developed and that you have graciously allowed them to use. Be attentive to your siblings' problems, the ones they solve may be your own. And do not dwell on their poor home life before they moved into the cognitive temple; they are probably already ashamed of their impoverished childhood.

Improvements to rooms are encouraged, but dont reinvent the wheel—at least try something more difficult (like the horse)

Rationale: It might be useful to learn more about the achievements and disasters of other scientific enterprises, lest the blinding insights that you discover at regular intervals turn out to be somebody else's old saws. Insights (blinding or otherwise) need to be examined carefully and should be given wide berths when they seem to be obvious—other people might think so too, and might have thought so for a long time. If you must (can) discover something new, try to challenge your own intellect and attempt something difficult. In particular, beware of insights into other people's disciplines; contrary to general belief not everybody who preceded you was an idiot.

Everybody is invited to the weekly cocktail party in the lounge, but tape recorders are prohibited

Rationale: Communication among people and disciplines is the *sine qua non* of a successful cognitive science. The more we talk to each other the more we can learn and the more likely it is that the millenium of cognitive integration will occur. In particular, undisciplined conversation is a useful devide for discovering common ideas and approaches. Conversely, lack of intellectual discipline does not make for scientific advance. Dont publish your loose talk, leave that for hard ideas and evidence.

Dont lock your door, other people may want to use your room

Rationale: Secrecy and privacy have no place in the creation of a new science. Be forthcoming in sharing your ideas, your offices, and your laboratories. Offer your friends in other disciplines the use of your tools and methods, make them feel comfortable in the strange surroundings of your own field.

Visits to other rooms are permitted at all hours.

Rationale: Try out your problems (and solutions) on your friends; share your puzzles with them. Look over their shoulders as they work on their problems—ask questions; it may not all be as obvious as it seems.

Children are permitted; boarders are encouraged to procreate

Rationale: Although there are some genuine cognitive scientists in our midst who span disciplines and knowledge domains with ease, it will be in succeeding intellectual generations that cognitive science will mature if it is at all viable. Teach your students (and your colleagues) the advantages of knowing about cognitive endeavors other than your own. Send them out into the cognitive world, try to develop interdisciplinary projects. Dont inhibit members of your department, laboratory, or research center from exploring other people's work; remember cognitive science is being made.

Envoi

The cognitive sciences today are full of the more enjoyable part of the scientific enterprise—fun and games. We are discovering new problems to conquer, re-discovering some old truths, and meeting a lot of smart and interesting people. We are not quite sure how it all came about and we sometimes forget that science (even cognitive science) is not only part of, but also an expression of the culture in which it flourishes. In return, we need to be responsible for the consequence of our science in the culture at large and consider the ends for which it is used. To the extent that we are successful, to that extent will we have wares to offer to society at large that will be powerful and that will have potentials for social good or evil? Our enthusiasm for our current enterprise should live side by side with some sense of social effectiveness.

Consider the moral of the story of the truck driver who was asked to take a load of penguins to the zoo and was found the next day taking them to the museum because they had such a good time at the zoo.

The Moral

Just because you are having a good time, don't assume that you are doing the right thing.

ACKNOWLEDGMENTS

Copyright © 1981 George Mandler. The preparation of this chapter was supported in part by National Science Foundation Grant BNS 79-15336.

REFERENCES

Mandler, G. *What is cognitive psychology? What isn't?* Invited address to the Division of Philosoph-ical Psychology, American Psychological Association, Los Angeles, August 1981.

Author Index

Italics denote pages with bibliographic information.

Subject Index

A

Algorithm, 19, 286
ambiguity, 225
anaphora, 207
arithmetic problem solving, 287
attention, 9, 247
automaticity, 222
autonomous language faculty, 211, 283

B

backward chaining, 53
belief, 152, 281
binary contrasts, 290, 292
black box, 237
BORIS, 35, 41, 46
bottom-up, 229, 231, 236
brute force, 283, 298
bugs, 246

C

chess programs, 19
chunks, 235
cogency condition, 97
color terms, 194
common sense, 264
communication situation, 288

competence, 35, 103, 107, 115, 199, 201,
    244, 258, 266, 287, 300
competitive argument, 238, 260, 295
complexity, 94, 121
comprehension, 4, 13, 15, 19, 201, 220, 237,
    274, 286
computational methodologies, 216
computational models, 7, 235, 238, 241, 260,
    279, 294
consciousness, 3, 224, 288
constituent structure, 112, 193
constraints, 102, 109, 120, 140, 192, 195,
    199, 200, 202, 204, 208, 206, 209, 221,
    224, 236, 286, 287, 289, 296
context, 12, 140, 165, 179, 180, 206, 220,
    221, 222, 224, 285, 288
contrasts, 86, 87, 88, 290
control structures, 248, 291
counterbalancing, 278
creativity, 122, 130, 202
cross-modal priming technique, 226
crucial facts, 242

D

deductive inference, 206
deep theories, 12, 240
degrees of freedom, 236